Scattering in Condensed Matter

Scattering in Condensed Matter

By

Marian Apostol

Cambridge
Scholars
Publishing

Scattering in Condensed Matter

By Marian Apostol

This book first published 2024

Cambridge Scholars Publishing

Lady Stephenson Library, Newcastle upon Tyne, NE6 2PA, UK

British Library Cataloguing in Publication Data
A catalogue record for this book is available from the British Library

ISBN (10): 1-5275-5835-5
ISBN (13): 978-1-5275-5835-9

Contents

1 Preface **1**

2 Scattering, Collisions **9**
2.1 Born scattering 9
2.2 An exact solution 13
2.3 Momentum transfer 15
2.4 Beam 17
2.5 Quantum transitions 19
2.6 Energy transfer 24
2.7 Phase factors 26

3 Electromagnetic Radiation **29**
3.1 Dipole field 29
3.2 Damping force 31
3.3 Line breadth 32
3.4 Scattering of radiation 33
3.5 Absorption of radiation 36
3.6 Maxwell equations 37
3.7 Energy, force and momentum 40
3.8 Longitudinal field 44
3.9 Photons 47
3.10 Charge in electromagnetic field 49
3.11 Scattering off a charge distribution 51
3.12 The atom 54
3.13 Quantum theory of radiation 59
3.14 Emission and absorption: free charges 62
3.15 Emission and absorption: bound charges 63
3.16 Propagation of light, interference 67
3.17 Doppler effect 70
3.18 Scattering: first-order perturbation 72
3.19 Scattering: second-order perturbation 74

4 X-Rays Diffraction **79**
4.1 X-rays . . . 79
4.2 Extinction . . . 81
4.3 Polarization . . . 82
4.4 Elastic scattering of X-rays . . . 87
4.5 Diffraction peaks in crystals . . . 88
4.6 Form-factor . . . 91
4.7 Energy and momentum . . . 97
4.8 Average total intensity . . . 100
4.9 Diffuse thermal scattering of X-rays . . . 102
4.10 Lattice distortion and phase defects . . . 103
4.11 Temperature dependence: X-rays scattering and phonons 106
4.12 Higher-order effects . . . 109

5 Neutrino Scattering **113**
5.1 Neutrinos . . . 113
5.2 Scattering parameters . . . 115
5.3 Classical scattering . . . 116
5.4 Form-factor . . . 120
5.5 Temperature dependence . . . 121
5.6 Coherent *vs* incoherent scattering . . . 126
5.7 Neutrino coherent scattering . . . 127
5.8 Coherent-incoherent interplay . . . 129
5.9 Quantum scattering . . . 133
5.9.1 Introduction . . . 133
5.9.2 Center of mass . . . 134
5.9.3 Crystal field . . . 138
5.9.4 Photon emission (absorption) . . . 139
5.9.5 Coherent scattering . . . 142
5.9.6 Cross-section . . . 144
5.10 Neutrino scattering . . . 146
5.11 Coherence domains . . . 151

6 Neutron Scattering **157**
6.1 Single-particle scattering . . . 157
6.2 Scattering parameters . . . 158
6.3 Macroscopic target. Quantum scattering . . . 159

6.4 Coherence domains. The principle of maximal cross-section 164
6.5 Classical scattering 166
6.6 Macroscopic quantum-mechanical scattering 169
6.6.1 Introduction 169
6.6.2 Macroscopic quantum-mechanical scattering 171
6.6.3 Neutrino scattering 174
6.6.4 Other projectiles. Coherence domains 176
6.6.5 Classical scattering 177

7 Moessbauer Effect 181
7.1 Resonance fluorescence 181
7.2 Photon emission 186

8 Light scattering 191
8.1 Sound 191
8.2 Scattering by sound 197

9 Scattering by laser pulses 203
9.1 Charge scattering by electromagnetic radiation 203
9.2 Cross-section 211
9.3 Quasi-classical approximation 218

10 Electron scattering 231

11 Appendix 235
11.1 Center of mass 235
11.2 δ-function and the Fourier transform 237
11.3 Electromagnetic field and polarization 238
11.4 Maxwell equations in matter 239

Index 241

1 Preface

Scattering with one-particle targets is well developed in atomic and nuclear physics, but its counterpart with macroscopic targets still exhibits insufficiencies, in spite of the large amount of spectroscopical techniques employed in condensed matter. This book is devoted to an exploration of scattering in condensed matter. By condensed matter we mean large macroscopic targets, consisting of a large number of scattering centres like atoms, molecules, and unit cells in crystals. Scattering by inhomogeneities in continuous media (*e.g.*, light sound) is briefly treated. Also, we include a recent development of charge scattering by laser pulses. Many other types of condensed matter, like plasmas, soft matter, superfluids, etc, are left aside. As regards the projectiles, we deal with pointlike charges like electrons or ions, light and X- and gamma rays, neutrinos, neutrons. Many other types of scattering in condensed matter are not included. The book focuses on basic problems raised by scattering in condensed matter.

The first problem concerns the classical or quantum-mechanical nature of the interaction of the projectile with a one-particle target. If the range of the interaction is larger than the variation range of the target wavefunction, the collision is classical. For instance, this may happen for radiation scattering, where we can derive the cross-section by classical electromagnetism. On the other hand, the interaction may vary abruptly over the range of the wavefunction, like for neutron contact interaction, where the motion of the target is quantum-mechanical. However, the corresponding energy transfer is thermalized, such that the scattering is classical, as in diffraction.

The scattering process in condensed matter is a local process: a local portion of the incoming wave (radiation, particle wave) interacts with a local portion of the macroscopic target. In this respect, the scattering in condensed matter may have a quantum-mechanical character. We may imagine that an incoming quantum particle is absorbed by the target and thereafter released, emitted by the target. Therefore, the scattering interaction may carry a quantum phase. If the phase is the same for all scatterers, the scattering is coherent, and the scattering cross-section is proportional to N^2, where N is the number of scatterers. If the phase varies randomly from scatterer to scatterer, the scattering is incoherent and the cross-section is proportional to N. The difference between coherent and incoherent scattering is well-known for multi-particle targets, but it is not sufficiently exploited in condensed matter.

Although the scattering in condensed matter may be ultimately quantum-mechanical, in some cases a quasi-classical description is valid. However, a phase carried by interaction may also be present in these cases. According to the interpretation of the scattering as an absorption/emission process, this phase has in fact a statistical character.

Another element which receives less attention, in spite of its importance, is the momentum and energy conservation in scattering in condensed matter. This conservation is emphasized throughout this book, together with the force acting upon the target. An estimation of this force, acting in the forward direction, is $F = \Phi p\sigma$, where Φ is the flux density of the projectile (number of particles per unit time and unit cross-section of the target), p is the momentum of the incident particle and σ is the scattering cross-section. The momentum conservation in coherent scattering may lead to diffraction peaks, infinitesimally extended for macroscopic targets; while the incoherent scattering does not exhibit peaks; it proceeds by the conservation of the momentum of individual scatterers, or elementary excitations like phonons. The infinitesimal extension of the coherent peaks diminishes the scattering cross-section from $\sim N^2$ to $\sim N^{4/3}$. We call this scattering a coherent diffraction. This is the usual point which receives the main attention in scattering in condensed matter. It seems that only the coherent diffraction and the incoherent scattering are known in condensed matter. We call these scatterings classical scatterings.

Here we encounter the first big problem. The cross-section of the coherent diffraction is $\sigma = N^{4/3}\sigma_0$, where σ_0 is the one-particle cross-section. For all scatterings in condensed matter, with one exception, this cross-section is larger than the cross-section A of the target. The exception is neutrino scattering, which, anyway, raises other problems.

Now, the number of scattered particles (per unit time) cannot exceed the number of incident particles, so the cross-section σ should always be smaller than A. This condition is usually overlooked. It seems that in X-ray diffraction it is implied that this condition would be satisfied by the so-called dynamic X-ray scattering. However, this dynamic scattering brings only small corrections to the cross-section.

The way out from this deadlock is provided by the coherence domains, which we introduce in this book. Usually, the macroscopic target is at thermal equilibrium. Under the action of the incident particles the macroscopic target is taken out of equilibrium. Consequently, the target tries to regain as much of its thermal equilibrium as possible. This can only be done by developing a disorder, which would increase the entropy. In these circumstances this disorder can only be associated to phase factors. Being still under the action of an external agent, the increase in entropy should be minimal, such that the cross-section is maximal: it is reduced to the target cross-section A. The target cross-section A is a macroscopic quantity in reasonable limits. If the target is very large, then radiation extinction may appear, as well as absorption, multiple scatterings, such that the description of the scattering is not so simple anymore. We assume that a number $n_d = N/N_d$ of (identical) domains appear in the target, where N_d is the number of particles in each domain, $1 < N_d < N$. In each domain the scattering phase is constant and the domain phases are randomly distributed. It follows that the cross-section becomes

$$\overline{\sigma} = n_d \sigma_d \ , \tag{1.1}$$

where σ_d is the cross-section of one domain. For a coherent scattering we have $\sigma_d = N_d^2 \sigma_0$, for a coherent diffraction we have $\sigma_d = N_d^{4/3} \sigma_0$, such that we have $\overline{\sigma} = N N_d \sigma_0$ for coherent scattering and $\overline{\sigma} = N N_d^{1/3} \sigma_0$ for coherent diffraction. We note that for an incoherent scattering $N_d = 1$ and both formulae give the same result $\overline{\sigma} = N \sigma_0$. It is convenient to introduce $f = 1/N_d$, where $1/N < f < 1$. We get

$$\overline{\sigma} = \frac{N \sigma_0}{f} \tag{1.2}$$

for coherent scattering and

$$\overline{\sigma} = \frac{N\sigma_0}{f^{1/3}} \tag{1.3}$$

for coherent diffraction. The condition

$$\overline{\sigma} < A \tag{1.4}$$

should be fulfilled for any scattering in condensed matter.

The above inequality should be satisfied for $1/N < f < 1$. Now, there is another circumstance which should be included in such an analysis. The domains are regions of continuity of the scattering phase. The phase is associated with the atomic constituents of the macroscopic target. Any macroscopic target has a discrete structure. Therefore, there should exist a critical extension of each continuity domain. We show in this book that the critical number of particles in each (three-dimensional) domain is given by $N_{dc} = (a/a_0)^9$, where a is the mean separation distance between the particles and a_0 is the atomic dimension of each particle. For instance, for $a = 4a_0$ we can take N_{dc} of the order $\simeq 10^6$. In two dimensions $N_{dc} = (a/a_0)^4$, which is of the order $\simeq 10^3$ for $a = 4a_0$. Therefore, f should satisfy the condition $f < 10^{-6}$ ($f < 10^{-3}$). For $f > 10^{-6}$ ($f > 10^{-3}$) the domains are not well defined; they are reduced to individual particles and the scattering is incoherent ($f = 1$).

On the other hand, the principle of maximal cross-section requires $\overline{\sigma} = A$, *i.e.* $f = N\sigma_0/A$ for coherent scattering and $f = (N\sigma_0/A)^3$ for coherent diffraction. For neutrino scattering, with $\sigma_0 = 10^{-44}cm^2$ and $N = 10^{22}$ ($0.1mol$) we get $f = 10^{-22}/A$ for coherent scattering, which is less than $1/N$ for a macroscopic target. It follows $N_d = N$, *i.e.* a coherent scattering may appear in the whole target. This is a novelty for the scattering in condensed matter, which needs to be accommodated by a theory. For $\sigma_0 = 10^{-24}cm^2$ (X-, gamma rays, neutrons) we get $f = 10^{-2}/A$, which is not acceptable; a tendency exists towards an incoherent scattering ($f \to 1$). Therefore, a coherent scattering cannot appear in this case. However, for a coherent diffraction we have $f = 10^{-6}/A^3$, which is acceptable, and $N_d = 10^6 A^3$: a coherent diffraction may appear in this case. The condition $\overline{\sigma} < A$ ($\overline{\sigma} = A$) is a necessary condition for scattering. It remains to differentiate between the coherent scattering and the coherent diffraction.

The above hypothesis of coherence domains leads, in certain conditions (neutrino scattering), to a coherent scattering in the whole macroscopic target. That means that the scattering phase is the same in the whole macroscopic target. A constant scattering phase may appear in small targets, with a small number of particles, for incident wavelengths much longer than the dimension of the target; indeed, the phase variation over the dimension of such a target is small, so the phase may be taken constant. In macroscopic targets the scattering phase can only be constant if the incoming particles see the target as a whole. This means that the macroscopic target, namely the constituents of the macroscopic target, should be described in this case by a quantum field (a distributed wavefunction). This is a macroscopic quantum scattering, *i.e.* a scattering which is quantum-mechanical at the macroscopic scale. As long as we introduce phase factors in describing the scattering we deal with a quantum scattering, which requires quantum fields. The coherent diffraction does not employ such fields, although a wavefunction of the whole target (center of mass) is necessary. We may call the coherent diffraction a classical scattering. The macroscopic quantum scattering is introduced in this book. For crystals, a field of the whole target implies the quasi-momentum of the crystal. This is a coherent field (wavefunction). In addition, for all types of targets, there exist fields (wavefunctions) associated with the thermal motion of the atomic constituents. Basically, they are reduced to the one-particle wavefunctions of the target. We call them incoherent fields (wavefunctions). The coherent field has a significant weight in comparison to the incoherent field for stiff crystals with a high Debye temperature. The coherent field leads to a coherent cross-section $\sim N^2$, while the incoherent fields give a cross-section $\sim N$. In the former case the momentum transfer is taken by the quasi-momentum, *i.e.* by the whole crystal, while in the latter case the momentum is transferred to collective excitations like phonons.

A macroscopic quantum scattering is only possible if the collision time τ is longer than the equilibrium time Δt_{eq},

$$\tau > \Delta t_{eq} \ , \tag{1.5}$$

such that the incident particle flux sees the target as a whole. The collision time is the mean time between two succesive collisions. It is given by $\tau = 1/\Phi\overline{\sigma}$, where Φ is the flux density of the incident particles. The equilibrium time is the mean time needed by a target particle to regain its equilibrium. It is given by $\Delta t_{eq} = \hbar/\Delta E$, where ΔE is the energy transfer to a particle. The energy transfer is estimated as $\Delta E = (v + p/M)p$, where v is the thermal velocity of the target particles, M is the mass of the target particles and p is the momentum of the incident particles (of the order of the momentum transfer). It is convenient to introduce a threshold given by $p = Mv$, such that for $p < Mv$ we may use $\Delta E = vp$ and for $p > Mv$ we may use $\Delta E = p^2/M$. The threshold energy of the incident particles with mass m is $E = p^2/m = \frac{M}{m}T$, while for radiation $E = cp = c\sqrt{MT}$, where T is the temperature. For instance, for $M = 10^5 m_e$, where m_e is the electron mass, and room temperature $T = 300K$ we get $E \simeq 1eV$ for neutrons, $E \simeq 1keV$ for electrons and $E \simeq 10keV$ for radiation (neutrino included).[1] If $\tau < \Delta t_{eq}$ the incident beam sees the target particles as individual scatterers. This is a coherent diffraction (classical scattering).

Let us apply inequality (1.5) for neutrino coherent scattering. Together with condition (1.4) we have

$$f > \frac{N\sigma_0}{A} \; , \; f > \frac{\hbar\Phi N\sigma_0}{\Delta E} \; . \tag{1.6}$$

For an incident flux density $\Phi = 10^{12}/cm^2 \cdot s$ we have $\Delta E/\hbar\Phi \simeq 10^3 \Delta E(eV)$; for $E = 1MeV$ the energy transfer is $\simeq 20eV$, such that $A \ll \Delta E/\hbar\Phi$. Consequently, only the first inequality (1.6) needs to be satisfied. This condition has been examined above, so we may conclude that neutrino coherent scattering appears in the whole macroscopic target (for the numerical data used here). The force acting upon the macroscopic target is a measurable force, due to the large cross-section of the sample. This may justify the claims of Weber and others, in spite of many negative opinions, of the possibility of neutrino detection by using a torsion balance with a stiff sapphire crystal.

---

[1] $m_e = 10^{-27}g$, neutron mass $m \simeq 2 \times 10^3 m_e$, $c = 10^{10}cm/s$, $1eV = 1.6 \times 10^{-12}erg$, $1K = 1.38 \times 10^{-16}erg$, $\hbar = 10^{-27}$erg$\cdot s$.

For a coherent diffraction we need

$$\frac{N\sigma_0}{A} < f^{1/3} < \frac{\hbar\Phi N\sigma_0}{\Delta E} . \tag{1.7}$$

First, we note that these inequalities imply a small energy transfer, such that the energy of the incident neutrons, or radiation is low ($\Delta E < \hbar\Phi A$). Since, according to the principle of maximal cross-section, we have $\frac{N\sigma_0}{A} = f^{1/3}$, it follows that this coherent diffraction is possible, with domain dimension as large as that found above ($N_d = 10^6 A^3$). However, the force acting upon the target is small, precisely due to the domain occurrence. For higher energies the energy transfer is larger and we need to apply the inequalities (1.6). Making use of $\sigma_0 = 10^{-24}cm^2$, we get $f = 10^{-2}/A$, which indicates an incoherent scattering ($f \to 1$). The force acting upon the target is very small.

This book deals with such considerations. Their results depend on the numerical data. Besides a large variety of the known input data, serious uncertainties exist also in some input data. Many types of scattering in condensed matter are not analyzed herein, both for various projectiles and various macroscopic targets. For instance, a special case is the electron scattering, which proceeds on small superficial layers; the coherence domains are two-dimensional. It is an incoherent scattering with a weak coherent diffraction.

2 Scattering, Collisions

2.1 Born scattering

A quantum particle with mass m comes from somewhere at minus infinity, where it starts at the moment of time $t = 0$, and collides with a particle with mass M. The two particles interact with the potential $U(\boldsymbol{r})$, where $\boldsymbol{r}$ is the relative position of the two particles. The M-particle is fixed and structureless. After collision the m-particle goes somewhere at plus infinity, where it arrives at the infinitely long time t. We assume that the potential U falls sufficiently rapidly at infinity, such that the incoming particle at minus infinity and the outgoing particle at plus infinity are free. The motion of the colliding particle obeys the Schroedinger equation

$$-\frac{\hbar^2}{2m}\Delta\psi + U(\boldsymbol{r})\psi = E\psi \ , \tag{2.1}$$

where ψ is its wavefunction, E is the particle energy (and $\hbar$ is Planck's constant). The solution of equation (2.1) is a stationary solution, with a constant energy, which means that the collision is elastic. We write this energy as $E = \hbar^2k^2/2m$, where $\boldsymbol{k}$ is the wavevector of the incoming particle at minus infinity (and $\boldsymbol{p} = \hbar\boldsymbol{k}$ is its momentum). The particle wavefunction is $\psi_i = \frac{1}{\sqrt{V}}e^{i\boldsymbol{kr}}$, where V denotes the volume. It satisfies the Schroedinger equation

$$-\frac{\hbar^2}{2m}\Delta\psi_i = E\psi_i \ , \tag{2.2}$$

written also as

$$\Delta\psi_i + k^2\psi_i = 0 \ . \tag{2.3}$$

Let us assume that the potential U is sufficiently small, in a sense which will be shortly specified, such that we seek a perturbation-theory solution of equation (2.1), written as $\psi = \psi_i + \psi_s$, where the small correction ψ_s satisfies the equation

$$\Delta\psi_s + k^2\psi_s = \frac{2m}{\hbar^2}U\psi_i + ... \ . \tag{2.4}$$

The solution is valid as long as $| \psi_s | \ll | \psi_i |$. ψ_s is the scattered wave. This is a Helmholtz equation. Its solution is given by

$$\psi_s(\boldsymbol{r}) = \frac{2m}{\hbar^2} \int d\boldsymbol{r}' U(\boldsymbol{r}')\psi_i(\boldsymbol{r}')G(\boldsymbol{r} - \boldsymbol{r}') \ , \tag{2.5}$$

where G is the Green function satisfying the equation

$$\Delta G + k^2 G = \delta(\boldsymbol{r}) \ . \tag{2.6}$$

By using Fourier transformations, we get

$$G(\boldsymbol{r}) = -\frac{1}{(2\pi)^3} \int d\boldsymbol{q} \frac{e^{i\boldsymbol{q}\boldsymbol{r}}}{q^2 - k^2} = -\frac{1}{4\pi^2 ir} \int_{-\infty}^{+\infty} dq \frac{qe^{iqr}}{q^2 - k^2} \ . \tag{2.7}$$

In order to perform this integration, we need to give a sense to the singularities $q = \pm k$. Now, we remember that $k^2 \sim E/\hbar = \omega$, and the integration over frequency ω of the phase factor $e^{-i\omega t}$ should be zero for $t < 0$, in order to satisfy the causality principle. Therefore, the integration should be performed in the upper half-plane (for $t < 0$) and the ω-poles must lie in the lower half plane, *i.e.* $\omega \to \omega + i\varepsilon$, $\varepsilon \to 0^+$. Therefore, the integral in equation (2.7) becomes

$$G(\boldsymbol{r}) = -\frac{1}{4\pi^2 ir} \int_{-\infty}^{+\infty} dq \frac{qe^{iqr}}{q^2 - k^2 - i\varepsilon} = -\frac{1}{4\pi}\frac{e^{ikr}}{r} \ , \tag{2.8}$$

and the solution of the Helmholtz equation (2.4) is[1]

$$\psi_s(\boldsymbol{r}) = -\frac{m}{2\pi\hbar^2\sqrt{V}} \int d\boldsymbol{r}' U(\boldsymbol{r}') e^{i\boldsymbol{k}\boldsymbol{r}'} \frac{e^{ik|\boldsymbol{r}-\boldsymbol{r}'|}}{| \boldsymbol{r} - \boldsymbol{r}' |} \ . \tag{2.9}$$

Now, we can estimate the magnitude of ψ_s for the condition $| \psi_s | \ll | \psi_i |$. For a short range potential with magnitude U, extending over a small range a, and low energies $ka \ll 1$ we get from equation (2.9)

$$\frac{m | U | a^2}{\hbar^2} \ll 1 \ ; \tag{2.10}$$

[1] M. Apostol, *Equations of Mathematical Physics*, Cambridge Scholars Publishing, Newcastle upon Tyne (2018).

the potential should be smaller than the localization energy $\hbar^2/ma^2$ over region of extension a. In the same conditions but for high energies $ka \gg 1$, the condition reads, from equation (2.9),

$$\frac{m \mid U \mid a^2}{\hbar^2 ka} \ll 1 \ , \tag{2.11}$$

or

$$\mid U \mid \ll \frac{\hbar v}{a} \ , \tag{2.12}$$

where $v = \hbar k/a$ is the velocity of the incoming particle; for a Coulomb potential $U = \alpha/r$, there is no a, and the evaluation of the integral in equation (2.9) leads to $m \mid \alpha \mid /\hbar^2 k \ll 1$, or $\mid \alpha \mid \ll \hbar v$. These are the validity conditions for the perturbation-theory treatment of this collision problem.[2]

We are interested in the scattered wave at plus infinity. For $r \gg r'$, $\mid \boldsymbol{r} - \boldsymbol{r}' \mid \simeq r - \boldsymbol{r}\boldsymbol{r}'/r$, such that $\boldsymbol{k}' = k\boldsymbol{r}/r$ is the wavevector of the scattered particle (Fraunhofer diffraction); the scattered wavefunction given by equation (2.9) becomes

$$\psi_s(\boldsymbol{r}) \simeq -\frac{m}{2\pi\hbar^2\sqrt{V}} \frac{e^{ikr}}{r} \int d\boldsymbol{r}' U(\boldsymbol{r}') e^{i(\boldsymbol{k}-\boldsymbol{k}')\boldsymbol{r}'} \ ; \tag{2.13}$$

it is a spherical wave e^{ikr}/r distorted by the scattering amplitude

$$F(\boldsymbol{k} - \boldsymbol{k}') = -\frac{m}{2\pi\hbar^2} \int d\boldsymbol{r}' U(\boldsymbol{r}') e^{i(\boldsymbol{k}-\boldsymbol{k}')\boldsymbol{r}'} \ ; \tag{2.14}$$

it is a scattering length. This is Born's formula.[3] The scattering amplitude is proportional to the Fourier transform of the potential. The number of particles scattered in the solid angle at large distances per unit time is

$$v' \mid \psi_s \mid^2 r^2 do = \frac{\hbar k}{mV} \mid F \mid^2 do \ , \tag{2.15}$$

---

[2] L. Landau and E. Lifshitz, *Course of Theoretical Physics*, vol. 3, *Quantum Mechanics*, Butterworth-Heinemann, Oxford (1993).

[3] M. Born, "Zur Quantenmechanik der Stossvorgaenge", Z. Phys. **37** 863 (1926); "Quantenmechanik der Stossvorgaenge", Z. Phys. **38** 803 (1926).

where $\boldsymbol{v}' = \hbar\boldsymbol{k}'/m$, $v' = \hbar k/m$; the flux density of the incoming particle is $v \mid \psi_i \mid^2 = \frac{\hbar k}{mV}$ (current density, number of particles per unit area and unit time). The ratio of these two quantities gives the (differential) cross-section

$$d\sigma = \mid F \mid^2 do \qquad (2.16)$$

(it has the dimension of an area).

We note that the total number of scattered particles $\sim \mid \psi_s \mid^2 r^3 \sim \sigma/r^2$ in the volume V is much smaller than the total number of incident particles (1), due to the factor $1/r$ in equation (2.13).

For a screened Coulomb potential $U = \alpha e^{-\gamma r}/r$ we get $F = (2m\alpha/\hbar^2)/(q^2+\gamma^2)$, where $\boldsymbol{q} = \boldsymbol{k} - \boldsymbol{k}'$. The cross-section is much smaller than the characteristic area $1/\gamma^2$ ($m\alpha/\hbar^2\gamma \ll 1$), precisely due to the condition (2.10).

For the Coulomb potential ($\gamma \rightarrow 0$) we get the Rutherford cross-section[4]

$$d\sigma = \frac{\alpha^2}{16E^2 \sin^4 \frac{\theta}{2}} do \ , \qquad (2.17)$$

where $E = \hbar^2 k^2/2m$ is the energy of the projectile and θ is the scattering angle (the angle between $\boldsymbol{k}$ and $\boldsymbol{k}'$). The total cross-section is infinite, due to the singularity at $\theta = 0$. The Coulomb potential is an unrealistic case (a "pathological" case). The physical electric potentials are screened.

For an atom with the nucleus charge $-Ze$, where e is the electron charge, the potential is given by the Poisson equation

$$\Delta\varphi = 4\pi Ze\delta(\boldsymbol{r}) - 4\pi en(\boldsymbol{r}) \ , \qquad (2.18)$$

where $n(\boldsymbol{r})$ is the electron density. The Fourier transformation gives $\varphi(\boldsymbol{q}) = -4\pi e[Z - n(\boldsymbol{q})]/q^2$, where $n(\boldsymbol{q}) = \int d\boldsymbol{r} n(\boldsymbol{r}) e^{-i\boldsymbol{q}\boldsymbol{r}}$ is the atomic form-factor. For an incident charge Q the potential energy is $U(\boldsymbol{q}) = -4\pi eQ[Z - n(\boldsymbol{q})]/q^2$ and the cross-section is $d\sigma = (2meQ/\hbar^2 q^2)^2 [Z - n(\boldsymbol{q})]^2 do$.[5]

[4] E. Rutherford, "The scattering of α and β particles by matter and the structure of the atom", Phil. Mag. **21** 669 (1911).

[5] N. F. Mott, "The scattering of electrons by atoms", Proc. Roy. Soc. **A127** 658 (1930).

In order to get an estimate of the Coulomb cross-section we assume an electron-electron collision with a screened Coulomb potential. The cross-section is

$$\sigma = \pi \left(\frac{2me^2}{\hbar^2}\right)^2 \frac{1}{k^2\gamma^2}\frac{4k^2/\gamma^2}{4k^2/\gamma^2+1} \ . \tag{2.19}$$

The parameter γ is of the order $\gamma = 1/a_H$, where $a_H = \frac{\hbar^2}{me^2} = 0.53\text{Å}$ is the Bohr radius ($\hbar = 10^{-27}erg\cdot s$, $m = 10^{-27}g$, $e = -4.8\times10^{-10}esu$, $1\text{Å} = 10^{-8}cm$). For $ka_H \ll 1$ the cross-section is of the order a_H^2, for $ka_H \gg 1$ the cross-section is of the order $\sigma = 4\pi/k^2 \ll a_H^2$ (in both these limiting cases the Born formula is valid).

The atomic structure factor can be written as

$$n(\boldsymbol{q}) = \int d\boldsymbol{r}n(\boldsymbol{r})e^{-i\boldsymbol{q}\boldsymbol{r}} = \sum_i \int d\boldsymbol{r}\delta(\boldsymbol{r}-\boldsymbol{r}_i)e^{-i\boldsymbol{q}\boldsymbol{r}} = \sum_i e^{i\boldsymbol{q}\boldsymbol{r}_i} \ , \tag{2.20}$$

where the summation is peformed over all the positions $\boldsymbol{r}_i$ of the electrons in the atom (these positions are meaningful for heavy atoms, where the Thomas-Fermi theory is valid). At the point $\boldsymbol{r}$ the incident wave and the scattered wave interfere, with a phase factor $e^{-i\boldsymbol{k}'(\boldsymbol{r}-\boldsymbol{r}_i)}e^{i\boldsymbol{k}(\boldsymbol{r}-\boldsymbol{r}_i)} = e^{i\boldsymbol{q}\boldsymbol{r}}e^{-i\boldsymbol{q}\boldsymbol{r}_i}$. We can see that the structure factor controls the interference of the two waves. The extension of the heavy atoms is of the order a_H (with most of the electrons localized around the distance $a_H/Z^{1/3}$). For $qa_H/Z^{1/3} \ll 1$ the form-factor is $n(\boldsymbol{q}) = Z$; all the electrons contribute with the same (vanishing) phase to the interference; we say that we have a coherent scattering. For $qa_H/Z^{1/3} \gg 1$ the form-factor decreases considerably as $Z^{5/3}/(qa_H)^2$; we say that we have an incoherent scattering. The Thomas-Fermi density is $n(r) = (\kappa^2/4\pi)\frac{Z}{r}e^{-\kappa r}$, where $\kappa \simeq 0.85Z^{1/3}/a_H$ is the Thomas-Fermi screening wavevector.[6]

2.2 An exact solution

Let us assume a potential U given by a δ-function of the form

$$U(\boldsymbol{r}) = a^3U_0\delta(\boldsymbol{r}) \ , \tag{2.21}$$

[6]M. Apostol, *Structure of Matter*, Nova, NY (2019).

where a is its range. Equation (2.1) reads

$$\Delta\psi + k^2\psi = \frac{2ma^3U_0}{\hbar^2}\delta(\boldsymbol{r})\psi \ , \tag{2.22}$$

with the solution $\psi = \psi_i + \psi_s$, where the scattered wave is given by

$$\Delta\psi_s + k^2\psi_s = \frac{2ma^3U_0}{\hbar^2}\delta(\boldsymbol{r})\left[\psi_i(0) + \psi_s(0)\right] \ . \tag{2.23}$$

The solution is

$$\psi_s(\boldsymbol{r}) = -\frac{ma^3U_0}{2\pi\hbar^2}\left[\psi_i(0) + \psi_s(0)\right]\frac{e^{ikr}}{r} \ , \tag{2.24}$$

whence

$$\psi_s(0) = -\frac{ma^2U_0/2\pi\hbar^2}{1 + ma^2U_0/2\pi\hbar^2}\psi_i(0) \tag{2.25}$$

and

$$\psi_s(\boldsymbol{r}) = -a\frac{ma^2U_0/2\pi\hbar^2}{1 + ma^2U_0/2\pi\hbar^2}\psi_i(0)\frac{e^{ikr}}{r} \ . \tag{2.26}$$

The scattering amplitude is

$$F = -a\frac{ma^2U_0/2\pi\hbar^2}{1 + ma^2U_0/2\pi\hbar^2} \ . \tag{2.27}$$

We can see that for a δ-potential the scattering is isotropic (it does not depend on $\boldsymbol{q} = \boldsymbol{k}' - \boldsymbol{k}$). For $ma^2\left|U_0\right|/2\pi\hbar^2 \ll 1$ we get the Born amplitude $F = -ma^3U_0/2\pi\hbar^2$ (equation (2.14)). For $ma^2\left|U_0\right|/2\pi\hbar^2 \gg 1$ the scattering amplitude is $F = -a$. It is worth noting that there exists a singularity in the scattering amplitude for a resonant scattering by attractive potentials.

For resonant scattering it is important to keep the phase factor e^{ika} in equation (2.25), especially for small values of ka. The scattering amplitude becomes

$$F = \frac{1}{-\frac{1+u}{u}a^{-1} - ik} \ , \tag{2.28}$$

where $u = ma^2U_0/2\pi\hbar^2$. This is the Wigner formula. $au/(1+u)$ is the scattering length.[7]

[7]See M. Apostol, *Theory of Quanta*, Nova, NY (2019) and the References therein.

2.3 Momentum transfer

The scattered particle receives an additional momentum $\hbar(\boldsymbol{k}' - \boldsymbol{k})$. Since the total momentum is conserved the target receives a momentum $\hbar(\boldsymbol{k} - \boldsymbol{k}')$. This is the momentum transfer to the target.

Let us assume two colliding particles, denoted by 1 and 2, with masses $m_{1,2}$, interacting with a potential energy $U(1-2)$. Their wavefunction satisfies the Schroedinger equation

$$\left(-\frac{\hbar^2}{2m_1}\Delta_1 - \frac{\hbar^2}{2m_2}\Delta_2\right)\psi + U(1-2)\psi = E\psi \ . \tag{2.29}$$

The initial wavefunction is the product of two free plane waves $\psi_{i1,2} = \frac{1}{\sqrt{v}}e^{i\boldsymbol{k}_{1,2}\boldsymbol{r}_{1,2}}$,

$$\psi_i = \psi_{i1}\psi_{i2} = \frac{1}{V}e^{i\boldsymbol{k}_1\boldsymbol{r}_1}e^{i\boldsymbol{k}_2\boldsymbol{r}_2} \ , \tag{2.30}$$

which satisfies the Schroedinger equation

$$\begin{aligned}&\left(-\tfrac{\hbar^2}{2m_1}\Delta_1 - \tfrac{\hbar^2}{2m_2}\Delta_2\right)\psi_{i1}\psi_{i2} = \\ &= E\psi_{i1}\psi_{i2} = (E_1+E_2)\psi_{i1}\psi_{i2} \ ,\end{aligned} \tag{2.31}$$

where $E_{1,2} = \hbar^2 k_{1,2}^2/2m_{1,2}$ are the energies of the two particles. We assume that the collision is elastic, *i.e.* each particle preserves its energy. In the first order of the perturbation theory the scattered wave satisfies the Schroedinger equation

$$\left(\frac{\hbar^2}{2m_1}\Delta_1 + \frac{\hbar^2}{2m_2}\Delta_2\right)\psi_s + (E_1+E_2)\psi_s = U(1-2)\psi_{i1}\psi_{i2} \ . \tag{2.32}$$

We are interested in a solution for a large separation distance between the particles, such that the scattered wave is separated, *i.e.* $\psi_s = \psi_{s1}\psi_{s2}$, where $\psi_{s1,2} = \frac{1}{\sqrt{V}}e^{i\boldsymbol{k}'_{1,2}\boldsymbol{r}_{1,2}}$ are two free plane waves. This approximation brings small errors of higher orders in the interaction. Let us assume that we are interested in the scattering of the particle 1 off the particle 2. Equation (2.32) becomes

$$\frac{\hbar^2}{2m_1}\Delta_1\psi_{s1}\cdot\psi_{s2} + E_1\psi_{s1}\cdot\psi_{s2} = U(1-2)\psi_{i1}\psi_{i2} \ , \tag{2.33}$$

or

$$\Delta_1\psi_{s1}+k_1^2\psi_{s1}=\frac{2m_1}{\hbar^2}\int d\boldsymbol{r}_2\psi_{2s}^*U(1-2)\psi_{i2}\cdot\psi_{i1}\ . \tag{2.34}$$

The solution of this equation is

$$\psi_{s1}(\boldsymbol{r}_1)=-\frac{m_1}{2\pi\hbar^2}\frac{e^{ik_1r_1}}{r_1}\cdot$$
$$\cdot\int d\boldsymbol{r}_1^{'}d\boldsymbol{r}_2\psi_{2s}^*U(\boldsymbol{r}_1^{'}-\boldsymbol{r}_2)\psi_{i2}\cdot\psi_{i1}(\boldsymbol{r}_1^{'})e^{-i\boldsymbol{k}_1^{'}\boldsymbol{r}_1^{'}}\ , \tag{2.35}$$

or

$$\psi_{s1}(\boldsymbol{r}_1)=-\frac{m_1}{2\pi\hbar^2V\sqrt{V}}\frac{e^{ik_1r_1}}{r_1}\cdot$$
$$\cdot\int d\boldsymbol{r}_1^{'}d\boldsymbol{r}_2e^{-i(\boldsymbol{k}_2^{'}-\boldsymbol{k}_2)\boldsymbol{r}_2}U(\boldsymbol{r}_1^{'}-\boldsymbol{r}_2)e^{-i(\boldsymbol{k}_1^{'}-\boldsymbol{k}_1)\boldsymbol{r}_1^{'}}\ , \tag{2.36}$$

which can also be written as

$$\psi_{s1}(\boldsymbol{r}_1)=-\frac{m_1}{2\pi\hbar^2V\sqrt{V}}\frac{e^{ik_1r_1}}{r_1}\cdot$$
$$\cdot\int d\boldsymbol{r}_1^{'}d\boldsymbol{r}_2e^{-i[(\boldsymbol{k}_2^{'}-\boldsymbol{k}_2)+(\boldsymbol{k}_1^{'}-\boldsymbol{k}_1)]\boldsymbol{r}_2}U(\boldsymbol{r}_1^{'})e^{-i(\boldsymbol{k}_1^{'}-\boldsymbol{k}_1)\boldsymbol{r}_1^{'}}\ . \tag{2.37}$$

The scattering amplitude is

$$F=-\frac{m_1}{2\pi\hbar^2}\delta_{\boldsymbol{k}_2^{'}-\boldsymbol{k}_2,\boldsymbol{k}_1-\boldsymbol{k}_1^{'}}\int d\boldsymbol{r}U(\boldsymbol{r})e^{-i(\boldsymbol{k}_1^{'}-\boldsymbol{k}_1)\boldsymbol{r}^{'}}\ . \tag{2.38}$$

We can see that the total momentum is conserved, *i.e.* $\boldsymbol{k}_1^{'}+\boldsymbol{k}_2^{'}=\boldsymbol{k}_1+\boldsymbol{k}_2$. The momentum transferred to the particle 2 (the target) is $\boldsymbol{q}=\boldsymbol{k}_2^{'}-\boldsymbol{k}_2=\boldsymbol{k}_1-\boldsymbol{k}_1^{'}$. Leaving aside the factor of momentum conservation, equation (2.38) gives the scattering amplitude of the particle 1. The deviation angle θ' of the particle 2 with respect to the collision direction, *i.e.* the angle between $\boldsymbol{q}$ and $\boldsymbol{k}_1$ ($\theta'=(\boldsymbol{q},\boldsymbol{k}_1)$) is given by $qk_1\cos\theta'=2k_1^2\sin^2(\theta/2)$, where $\theta=(\boldsymbol{k}_1^{'},\boldsymbol{k}_1)$ is the deviation angle of the scattered particle. From $\boldsymbol{q}=\boldsymbol{k}_1-\boldsymbol{k}_1^{'}$ we get $q=2k_1\sin(\theta/2)$, such that $\cos\theta'=\sin(\theta/2)$ and $\theta'=\frac{1}{2}(\pi-\theta)$.

The momentum transfer to the target is $\hbar(\boldsymbol{k}-\boldsymbol{k}')$. The number of collisions per unit time in the solid angle do is $\frac{\hbar k}{mV}\mid F\mid^2 do$. Therefore, the force acting upon the target is

$$\boldsymbol{f}=\frac{\hbar^2k}{mV}\int(\boldsymbol{k}-\boldsymbol{k}')\mid F\mid^2do\ . \tag{2.39}$$

If the scattering amplitude depends only on the angle θ between $\boldsymbol{k}$ and $\boldsymbol{k}'$ (and k), then the force has only one component along the forward direction, say the z-direction, given by

$$f_z = \frac{\hbar^2 k^2}{mV} \int (1 - \cos\theta) \mid F \mid^2 do \ , \tag{2.40}$$

or

$$f_z = \frac{\hbar^2 k^2}{mV} \int (1 - \cos\theta) d\sigma \ . \tag{2.41}$$

The integral in this equation is called the transport cross-section. For a δ-potential $U = a^3 U_0 \delta(\boldsymbol{r})$ the force is

$$f_z = \frac{\hbar^2 k^2 a^2}{\pi mV} \left(\frac{ma^2 U_0}{\hbar^2} \right)^2 \ ; \tag{2.42}$$

f_z/a^2 is the pressure exerted by the projectile upon the target. The volume V in the denominator of this equation arises from the normalization of one particle in the volume V. If we have N_i incoming particles in the volume V the factor $n_i = N_i/V$ appears in equation (2.42) instead of $1/V$. The flux density of the incoming particles is $S = vn_i$, where v is the particle velocity. Usual values of the flux density are of the order $10^{12}/cm^2 \cdot s$. For the collision of an electron by an atom the force is $f_z \simeq 4\pi n_i (\alpha^2/E) \ln(2\hbar v/e^2)$, where $\alpha = Ze^2$, E is the energy of the electron and v is its velocity ($\hbar v \gg e^2$); the screening coefficient is $\gamma = 1/a_H$. This is related to the stopping power of charged particles in matter.[8]

2.4 Beam

Usually, the particle beams are prepared as a set of identical, independent particles, propagating along the beam direction, with the same velocity for monoenergetic beams. Each particle has an uncertainty along the transverse direction of the order of the atomic distance a, at most, since they are extracted from solid targets. The dispersion

[8] H. Bethe, "Zur Theorie des Durchgangs schneller Korpuskularstrahlen durch Materie", Ann. Phys. **5** 325 (1930).

angle of the beam is $\theta \simeq \lambda/a \ll 1$, at most, where λ is the wavelength along the beam direction. Being localized along the transverse direction the statistics of the beam particles is irrelevant. Indeed, the wavevector $\boldsymbol{k}$ in the wavefunction

$$e^{-i\frac{\hbar k^2}{2m}t+i\boldsymbol{k}\boldsymbol{r}} \tag{2.43}$$

has an uncertainty $\boldsymbol{q}$, such that $\boldsymbol{k} = \boldsymbol{k}_0 + \boldsymbol{q}$, where $\boldsymbol{k}_0$ is the central wavevector of the beam along the beam direction. The wavefunction reads

$$e^{-i\frac{\hbar k_0^2}{2m}t+i\boldsymbol{k}_0\boldsymbol{r}}e^{-i(\boldsymbol{v}_0 t-\boldsymbol{r})\boldsymbol{q}-i\frac{\hbar t}{2m}q^2} \quad , \tag{2.44}$$

where $\boldsymbol{v}_0 = \hbar\boldsymbol{k}_0/m$ is the group velocity. For short times we may leave aside the quadratic q^2-term and integrate with respect to all $\boldsymbol{q}$; the wavefunction is proportional to

$$\delta(\boldsymbol{r} - \boldsymbol{v}_0 t) \ ; \tag{2.45}$$

we can see that the particle is localized and propagates with the velocity $\boldsymbol{v}_0$. This is a wavepacket. The δ-function in equation (2.45) should be read as $1/a^3$ around the point $\boldsymbol{r} = \boldsymbol{v}_0 t$, where a is a distance of the order of the atomic distances. Actually, for longer times we cannot neglect the quadratic term, so the integration reads in fact

$$\int d\boldsymbol{q}_\perp e^{-i\frac{\hbar t}{2m}q_\perp^2} \cdot e^{i\frac{mR^2}{2\hbar t}} \cdot \int dq_\parallel e^{-i\frac{\hbar t}{2m}(q_\parallel - mR/\hbar t)^2} \quad , \tag{2.46}$$

where $\boldsymbol{R} = \boldsymbol{r} - \boldsymbol{v}_0 t$, $q_\parallel$ is the longitudinal component of $\boldsymbol{q}$ and $q_\perp$ is the transverse component of $\boldsymbol{q}$ with respect to $\boldsymbol{R}$. This wavefunction is proportional to

$$(m/\hbar t)^{3/2} e^{i\frac{mR^2}{2\hbar t}} \ ; \tag{2.47}$$

it goes to zero for long times and has a relevant oscillation for distances of the order $\Delta R = (\hbar t/m)^{1/2}$. We can see that the localization is lost, and the particle position exhibits a dispersion. The distance ΔR is very small for collision durations. The integration over the whole range of $\boldsymbol{q}$ in the above equations brings small errors, since the imaginary exponentials with quadratic exponents in $q_{\perp,\parallel}$ vary very rapidly.

The wavefunction of the beam is a multiparticle wavefunction $\psi = \varphi(\boldsymbol{r}_1)\varphi(\boldsymbol{r}_2)...$, with the same φ for each particle (monoenergetic beam) where $\boldsymbol{r}_i$, $i = 1, 2, ...N_i$ are the positions of the N_i particles in the beam. The potential energy is $\sum_i U(\boldsymbol{r}_i)$. The Schroedinger equation

$$\begin{gathered}\left(-\tfrac{\hbar^2}{2m}\Delta_1 - \tfrac{\hbar^2}{2m}\Delta_2 - ...\right)\varphi(\boldsymbol{r}_1)\varphi(\boldsymbol{r}_2)...+ \\ +\sum_i U(\boldsymbol{r}_i)\varphi(\boldsymbol{r}_1)\varphi(\boldsymbol{r}_2)... = N_i E\varphi(\boldsymbol{r}_1)\varphi(\boldsymbol{r}_2)...\end{gathered} \tag{2.48}$$

is reduced to a one-particle Schroedinger equation

$$-\frac{\hbar^2}{2m}\Delta\varphi(\boldsymbol{r}) + U(\boldsymbol{r})\varphi(\boldsymbol{r}) = E\varphi(\boldsymbol{r}) \ . \tag{2.49}$$

Alternatively, instead of multi-particle wavefunctions it is convenient to use the field

$$\psi(\boldsymbol{r}) = \sqrt{N_i}\varphi(\boldsymbol{r}) \tag{2.50}$$

normalized to the number of particles in the beam. The incident field is

$$\psi(\boldsymbol{r}) = \sqrt{\frac{N_i}{V}}e^{i\boldsymbol{kr}} \ . \tag{2.51}$$

The field satisfies the Schroedinger equation (2.49). The one-particle operators $\sum_i O(\boldsymbol{r}_i)$ are written as $\int d\boldsymbol{r}\psi^*(\boldsymbol{r})O(\boldsymbol{r})\psi(\boldsymbol{r})$. The scattered field acquires a factor $\sqrt{N_i/V}$, the scattered current acquires a factor $n_i = N_i/V$, which is the density of particles in the beam. The same factor appears in the incident current, such that the cross-section remains unchanged, as expected. The number of collisions per unit time acquires a factor n_i, which appears, for instance, in equation (2.39) (the momentum transfer and the force).

2.5 Quantum transitions

Apart from a momentum transfer there exists also an energy transfer to the target. In order to account for the energy transfer we need the time-dependent Schroedinger equation. Let us assume a target and an incident beam, described by a field ψ_i (or a one-particle wavefunction) which satisfies the Schroedinger equation

$$i\hbar\frac{\partial\psi_i}{\partial t} = (H_0 + H_1)\psi_i \ ; \tag{2.52}$$

this equation includes the degrees of freedom of both the projectile and the target, including the internal degrees of freedom. If the internal structure of these entities is not changed, the collision is elastic; if their internal structure is changed, the collision is inelastic. The hamiltonian H_0 is the free hamiltonian, while H_1 is the interaction hamiltonian; H_1 may depend on time. The label i stands for the initial state. We can check immediately that the solution of this equation is

$$\psi_i = \psi_i^0 - \frac{i}{\hbar}e^{-\frac{i}{\hbar}H_0 t}\int_0^t dt_1 e^{\frac{i}{\hbar}H_0 t_1}H_1\psi_i \ , \tag{2.53}$$

where ψ_i^0 is the initial field without interaction, satisfying the equation

$$i\hbar\frac{\partial\psi_i^0}{\partial t} = H_0\psi_i^0 \ ; \tag{2.54}$$

we write this field as

$$\psi_i^0 = e^{-\frac{i}{\hbar}E_i t}\varphi_i^0 \ , \tag{2.55}$$

where E_i is the initial energy of the assembly and φ_i^0 is the free initial field without time dependence. The lower limit $t = 0$ in equation (2.53) is misleading, because we still preserve the time dependence in ψ_i^0. However, for the upper limit going to infinity ($t \to \infty$) this is a correct assumption. The second term on the right in equation (2.53) is the scattered wave. We are interested in the content of a final free state ψ_f, with an energy E_f at $t \to \infty$,

$$\psi_f = e^{-\frac{i}{\hbar}E_f t}\varphi_f \ , \tag{2.56}$$

in the scattered state, *i.e.* we are interested in the matrix element

$$c_{fi} = -\frac{i}{\hbar}\left(\psi_f, e^{-\frac{i}{\hbar}H_0 t}\int_0^{t\to\infty} dt_1 e^{\frac{i}{\hbar}H_0 t_1}H_1\psi_i\right) \ . \tag{2.57}$$

This is the scattering matrix.[9] We write iteratively the first two contributions to $c_{fi} = c_{fi}^{(1)} + c_{fi}^{(2)}$:

$$c_{fi}^{(1)} = -\frac{i}{\hbar}\left(\psi_f, e^{-\frac{i}{\hbar}H_0 t}\int_0^{t\to\infty} dt_1 e^{\frac{i}{\hbar}H_0 t_1} H_1 \psi_i^0\right) \tag{2.58}$$

and

$$c_{fi}^{(2)} = \left(-\tfrac{i}{\hbar}\right)^2 \left(\psi_f, e^{-\frac{i}{\hbar}H_0 t}\int_0^{t\to\infty} dt_1 e^{\frac{i}{\hbar}H_0 t_1} H_1 e^{-\frac{i}{\hbar}H_0 t_1}\cdot \right. \\ \left. \cdot \int_0^{t_1} dt_2 e^{\frac{i}{\hbar}H_0 t_2} H_1 \psi_i^0\right) ; \tag{2.59}$$

by introducing the temporal factors we get

$$c_{fi}^{(1)} = -\frac{i}{\hbar}(H_1)_{fi}\int_0^{t\to\infty} dt_1 e^{\frac{i}{\hbar}(E_f-E_i)t_1} \tag{2.60}$$

and

$$c_{fi}^{(2)} = \left(-\tfrac{i}{\hbar}\right)^2 \sum\nolimits_n (H_1)_{fn}(H_1)_{ni}\cdot \\ \cdot \int_0^{t\to\infty} dt_1 e^{\frac{i}{\hbar}(E_f-E_n)t_1}\int_0^{t_1} dt_2 e^{\frac{i}{\hbar}(E_n-E_i)t_2} , \tag{2.61}$$

where n denotes the intermediate states. The energies include all the contributions of the assembly. If H_1 depends on the time through factors like $e^{\pm i\omega t}$, these factors appear in the time exponentials, which read

$$e^{\frac{i}{\hbar}(E_f-E_i\mp\hbar\omega)t_1} , \ e^{\frac{i}{\hbar}(E_f-E_n\mp\hbar\omega)t_1} , \ e^{\frac{i}{\hbar}(E_n-E_i\mp\hbar\omega)t_2} . \tag{2.62}$$

The squared coefficients $\mid c_{fi} \mid^2$ are the number of transitions from the initial state i to the final state f. The time factor in equation (2.60) is

$$\int_0^t dt_1 e^{\frac{i}{\hbar}(E_f-E_i)t_1} = \frac{e^{i\Delta\omega t}-1}{i\Delta\omega} , \tag{2.63}$$

where $\Delta\omega = (E_f - E_i)/\hbar$; its square is

$$\frac{\sin^2(\Delta\omega t/2)}{(\Delta\omega t/2)^2} \to_{t\to\infty} \pi t\delta(\Delta\omega/2) = 2\pi\hbar t\delta(E_f - E_i) , \tag{2.64}$$

[9] J. A. Wheeler, "On the mathematical description of light nuclei by the method of resonating group structure", Phys. Rev. **52** 1107 (1937); W. Heisenberg, "Die "beobachtbaren Grossen" in der Theorie der Elementarteilchen", Z. Phys. **120** 513 (1943).

which leads to a number of transitions per unit time

$$w_{fi}^{(1)} = \left| c_{fi}^{(1)} \right| /t = \frac{2\pi}{\hbar} \left|(H_1)_{fi}\right|^2 \delta(E_f - E_i) \ . \tag{2.65}$$

This is Fermi's golden rule.

The matrix element $(H_1)_{fi}$ is of the order $u\delta E/V$, where u is the volume over which the perturbation δE is active. For a discrete spectrum $\delta(E_f - E_i)$ is of the order of the inverse spacing $1/\Delta E$ of the energy levels. The total number of transitions given by equation (2.65) is of the order $\frac{u^2}{V^2}\frac{\delta E}{\hbar}\frac{\delta E}{\Delta E} \cdot \frac{l}{v}$, where $V = l^3$ and v is the velocity of the scattered particles. The ratio $\delta E/\Delta E$ is much smaller than unity, because the interaction is a small perturbation; the ratio $\delta E/\hbar$ is the inverse of the time uncertainty $1/\delta t$ produced by interaction, which is much smaller than the inverse time v/l, if the measurement of the collision is realized. We can see that the total number of scattered particles is much smaller than the total number of incident particles; mainly, this is due to the very small factor u/V. For a continuous spectrum we need to multiply by $Vk^2\Delta k$, such that, instead of $\frac{\delta E}{\Delta E}$, we get $\frac{V}{\lambda^3}(\delta E/E)$, where λ is the particle wavelength and E is the particle energy; we should have $\delta E/E \ll 1$, because the interaction is a small perturbation. The inequality loses a factor $1/V$, but still it is fulfilled. If the wavefunctions are normalized to the number of particles N, the above inequality reads $\frac{N^2u^2}{V^2}\frac{\delta E}{\hbar}\frac{\delta E}{\Delta E} \cdot \frac{l}{N^{1/3}v} \ll N$, which is again fulfilled, in the limit $N,\ V \to \infty$, $N/V = const.$

For the second contribution $c_{fi}^{(2)}$ the integration with respect to t_2 is over finite times, so the lower limit $t_2 = 0$ is improper. Actually, the interaction is vanishing at $t \to -\infty$, which requires a factor $e^{\alpha t}$, where $\alpha \to 0^+$. This is the adiabatic hypothesis. Equation (2.63) reads

$$\int_{-\infty}^{t} dt_1 e^{\frac{i}{\hbar}(E_f - E_i)t_1 + \alpha t_1} = \frac{e^{i\Delta\omega t + \alpha t}}{i\Delta\omega + \alpha} \ , \tag{2.66}$$

whose square is

$$\frac{e^{2\alpha t}}{\Delta\omega^2 + \alpha^2} \ . \tag{2.67}$$

In order to get the number of transitions per unit time we need to take the time derivative of this quantity, which gives

$$\frac{2\alpha e^{2\alpha t}}{\Delta\omega^2+\alpha^2} \rightarrow_{\alpha\rightarrow 0^+} 2\pi\delta(\Delta\omega) \ , \tag{2.68}$$

as in equation (2.64). By doing so, we get from equation (2.61)

$$c_{fi}^{(2)} = \frac{i}{\hbar}\sum_n \frac{(H_1)_{fn}(H_1)_{ni}}{E_n-E_i-i0^+}\cdot\frac{e^{i\omega_{fi}t+\alpha}}{i\omega_{fi}+\alpha} \ , \tag{2.69}$$

where $\omega_{fi}=(E_f-E_i)/\hbar$, which leads to

$$w_{fi}^{(2)} = \frac{2\pi}{\hbar}\left|\sum_n \frac{(H_1)_{fn}(H_1)_{ni}}{E_n-E_i-i0^+}\right|^2 \delta(E_f-E_i) \ ; \tag{2.70}$$

we can see that the adiabatic hypothesis gives the correct causality pole $E_n-E_i-i0^+$, where $-i0^+$ should be read $-i\alpha$, $\alpha\rightarrow 0^+$. Usually, the coefficients $c_{fi}^{(1,2)}$ describe distinct processes, so they do not interfere. The total number of transitions per unit time is obtained by multiplying these expressions by the number of final states $d\nu$,

$$dw_{fi} = w_{fi}d\nu \ . \tag{2.71}$$

Let us assume that the incoming particle has the momentum $\hbar\boldsymbol{k}$, the energy $E_i=\hbar^2k^2/2m$ and the wavefunction $\varphi_i^0=\frac{1}{\sqrt{V}}e^{i\boldsymbol{kr}}$; the outgoing particle has the momentum $\hbar\boldsymbol{k}'$, the energy $E_f=\hbar^2k'^2/2m$ and the wavefunction $\varphi_f=\frac{1}{V}e^{j\boldsymbol{k}'\boldsymbol{r}}$; the interaction hamiltonian is the potential energy $H_1=U(\boldsymbol{r})$. The number of final states in the solid angle do and the wavevector dk' is $d\nu=Vk'^2dk'do/(2\pi)^3$. From equation (2.65) we get the number of transitions per unit time

$$\begin{gathered} dw_{fi} = \tfrac{2\pi}{\hbar}\left|\tfrac{1}{V}\int d\boldsymbol{r}U(\boldsymbol{r})e^{i(\boldsymbol{k}-\boldsymbol{k}')\boldsymbol{r}}\right|^2 \cdot \\ \cdot\delta\left(\tfrac{\hbar^2k'^2}{2m}-\tfrac{\hbar^2k^2}{2m}\right)\tfrac{Vk'^2dk'do}{(2\pi)^3} \ , \end{gathered} \tag{2.72}$$

or

$$dw_{fi} = \frac{mk}{4\pi^2\hbar^3V}\left|\int d\boldsymbol{r}U(\boldsymbol{r})e^{i(\boldsymbol{k}-\boldsymbol{k}')\boldsymbol{r}}\right|^2 \delta\left(k'-k\right)dk'do \ ; \tag{2.73}$$

$\delta(k'-k)$ is $L/2\pi$, where L is the linear dimension of the region, and $dk' = 2\pi/L$; the same result is obtained by integrating over k'; we get

$$dw_{fi} = \frac{mk}{4\pi^2\hbar^3 V}\left|\int d\boldsymbol{r} U(\boldsymbol{r}) e^{i(\boldsymbol{k}-\boldsymbol{k}')\boldsymbol{r}}\right|^2 do \ , \tag{2.74}$$

which coincides with equation (2.15). Dividing by the incident current density $\hbar k/mV$, we get the Born cross-section

$$d\sigma = \frac{m^2}{4\pi^2\hbar^4}\left|\int d\boldsymbol{r} U(\boldsymbol{r}) e^{i(\boldsymbol{k}-\boldsymbol{k}')\boldsymbol{r}}\right|^2 do \ . \tag{2.75}$$

If we allow for a change in energy R of the target (recoil energy), then the final energy of the assembly is $E_f = R + \hbar^2 k'^2/2m$ and the initial energy is $E_i = \hbar^2 k^2/2m$; the δ-function in equation (2.72) gives the energy conservation $R = \hbar^2 k^2/2m - \hbar^2 k'^2/2m$. It is the mechanical work done by the force f (momentum transfer per unit time) along the finite short-range a of the potential; thereby, we can have an estimation of the variation of the momentum $\hbar\Delta k$. For instance, for a δ-potential we get $\Delta k = \frac{ka^3}{\pi V}(ma^2U_0/\hbar^2)^2$ from equation (2.42).

2.6 Energy transfer

Let us consider a collision between a projectile with mass m and momentum $\hbar\boldsymbol{k}$ and a target with mass M initially at rest. We assume the most common case $m < M$. The momentum and energy conservation reads

$$\boldsymbol{q} = \boldsymbol{k} - \boldsymbol{k}' \ , \ R = \frac{\hbar^2 q^2}{2M} = \frac{\hbar^2 k^2}{2m} - \frac{\hbar^2 k'^2}{2m} \ , \tag{2.76}$$

where $\hbar\boldsymbol{k}'$ is the final momentum of the projectile and $R > 0$ is the recoil energy of the target. The magnitude of the momentum $\hbar k'$ is given by the equation

$$(1+x)k'^2 - 2xk\cos\theta\cdot k' - (1-x)k^2 = 0 \ , \tag{2.77}$$

where $x = m/M$ (< 1) and θ is the angle between $\boldsymbol{k}$ and $\boldsymbol{k}'$. The solutions of this equation are

$$k' = \frac{k}{1+x}\left(x\cos\theta \pm \sqrt{1 - x^2\sin^2\theta}\right) \ ; \tag{2.78}$$

we note that $\sin\theta < M/m < 1$, *i.e.* the scattering is produced for all angles. These solutions should satisfy the inequalities

$$0 < k' < k \ . \tag{2.79}$$

It is easy to verify that only the solution

$$k' = \frac{k}{1+x}\left(x\cos\theta + \sqrt{1 - x^2\sin^2\theta}\right) \tag{2.80}$$

satisfies these inequalities. The recoil energy is

$$R = \frac{\hbar^2 k^2 x}{m(1+x)^2}\left(1 - \cos\theta\sqrt{1 - x^2\sin^2\theta} + x\sin^2\theta\right) \ , \tag{2.81}$$

or

$$R = 2E\frac{x}{(1+x)^2}\left(1 - \cos\theta\sqrt{1 - x^2\sin^2\theta} + x\sin^2\theta\right) \ , \tag{2.82}$$

where $E = \hbar^2 k^2/2m$ is the energy of the incident particle. (These equations are also valid for $x = 1$[10]). The number of collisions per unit time in the solid angle do is $\frac{\hbar k'}{mV} \mid F \mid^2 do$, where F is the scattering amplitude (equation (2.74)), so the rate of energy transfer to the target is

$$\frac{dE_t}{dt} = \int R\frac{\hbar k'}{mV} \mid F \mid^2 do \ . \tag{2.83}$$

For $x = m/M \ll 1$ we may use the approximations

$$\begin{aligned} k' &\simeq k[1 - x(1 - \cos\theta)] \ , \\ R &\simeq 2Ex(1 - \cos\theta) \ , \end{aligned} \tag{2.84}$$

which lead to

$$\frac{dE_t}{dt} = \frac{2E\hbar kx}{mV}\int (1 - \cos\theta) \mid F \mid^2 do \ ; \tag{2.85}$$

---

[10]For identical particles the wavefunction should be properly symmetrized, N. F. Mott and H. S. W Massey, *Theory of Atomic Collisions*, Oxford University Press, Oxford (1965).

we can see that this equation implies the transport cross-section and the force f_z acting upon the target (equation (2.41)), such that we can write

$$\frac{dE_t}{dt} = \frac{2Ex}{\hbar k} f_z \; ; \tag{2.86}$$

indeed, $E/\hbar k = v$ is the velocity of the incident particle and $2Ex/\hbar k$ is the velocity of the target (for backward scattering). We may estimate the duration of the collision as $\tau = a/v$, where a is the short range of the potential, such that we get the energy transfer in a collision act $\Delta E_t = 2axf_z$.

2.7 Phase factors

The derivation of the scattered wave given in the Born scattering section has certain insufficiencies. The solution of the scattering problem needs to be resumed from the beginning.

The problem is to solve the equation

$$\Delta\psi + k^2\psi = \frac{2m}{\hbar^2} U(\boldsymbol{r})\psi \tag{2.87}$$

for the wave ψ with the initial condition $\psi = \psi_i$, when $U = 0$ and the initial (incident) wavefunction ψ_i is a free wave (the boundary conditions are vanishing of the solution and its derivatives at infinity). Obviously, the solution ψ is defined up to a phase factor $e^{i\chi}$. It is convenient to write $\psi = \psi_i + \psi_s$ in equation (2.87), where $\psi_s \to 0$ for $t \to 0$ (the initial time) and to assign the phase factor to ψ_i, such that equation (2.87) becomes

$$\Delta\psi_s + k^2\psi_s = \frac{2m}{\hbar^2} e^{i\chi} U(\boldsymbol{r})\psi_i + ... \tag{2.88}$$

In fact, the potential U vanishes both at the initial and the final time. If we associate these moments with spatial plus and minus infinity, the equation to be solved (equation (2.87)) is

$$\Delta\psi + k^2\psi = 0 \;\;, \tag{2.89}$$

with boundary conditions produced by U in the region of action of the potential. This equation is solved by using the Green function (equations (2.5) and (2.6)). Also, the solution of this equation is defined up to a phase factor $e^{i\chi}$. We can write the solution of equation (2.87) as

$$\psi(\boldsymbol{r}) = \frac{2m}{\hbar^2}\int d\boldsymbol{r}' U(\boldsymbol{r}')G(\boldsymbol{r}-\boldsymbol{r}')\psi(\boldsymbol{r}') \ , \tag{2.90}$$

where G is the Green function. We can see that ψ is defined up to a phase factor $e^{i\chi}$. If we write $\psi = \psi_i + \psi_s + ...$ in this equation, we need to retain the phase factor in ψ_i. Therefore, the asymptotic scattered wave is

$$\psi_s(\boldsymbol{r}) = e^{i\chi}F(\boldsymbol{k}-\boldsymbol{k}')\frac{e^{ikr}}{\sqrt{V}r} \ , \tag{2.91}$$

where F is the form-factor given by equation (2.14); the scattered wave should carry a label $\boldsymbol{k}'$.

Writing explicitly the phase factor in the solution, we get

$$\psi_s(\boldsymbol{r}) = \frac{2m}{\hbar^2}\int d\boldsymbol{r}' e^{i\chi} U(\boldsymbol{r}')G(\boldsymbol{r}-\boldsymbol{r}')\psi_i(\boldsymbol{r}') + ... \ ; \tag{2.92}$$

it is as if any point in the potential generates a phase factor (which in this case is constant). If $U(\boldsymbol{r})$ is generated by several sources, *i.e.* $U = \sum_{i=1}^{N} U_i$, the solution of equation (2.88) is a superposition of terms corresponding to each potential U_i, such that the corresponding factors will depend on i. The asymptotic scattered wave is

$$\psi_s(\boldsymbol{r}) = \sum_i e^{i\chi_i}F_i(\boldsymbol{k}-\boldsymbol{k}')\frac{e^{ikr}}{\sqrt{V}r} \ , \tag{2.93}$$

where

$$F_i(\boldsymbol{k}-\boldsymbol{k}') = -\frac{m}{2\pi\hbar^2}\int d\boldsymbol{r}' U_i(\boldsymbol{r}')e^{i(\boldsymbol{k}-\boldsymbol{k}')\boldsymbol{r}'} \ . \tag{2.94}$$

For $U_i(\boldsymbol{r}) = U_i\delta(\boldsymbol{r}-\boldsymbol{r}_i)$ the form-factors become

$$F_i(\boldsymbol{k}-\boldsymbol{k}') = -\frac{m}{2\pi\hbar^2}U_i e^{i(\boldsymbol{k}-\boldsymbol{k}')\boldsymbol{r}_i} \tag{2.95}$$

and the scattered wave is

$$\psi_s(\boldsymbol{r}) = -\frac{m}{2\pi\hbar^2}\sum_i e^{i\chi_i}U_i e^{i(\boldsymbol{k}-\boldsymbol{k}')\boldsymbol{r}_i}\frac{e^{ikr}}{\sqrt{V}r} \ ; \tag{2.96}$$

finally, for $U_i = U$ and $\boldsymbol{q} = \boldsymbol{k} - \boldsymbol{k}'$ we get

$$\psi_s(\boldsymbol{r}) = -\frac{mU}{2\pi\hbar^2}\sum_i e^{i\chi_i} e^{i\boldsymbol{q}\boldsymbol{r}_i}\frac{e^{ikr}}{\sqrt{V}r} , \tag{2.97}$$

where we can define a form-factor

$$F(\boldsymbol{q}) = -\frac{mU}{2\pi\hbar^2}\sum_i e^{i\chi_i} e^{i\boldsymbol{q}\boldsymbol{r}_i} . \tag{2.98}$$

We can see that every scattering centre brings its own phase factor to the amplitude of the cross-section. If these phase factors are the same, we have a coherent scattering, with the cross-section $\sim N^2$ $(\left|\sum_i e^{i\boldsymbol{q}\boldsymbol{r}_i}\right|^2)$; if they are randomly distributed, we have an incoherent cross-section $\sim N$.

According to equation (2.97), the scattered wave is a free wave $e^{ikr}/\sqrt{V}r$, affected by the phase jumps $e^{i\boldsymbol{q}\boldsymbol{r}_i}$ from the incident $\boldsymbol{k}$-wave to the scattered $\boldsymbol{k}'$-wave, suffered at the source position $\boldsymbol{r}_i$, and an additional phase factor $e^{i\chi_i}$, arising from distinct scattered waves produced at every source position. At first sight, the phase factor is a quantum-mechanical property; for $\chi_i = 0$ the scattering is a classical scattering. However, the way described above of their occurrence indicates rather an arbitrary origin for them. We show in the next chapters that they have in fact a statistical origin.

3 Electromagnetic Radiation

3.1 Dipole field

The electromagnetic wave equations are

$$\frac{1}{c^2}\frac{\partial^2\Phi}{\partial t^2}-\Delta\Phi=4\pi\rho \ ,$$
$$\frac{1}{c^2}\frac{\partial^2\boldsymbol{A}}{\partial t^2}-\Delta\boldsymbol{A}=\frac{4\pi}{c}\boldsymbol{j} \ , \tag{3.1}$$

where Φ is the scalar potential, $\boldsymbol{A}$ is the vector potential, ρ is the charge density, $\boldsymbol{j}$ is the current density and c denotes the speed of light in vacuum; the potentials satisfy the Lorenz gauge $\frac{1}{c}\frac{\partial\Phi}{\partial t}+div\boldsymbol{A}=0$, the charge density and the current density satisfy the continuity equation $\frac{\partial\rho}{\partial t}+div\boldsymbol{j}=0$. The retarded solutions of these wave equations are given by the Kirchhoff formulae

$$\Phi(t,\boldsymbol{r})=\int d\boldsymbol{r}'\frac{\rho(\boldsymbol{r}-\boldsymbol{r}',t-|\boldsymbol{r}-\boldsymbol{r}'|/c)}{|\boldsymbol{r}-\boldsymbol{r}'|} \ ,$$
$$\boldsymbol{A}(t,\boldsymbol{r})=\frac{1}{c}\int d\boldsymbol{r}'\frac{\boldsymbol{j}(\boldsymbol{r}-\boldsymbol{r}',t-|\boldsymbol{r}-\boldsymbol{r}'|/c)}{|\boldsymbol{r}-\boldsymbol{r}'|} \ . \tag{3.2}$$

Let us assume that the charge and current distribution is confined to a limited region with dimension l. At large distances r from the distribution we may omit $\boldsymbol{r}'$ in the Kirchhoff formulae in comparison with $\boldsymbol{r}$, such that we get

$$\Phi=\frac{q}{r}+... \ , \ \boldsymbol{A}=\frac{\dot{\boldsymbol{d}}}{cr}\mid_{t-r/c}+... \ , \tag{3.3}$$

where $\boldsymbol{d}$ is the dipole moment; for a charge placed at $\boldsymbol{r}_0$ the dipole moment is $\boldsymbol{d}=q\boldsymbol{r}_0$. Since $div\boldsymbol{A}=-\ddot{\boldsymbol{d}}\boldsymbol{r}/c^2r^2+...=-\frac{1}{c}\frac{\partial}{\partial t}\frac{\dot{\boldsymbol{d}}\boldsymbol{r}}{cr^2}+...$, the scalar potential has an additional contribution:

$$\Phi=\frac{q}{r}+\frac{\dot{\boldsymbol{d}}\boldsymbol{r}}{cr^2}\mid_{t-r/c}+...=\frac{q}{r}-div\frac{\boldsymbol{d}}{r}\mid_{t-r/c}+... \ , \tag{3.4}$$

in order to satisfy the Lorenz gauge. We note that in the limit of large distances r the derivatives arise only from the retarded time $t - r/c$. Also, we note that the static Coulomb potential is present in this limit. This is the dipolar approximation. Higher-order terms in powers of $\boldsymbol{r}'$ in the Kirchhoff formulae give quadrupole terms and magnetic dipole terms.

Within the dipole approximation the fields are given by

$$\begin{gathered}\boldsymbol{E} = -\tfrac{\ddot{\boldsymbol{d}}}{c^2 r} + grad\, div \tfrac{\boldsymbol{d}}{r} - grad \tfrac{q}{r} = \\ = -\Delta \tfrac{\boldsymbol{d}}{r} + grad\, div \tfrac{\boldsymbol{d}}{r} - grad \tfrac{q}{r} = curl\, curl \tfrac{\boldsymbol{d}}{r} - grad \tfrac{q}{r}\end{gathered} \tag{3.5}$$

and

$$\boldsymbol{H} = curl \frac{\dot{\boldsymbol{d}}}{cr} \ , \tag{3.6}$$

at the retarded time $t - r/c$. At large distances the last term in equation (3.5) may be neglected.

A finite motion confined to a region of dimension l changes over a characteristic time T, such that this motion generates waves with period of the order T and wavelengths of the order $\lambda = cT = \frac{c}{v} l$, at least, where v is the charge velocity. Since $v < c$, it follows $\lambda > l$. Therefore, wavelengths shorter than the dimension of the charge and current distribution are meaningless. The fields vary slowly over the region of the charge and current distribution. Indeed, the main contribution of the Fourier components $e^{i\boldsymbol{kr}}$ comes from $kr < 1$, *i.e.* $l < \lambda$. If we preserve in the above formulae the leading contributions for $l \ll \lambda \simeq r$, we are in the near-field region. It is worth noting that the condition $l \ll \lambda$ amounts to $v \ll c$. In fact, for a finite motion we have always $v \ll c$. If we limit ourselves to the leading contributions for $l, \lambda \ll r$, we are in the far-field region (wave zone). This is the radiation field of the dipole (dipole field). It is given by

$$\begin{gathered}\boldsymbol{E} = \tfrac{\boldsymbol{r} \times (\boldsymbol{r} \times \ddot{\boldsymbol{d}})}{c^2 r^3} = -\tfrac{\boldsymbol{r}}{r} \times \boldsymbol{H} \ , \\ \boldsymbol{H} = -\tfrac{\boldsymbol{r} \times \ddot{\boldsymbol{d}}}{c^2 r^2} \ .\end{gathered} \tag{3.7}$$

We can see that the radiation field is transverse. The Poynting vector is

$$\boldsymbol{S} = \frac{c}{4\pi} \boldsymbol{E} \times \boldsymbol{H} = \frac{c\boldsymbol{r}}{4\pi r} H^2 = \frac{\ddot{d}^2}{4\pi c^3 r^2} \sin^2\theta \cdot \frac{\boldsymbol{r}}{r} \ , \tag{3.8}$$

where θ is the angle between $\ddot{\boldsymbol{d}}$ and $\boldsymbol{r}$; $\boldsymbol{S}ds = cr^2H^2do/4\pi$ is the radiation intensity through the area $ds = r^2do$, corresponding to the solid angle do. The total intensity is

$$I = \frac{2\ddot{d}^2}{3c^3} \; ; \tag{3.9}$$

this is the total energy radiated by the dipole per unit time.

3.2 Damping force

The momentum carried per unit time and per unit area is $c\boldsymbol{g}$, where $\boldsymbol{g} = \boldsymbol{S}/c^2$ is the momentum density of the field; therefore, the total radiated momentum per unit time is $\oint ds c\boldsymbol{g}$, where the integration is performed over a surface placed at large distances. In this case equation (3.8) is valid, and we can see that the total radiated momentum is zero, because $\boldsymbol{S} \sim \boldsymbol{r}$. It follows that the total momentum of the field does not change ($\Delta\boldsymbol{G} = 0$) and the total force is reduced to $\int d\boldsymbol{r}\boldsymbol{f} = \int d\boldsymbol{r}\boldsymbol{t}$, where $t_i = \partial_j\sigma_{ij}$ are the components of the stress forces.[1] It is easy to see that the i-th component of the total stress force

$$\int d\boldsymbol{r}t_i = \tfrac{1}{4\pi}\oint \left[E_i\boldsymbol{E}d\boldsymbol{s} + H_i\boldsymbol{H}d\boldsymbol{s} - \tfrac{1}{2}(E^2+H^2)ds_i\right] \tag{3.10}$$

is zero for the radiation field given by equations (3.7). Therefore, the Lorentz force acting between the dipole and its own field, *i.e.* the self-force, is zero.

However, the energy radiated per unit time by a charge q can be written as

$$\dot{\mathcal{E}} = I = \frac{2q^2\dot{v}^2}{3c^3} = \frac{2q^2}{3c^3}\frac{d}{dt}(\boldsymbol{v}\dot{\boldsymbol{v}}) - \frac{2q^2}{3c^3}\boldsymbol{v}\ddot{\boldsymbol{v}} \; , \tag{3.11}$$

and the first term on the right in this equation, averaged over a finite motion, is zero. It follows that a force

$$\boldsymbol{f} = \frac{2q^2\ddot{\boldsymbol{v}}}{3c^3} \tag{3.12}$$

[1] M. Apostol, *Essays in Electromagnetism and Matter*, Lambert, Saarbruecken (2013).

is exerted by the field upon the charge. This is the Lorentz damping force, or the Lorentz frictional force. It is a reaction self-force. Besides the energy loss given by equation (3.11), the dipole suffers also a loss of angular momentum, produced by the damping force.

For a finite motion with a characteristic frequency ω_0 (*e.g.*, a harmonic oscillator, or a rotation) $\ddot{\boldsymbol{v}} = -\omega_0^2 \boldsymbol{v}$ and the damping force is $\boldsymbol{f} = -\frac{2q^2}{3c^3}\omega_0^2 \boldsymbol{v}$. The equation of motion $m\dot{\boldsymbol{v}} = \boldsymbol{f}$ for a charge with mass m leads to a very slow damping $v \sim e^{-\frac{2a\omega_0^2}{3c}t}$, where $a = q^2/mc^2$ is the classical radius of the charge. For an electron $a = 2.8 \times 10^{-13} cm$ ($q = e = -4.8 \times 10^{-10} esu$, $m = 0.9 \times 10^{-27} g$, $c = 3 \times 10^{10} cm/s$; typical frequencies $\omega_0 = 10^{15} s^{-1}$). This damping decreases the frequency of the emitted radiation, both for positive and negative frequencies, which amounts to both higher and lower frequencies for a positive central frequency; this is the natural line breadth of the emitted radiation.

For higher velocities (in an infinite motion) the damping force (and the equation of motion) should acquire a relativistic form. The relativistic motion of the charge is damped.[2]

3.3 Line breadth

A pointlike charge q with mass m, oscillating with frequency ω_0 along the z-axis, obeys the equation of motion

$$\ddot{z} = -\omega_0^2 z - \gamma \dot{z} \ , \tag{3.13}$$

where

$$\gamma = \frac{2q^2\omega_0^2}{3c^3 m} = \frac{2\omega_0^2}{3c} a \ll \omega_0 \tag{3.14}$$

is the damping coefficient ($a = q^2/mc^2$ is the classical radius). The solution of this equation is $z \sim e^{-\gamma t/2} e^{-i\omega_0 t}$; γ^{-1} is the (very long) lifetime of the damped oscillator. The field radiated by the dipole is also damped, E, $H \sim \ddot{z} \sim e^{-\gamma t/2} e^{-i\omega_0 t}$, and has a (long) lifetime. Its

[2] L. Landau and E. Lifshitz, *Course of Theoretical Physyics*, vol. 2, *The Classical Theory of Fields*, Butterworth-Heinemann, Oxford (1987).

Fourier transform goes like

$$E(\omega) \sim \int_0 dte^{i\omega t}e^{-i\omega_0 t}e^{-\gamma t/2} = \frac{1}{i(\omega_0 - \omega) + \gamma/2} \; ; \qquad (3.15)$$

the total energy of the emitted radiation

$$U \sim \int E^2(t)dt = \frac{1}{2\pi}\int d\omega \left|E(\omega)\right|^2 \qquad (3.16)$$

gives a spectral intensity (energy)

$$I(\omega) \sim \left|E(\omega)\right|^2 = I_0 \frac{1}{2\pi}\frac{\gamma}{(\omega_0 - \omega)^2 + \gamma^2/4} \; , \qquad (3.17)$$

where I_0 is the total intensity ($\omega > 0$). The ω-function in equation (3.17) is the Lorentz function, or dispersion function. At its half height $\omega = \omega_0 \pm \gamma/2$, so γ is also the breadth $\Delta\omega$ of the dispersion function (the reciprocal of the lifetime). It corresponds to a change

$$\Delta\lambda = 2\pi\frac{c\gamma}{\omega_0^2} = \frac{4\pi}{3}a \simeq 10^{-12}cm = 10^{-4}\text{Å} \qquad (3.18)$$

in the wavelength (for electrons). This is the natural breadth of the spectroscopic line.[3]

3.4 Scattering of radiation

Let a monochromatic wave with the electric field $E_0 \cos\omega t$ fall upon a point charge q with mass m at rest. The equation of motion is

$$m\ddot{z} = qE_0 \cos\omega t \; , \qquad (3.19)$$

whose solution is $z = -(qE_0/m\omega^2)\cos\omega t$. This dipole $d = qz$ generates a secondary radiation with the same frequency: the radiation is scattered. The energy per unit area scattered in the solid angle do

[3] W. Heitler, *The Quantum Theory of Radiation*, Oxford University Press, London (1954).

per unit time is $d\dot{\mathcal{E}} = Ido$, where I is the intensity per unit area; it is given by

$$I = |\mathbf{S}| = \frac{\ddot{d}^2}{4\pi c^3 r^2}\sin^2\theta = \frac{q^4}{4\pi m^2 c^3 r^2}E_0^2\cos^2\omega t\cdot\sin^2\theta \ , \tag{3.20}$$

where $\mathbf{S}$ is the Poynting vector and θ is the angle between $\mathbf{E}_0$ and $\boldsymbol{r}$ (equation (3.8)). The intensity per unit area of the incident radiation is $I_0 = (cE_0^2/4\pi)\cos^2\omega t$, so we can write

$$I = \frac{q^4}{m^2c^4}\sin^2\theta\left(\frac{I_0}{r^2}\right) \ . \tag{3.21}$$

We note that the scattered radiation energy is much smaller than the total incident energy. The quantity

$$\frac{q^4}{m^2c^4}\sin^2\theta = a^2\sin^2\theta \ , \tag{3.22}$$

where $a = q^2/mc^2$ is the classical radius of the charge, is the scattering cross-section (actually $d\sigma/do = a^2\sin^2\theta$ is the differential cross-section). We can see that it is of the order of the square of the classical electromagnetic radius of the charge. If the original wave is unpolarized, *e.g.* $\mathbf{E}_0 \sim (\cos\varphi', \sin\varphi', 0)$ (and propagates along the z-direction), then we average over the polarizations and get

$$\overline{\sin^2\theta} = \frac{1}{2}(1+\cos^2\Theta) \ , \tag{3.23}$$

where Θ is the angle of scattering (the angle between $\mathbf{r}$ - the propagation direction of the scattered radiation - and the propagation direction z of the incident radiation). The total scattering cross-section is

$$\sigma = \frac{8\pi}{3}a^2 \ ; \tag{3.24}$$

for electrons $\sigma = 6.65\times 10^{-25}cm^2$; this is the Thomson cross-section.

Let us consider the scattering of radiation by a linear harmonic oscillator. With usual notations, the equation of motion is

$$\ddot{\mathbf{r}} + \omega_0^2\mathbf{r} + \gamma\dot{\mathbf{r}} = \frac{q}{m}\mathbf{E}_0 e^{-i\omega t} \ , \tag{3.25}$$

where the damping is included (equation (3.14)). The solution of this equation is

$$\mathbf{r} = \frac{q}{m}\mathbf{E}_0 \frac{1}{\omega_0^2 - \omega^2 - i\omega\gamma} e^{-i\omega t} \tag{3.26}$$

and the intensity (per unit area) is given by the real part of the solution,

$$I = \frac{q^4 E_0^2}{4\pi m^2 c^3 r^2} \sin^2\theta \frac{\omega^4}{(\omega_0^2 - \omega^2)^2 + \omega^2\gamma^2} \cos^2(\omega t - \delta) \ , \tag{3.27}$$

where

$$\tan\delta = \frac{\gamma\omega}{\omega_0^2 - \omega^2} \ . \tag{3.28}$$

This is the Rayleigh scattering. We note that the intensity of the scattered radiation depends on the incident frequency, the maximum scattering occurring for the resonance $\omega = \omega_0$. The air molecules have a resonance frequency in the blue region, hence the blue colour of the sky. Also, we note that there appears a dephasing δ in the scattered radiation (which is maximal at resonance). Introducing the average intensity of the primary radiation $cE_0^2/8\pi$ and the classical radius a we get the differential cross-section

$$\frac{d\sigma}{do} = a^2 \sin^2\theta \frac{\omega^4}{(\omega_0^2 - \omega^2)^2 + \omega^2\gamma^2} \tag{3.29}$$

and the total cross-section

$$\sigma = \frac{8\pi}{3} a^2 \frac{\omega^4}{(\omega_0^2 - \omega^2)^2 + \omega^2\gamma^2} \ . \tag{3.30}$$

For ω_0, $\gamma = 0$ equation (3.22) for free charges is recovered. Close to the resonance, this formula becomes

$$\sigma = \frac{2\pi}{3} a^2 \frac{\omega_0^2}{(\omega_0 - \omega)^2 + \gamma^2/4} \ . \tag{3.31}$$

This is the resonance fluorescence: if the frequency of the incident radiation is close to the dipole frequency, the cross-section and the scattered radiation are large.

3.5 Absorption of radiation

With the notations used above, a harmonic oscillator subjected to a monochromatic wave,

$$\ddot{\mathbf{r}} + \omega_0^2 \mathbf{r} = \frac{q}{m} \mathbf{E}_0 \cos(\omega t + \delta_\omega) \ , \tag{3.32}$$

exchanges energy with the wave. The general solution of equation (3.32) is a particular solution (frequency ω) plus the solution of the homogeneous equation (frequency ω_0). Let us assume that the interaction begins at $t = 0$. Then, the solution is

$$\begin{gathered} \mathbf{r} = \tfrac{q}{m}\mathbf{E}_0 \tfrac{1}{\omega_0^2-\omega^2}\left[\cos(\omega t + \delta_\omega) - \cos(\omega_0 t + \delta_\omega)\right] + \\ + \mathbf{A}\cos(\omega_0 t + \theta) \ , \end{gathered} \tag{3.33}$$

where $\mathbf{A}$ and θ are the amplitude and the phase of the free oscillator. The interaction energy per unit time (mechanical work) is given by

$$e_m = q\dot{\mathbf{r}}\mathbf{E}_0 \cos(\omega t + \delta_\omega) \ , \tag{3.34}$$

the total interaction energy in the time τ is

$$\begin{gathered} \int_0^\tau dt e_m = \tfrac{q^2}{m} E_0^2 \tfrac{\omega_0}{\omega_0^2-\omega^2} \int_0^\tau dt \sin(\omega_0 t + \delta_\omega)\cos(\omega t + \delta_\omega) - \\ - q\mathbf{A}\mathbf{E}_0 \omega_0 \int_0^\tau dt \sin(\omega_0 t + \theta)\cos(\omega t + \delta_\omega) \end{gathered} \tag{3.35}$$

(the term with $\cos(\omega t + \delta_\omega)$ in equation (3.33) disappears by integration). This integral may have either sign, depending on the phases δ_ω and θ (the oscillator may give energy to the wave, which is an induced emission of radiation). We assume that the phases δ_ω are randomly distributed (as for incoherent radiation), and take the average: the second term on the right of equation (3.35) disappears. We get

$$\int_0^\tau dt \overline{e}_m = \frac{q^2}{2m} E_0^2 \frac{\omega_0}{\omega_0^2 - \omega^2} \frac{1 - \cos(\omega_0 - \omega)\tau}{\omega_0 - \omega} \ , \tag{3.36}$$

which is positive: this is the absorbed energy in the time τ. The intensity (per unit area) of the incident wave is

$$I_0(\omega) = \frac{c}{4\pi} E_0^2 \overline{\cos^2(\omega t + \delta_\omega)} = \frac{c}{8\pi} E_0^2 \ , \tag{3.37}$$

so we may write for the absorbed energy

$$\Delta E(\omega) = I_0(\omega)\frac{4\pi q^2}{mc}\frac{\omega_0}{\omega_0^2-\omega^2}\frac{1-\cos(\omega_0-\omega)\tau}{\omega_0-\omega} \ . \tag{3.38}$$

For large τ the function in equation (3.38) approaches $\tau\delta(\omega_0-\omega)$. The energy absorbed per unit time is

$$\Delta E(\omega)/\tau = I_0(\omega_0)\frac{2\pi^2 q^2}{mc}\delta(\omega-\omega_0) \ . \tag{3.39}$$

This is a resonance phenomenon; we may replace the function δ by a dispersion function,

$$\Delta E(\omega)/\tau = I_0(\omega)\frac{2\pi q^2}{mc}\frac{\gamma}{(\omega-\omega_0)^2+\gamma^2} \ , \tag{3.40}$$

where γ is the damping coefficient. The total energy absorbed per unit time is $\Delta E_{ab} \simeq I_0 ac/\gamma$, while the total energy emitted per unit time is $\Delta\dot{E}_{em} \sim I_0 a^2$ (equation (3.31)). We can see that $\Delta\dot{E}_{em} \ll \Delta E_{ab}$, since $a \ll c/\gamma$; the difference is spent for the energy dissipated by the charge.

3.6 Maxwell equations

E and **H** denote the electric field and the magnetic field in vacuum, ρ and **j** are the densities of electric charge and electric current, all real functions of the time t and the position **r**. The Maxwell equations are

$$\begin{aligned} div\mathbf{E} &= 4\pi\rho \ , \ div\mathbf{H} = 0 \ , \\ curl\mathbf{E} &= -\tfrac{1}{c}\tfrac{\partial\mathbf{H}}{\partial t} \ , \ curl\mathbf{H} = \tfrac{1}{c}\tfrac{\partial\mathbf{E}}{\partial t} + \tfrac{4\pi}{c}\mathbf{j} \ , \end{aligned} \tag{3.41}$$

where $c \simeq 3\times10^{10} cm/s$ is the speed of light in vacuum. The c in these equations is a universal constant. The first two equations are Gauss's laws, the third equation is Faraday's law, and the fourth equation is Maxwell-Ampere's equation. The charges and the currents satisfy the continuity equation

$$\frac{\partial\rho}{\partial t} + div\mathbf{j} = 0 \ . \tag{3.42}$$

By taking the *div* of the *curl* in equations (3.41) and using the continuity equation, the Gauss laws follow.

The Maxwell equations and the Theory of Electricity and Magnetism (Electromagnetism) were established in the 18th and 19th centuries and the beginning of the 20th century, by Coulomb, Gauss, Oersted, Ampere, Ohm, Fresnel, Faraday, Maxwell, Heaviside, Kirchhoff, Hertz, Lorentz, Einstein, etc.[4]

By using $curl\,curl = -\Delta + grad\,div$ in equations above, we get two equivalent wave equations with sources

$$\begin{aligned} \frac{1}{c^2}\frac{\partial^2 \boldsymbol{E}}{\partial t^2} - \Delta \boldsymbol{E} &= -4\pi grad\rho - \frac{4\pi}{c^2}\frac{\partial \boldsymbol{j}}{\partial t} \ , \\ \frac{1}{c^2}\frac{\partial^2 \boldsymbol{H}}{\partial t^2} - \Delta \boldsymbol{H} &= \frac{4\pi}{c} curl \boldsymbol{j} \ , \end{aligned} \tag{3.43}$$

which show that $\boldsymbol{E}$ and $\boldsymbol{H}$ satisfy two decoupled equations (with sources related by the continuity equation). According to these equations, the fields are electromagnetic waves propagating with velocity c. Of course, there exist also static solutions. The solution of equations (3.43) is provided by a particular field produced by sources plus a superposition of free waves, *i.e.* solutions of the homogeneous equation. Besides the equations, the solution is determined by boundary conditions and initial conditions.

The fields can be obtained from the vector potential $\mathbf{A}$ and the scalar potential Φ,

$$\mathbf{E} = -\frac{1}{c}\frac{\partial \mathbf{A}}{\partial t} - grad\Phi \ , \ \mathbf{H} = curl\mathbf{A} \ , \tag{3.44}$$

defined up to the gauge transformation

$$\mathbf{A} = \mathbf{A}' + grad\chi \ , \ \Phi = \Phi' - \frac{1}{c}\frac{\partial \chi}{\partial t} \ , \tag{3.45}$$

which leaves the fields unchanged; χ is an arbitrary function.

[4] J. C. Maxwell, *Lehrbuch der Elektrizitaet und der Magnetismus*, Berlin (1983); H. A. Lorentz, *The Theory of Electrons*, Teubner (Leipzig, 1916); see also L. Landau and E. Lifshitz, *Course of Theoretical Physics*, vol. 2 (*Classical Theory of Fields*), vol. 8 (*Electrodynamics of Continuous Media*), Elsevier, Oxford (2004).

The representation given by equation (3.44) satisfies the homogeneous Maxwell equations (the Faraday equation and the Gauss equation for magnetic field). By using the same representation in the Gauss equation for electric field and the Maxwell-Ampere equation we get[5]

$$\begin{aligned} &\tfrac{1}{c^2}\tfrac{\partial^2\Phi}{\partial t^2} - \Delta\Phi = 4\pi\rho + \tfrac{1}{c}\tfrac{\partial}{\partial t}\left(\tfrac{1}{c}\tfrac{\partial\Phi}{\partial t} + div\boldsymbol{A}\right) \ , \\ &\tfrac{1}{c^2}\tfrac{\partial^2\boldsymbol{A}}{\partial t^2} - \Delta\boldsymbol{A} = \tfrac{4\pi}{c}\boldsymbol{j} - grad\left(\tfrac{1}{c}\tfrac{\partial\Phi}{\partial t} + div\boldsymbol{A}\right) \ , \end{aligned} \tag{3.46}$$

which, by imposing the Lorenz ("Lorentz") gauge

$$\frac{1}{c}\frac{\partial\Phi}{\partial t} + div\boldsymbol{A} = 0 \ , \tag{3.47}$$

become wave equations with sources

$$\frac{1}{c^2}\frac{\partial^2\mathbf{A}}{\partial t^2} - \Delta\mathbf{A} = \frac{4\pi}{c}\mathbf{j} \ , \ \frac{1}{c^2}\frac{\partial^2\Phi}{\partial t^2} - \Delta\Phi = 4\pi\rho \ . \tag{3.48}$$

The gauge condition is necessary in order to preserve three independent unknowns (the components of a field) out of the four unknowns introduced by the potentials (three components of $\boldsymbol{A}$ and the scalar potential Φ). Subjected to the Lorenz gauge, the potentials are still defined up to a χ-gauge which satisfies the free wave equation $\frac{1}{c^2}\frac{\partial^2\chi}{\partial t^2} - \Delta\chi = 0$. The two wave equations (3.48) are not independent: they are coupled by the Lorenz gauge (and the continuity equation).

The charges are pointlike particles with mass. A pointlike charge q placed at $\boldsymbol{r}_0(t)$ has a charge density $\rho = q\delta(\boldsymbol{r} - \boldsymbol{r}_0)$; since

$$\dot{\rho} = -q(\boldsymbol{v}grad)\delta(\boldsymbol{r} - \boldsymbol{r}_0) = -qdiv[\boldsymbol{v}\delta(\boldsymbol{r} - \boldsymbol{r}_0)] \ , \tag{3.49}$$

where $\boldsymbol{v} = \dot{\boldsymbol{r}}_0$ is the charge velocity, from the continuity equation we get the current density $\boldsymbol{j} = q\boldsymbol{v}\delta(\boldsymbol{r} - \boldsymbol{r}_0)$.

[5]Various vectorial identities can be obtained by using $(\mathbf{a}\times\mathbf{b})_i = \varepsilon_{ijk}a_jb_k$ and $(curl\mathbf{a})_i = \varepsilon_{ijk}\partial_j a_k$, where ε_{ijk} is the totally antisymmetric tensor of rank 3 (Levi-Civita symbol, $\varepsilon_{ijk}\varepsilon_{lmk} = \delta_{il}\delta_{jm} - \delta_{im}\delta_{jl}$); $\delta_{ij} = 1$ for $i = j$, $\delta_{ij} = 0$ for $i \neq j$ is the Kronecker symbol.

3.7 Energy, force and momentum

From the Faraday equation and the Maxwell-Ampere equation (equations (3.41)) multiplied by $\mathbf{H}$ and $\boldsymbol{E}$, respectively, we get

$$\frac{1}{8\pi}\frac{\partial}{\partial t}(E^2+H^2)+\frac{c}{4\pi}div(\mathbf{E}\times\mathbf{H})+\mathbf{jE}=0\ . \tag{3.50}$$

This equation suggests that $u=(E^2+H^2)/8\pi$ is a density of electromagnetic energy,

$$\mathbf{S}=\frac{c}{4\pi}\mathbf{E}\times\mathbf{H} \tag{3.51}$$

is an energy flux density (*i.e.* energy per unit time per unit cross-sectional area) and $\mathbf{jE}$ is the rate of change in time of the mechanical work done by the field on the charges (or by charges on the field). When the field energy u decreases, part of it is lost through the surrounding surface and part of it is transferred by mechanical work to charges. When the field energy u increases, part of it comes through the surrounding surface and part of it comes by mechanical work done by charges upon the field. $\mathbf{S}$ is the Poynting vector. For a charge density ρ the current density is $\mathbf{j}=\rho\mathbf{v}$, where $\mathbf{v}$ is the velocity of the charge; therefore, $\rho\mathbf{E}$ is a force density $\mathbf{f}$ (of electromagnetic origin), and $\mathbf{jE}=\rho\boldsymbol{v}\boldsymbol{E}=\boldsymbol{v}\mathbf{f}$ is indeed the rate of change in time of the density of a mechanical (kinetic) energy E_m under the action of that force (if w is the mechanical work, then $dw/dt=\mathbf{f}\boldsymbol{v}$). Denoting

$$U=\frac{1}{8\pi}\int d\mathbf{r}(E^2+H^2) \tag{3.52}$$

we get

$$\frac{\partial}{\partial t}(U+E_m)+\int d\mathbf{s}\mathbf{S}=0\ , \tag{3.53}$$

where the integration is performed over the surface $\mathbf{s}$ enclosing the volume where the fields and charges are placed. Equations (3.50) and (3.53) tell that the volume electromagnetic energy (U) of the electromagnetic fields, the electromagnetic part of the mechanical energy (E_m) of the charges (*i.e.* that part which is changed by the action of electromagnetic forces) and the flux of the electromagnetic energy through the surface enclosing the volume (*i.e.* the surface integral in equation (3.53) multiplied by time) are all balanced against each other. These equations represent the energy conservation. For fields vanishing at infinity ($\mathbf{S}=0$) the energy $U+E_m$ is conserved (as for any closed motion). The density of the electromagnetic energy u is the pressure of the electromagnetic field. It is worth noting that these equations may include both an external field (which satisfies the homogeneous Maxwell equations) and the field generated by the charges and the currents.

According to Maxwell equations, the charges and the currents generate an electromagnetic field; we can see that they do that by mechanical work. Also, we see that the electromagnetic field acts by a force upon charges and currents. Of course, the self-force, *i.e.* the force acting upon a charge by that charge itself should be avoided (similarly, the force acting upon a field on behalf of that very field should be avoided). We may view the charges and the currents, on one side, and the field, on the other side, as distinct entities, interacting by the energy density $\int dt\boldsymbol{j}\boldsymbol{E}$. Making use of $\boldsymbol{E}=-\frac{1}{c}\frac{\partial\boldsymbol{A}}{\partial t}-grad\Phi$, we get the interaction energy

$$V=\int d\boldsymbol{r}\left(\rho\Phi-\frac{1}{c}\boldsymbol{j}\boldsymbol{A}\right) . \tag{3.54}$$

Indeed, $\int dt\boldsymbol{j}\left(-\frac{1}{c}\frac{\partial\boldsymbol{A}}{\partial t}\right)=-\frac{1}{c}\boldsymbol{j}\boldsymbol{A}$ (for fixed $\boldsymbol{j}$); as regards $-\int dt\boldsymbol{j}grad\Phi$, we note that $\boldsymbol{j}=\rho\boldsymbol{v}=-\rho\boldsymbol{v}'$ for a pointlike charge density $\rho=q\delta(\boldsymbol{r}-\boldsymbol{r}_0)$, where $\boldsymbol{v}=d\boldsymbol{r}_0/dt$ and $\boldsymbol{v}'=d\boldsymbol{r}/dt$, such that

$$-\int dt\,(\boldsymbol{j}grad\Phi)=\int dt\rho\frac{d\boldsymbol{r}}{dt}grad\Phi=\int dt\rho\frac{d\Phi}{dt}=\rho\Phi \tag{3.55}$$

It is worth noting that this interaction is not applicable to a self-interaction: it requires the charges and the currents, on one side, and the field, on the other side, to be viewed as distinct entities. On the other hand, the self-interaction is unavoidable, since the field moves everywhere in space. It requires the pointlike charge to move indefinitely within a small region of a finite radius equal to its classical radius q^2/mc^2, where m is the mass of the charge (for a quantum charge the Compton wavelength $\hbar/mc$, where $\hbar$ is Planck's constant). Within this region there exists no field.[6]

[6] M. Apostol, *Matter and Radiation*, Nova, NY (2022); "Relativistic uncertainties", J. Theor. Phys. **331** (2021).

Let us form the combination $\rho\mathbf{E}+\mathbf{j}\times\mathbf{H}/c$ from the Maxwell equations (3.41). We get

$$\begin{gathered}4\pi\rho\mathbf{E}+\tfrac{4\pi}{c}\mathbf{j}\times\mathbf{H}=\\=\mathbf{E}div\mathbf{E}+\left(curl\mathbf{H}-\tfrac{1}{c}\tfrac{\partial\mathbf{E}}{\partial t}\right)\times\mathbf{H}=\\=\mathbf{E}div\mathbf{E}+\mathbf{H}div\mathbf{H}-(\mathbf{E}\times curl\mathbf{E}+\mathbf{H}\times curl\mathbf{H})-\\-\tfrac{1}{c}\tfrac{\partial}{\partial t}(\mathbf{E}\times\mathbf{H})\ ;\end{gathered}\tag{3.56}$$

here we write for components $(\mathbf{E}div\mathbf{E})_i=\partial_j(E_iE_j)-(\mathbf{E}grad)E_i$ and use $grad(E^2)=2(\mathbf{E}grad)\mathbf{E}+\mathbf{2E}\times curl\mathbf{E}$ (and similarly for the magnetic field). Equations (3.56) can be cast in the form

$$\mathbf{f}=\mathbf{t}-\frac{\partial\mathbf{g}}{\partial t}\ ,\tag{3.57}$$

where

$$\mathbf{f}=\rho\mathbf{E}+\frac{1}{c}\mathbf{j}\times\mathbf{H}\tag{3.58}$$

is the density of the Lorentz force (acting upon charges and currents on behalf of the electromagnetic field), $t_i=\partial_j\sigma_{ij}$,

$$\sigma_{ij}=\frac{1}{4\pi}\left[E_iE_j+H_iH_j-\frac{1}{2}\delta_{ij}(E^2+H^2)\right]\ ,\tag{3.59}$$

are the stress forces (σ_{ij} is the electromagnetic stress tensor) and

$$\mathbf{g}=\frac{1}{4\pi c}\mathbf{E}\times\mathbf{H}=\frac{1}{c^2}\mathbf{S}\tag{3.60}$$

is the density of the electromagnetic momentum. We can see that the electromagnetic field acts upon charges and currents with the Lorentz force which contains $\rho\mathbf{E}$; the Lorentz term $\mathbf{j}\times\mathbf{H}/c$ does not change the motion energy and does not do mechanical work (since $\mathbf{v}(\rho\mathbf{v}\times\mathbf{H})/c=0$).

Equation (3.57) is the momentum conservation. The Lorentz force $\boldsymbol{f}$ acting upon charges and currents arises from the field stress force $\boldsymbol{t}$ minus the time variation of the field momentum $\boldsymbol{g}$. Also, the Lorentz force $-\boldsymbol{f}$ force acting upon the field is the time variation of the field momentum $\boldsymbol{g}$ minus the field stress force $\boldsymbol{t}$. We note that the stress forces are surface forces, *i.e.*

$$\int d\mathbf{r} f_i = \int ds_j \sigma_{ij} - \frac{\partial}{\partial t} \int d\mathbf{r} g_i \ . \tag{3.61}$$

The tensor σ_{ij} is the flux of forces (*i.e.* forces divided by the cross-sectional area), or the flow of momentum, *i.e.* the rate of change in time of the field momentum per unit cross-sectional area. There is a state of stress in the field, as in an elastic body, suggesting an aether. In contrast to elastic bodies, the aether has only transverse waves in the far-field region. For a pure radiation field, *i.e.* a free field, the rate of change in time of the electromagnetic momentum is compensated by the stress forces. For fields vanishing at infinity (a pulse for instance), the stress forces are zero, and the electromagnetic momentum is conserved. Similarly, the electromagnetic energy is then conserved. The existence of energy and momentum for radiation field indicates that the radiation field may be viewed as an independent physical entity. Let us assume that charges and currents act by a force $-\boldsymbol{f}$ and generate an electromagnetic field, *i.e.* they give to the field a change of momentum $\partial \boldsymbol{g}/\partial t$ and an additional stress force $-\boldsymbol{t}$; according to the law of action and reaction, and the momentum conservation, the newly created field acts upon the charges which generated it by an equal and opposite force $\boldsymbol{f}$. We can see that the field generated by charges and currents reacts back upon these charges and currents with a reaction force.

It is tempting, by the equations above, to give charges and currents a full electromagnetic nature.[7] Indeed, from equations (3.50) and (3.57) we may write

$$\begin{aligned} e_m &= -u - \int dt \cdot div\mathbf{S} \ , \\ \mathbf{p} &= \int \mathbf{f} dt = \int dt \cdot \mathbf{t} - \mathbf{g} \ , \end{aligned} \tag{3.62}$$

where e_m is the mechanical energy density of charges and currents and $\mathbf{p}$ denotes the density of their momentum. Then, it may appear that the energy and the momentum of particles (charges) are

[7] M. Abraham, *Theorie der Elektrizitaet*, Leipzig (1923) (ed. by R. Becker, Leipzig (1933)).

completely of electromagnetic nature, arising from their own electromagnetic field. It is worth noting that the integration over time makes these expressions relativistically non-invariant, even by an additional space integration.[8]

We multiply equation (3.57) by $\mathbf{r}$:

$$\mathbf{rf} = \partial_j(x_i\sigma_{ij}) - \mathbf{u} - \frac{\partial}{\partial t}(\mathbf{rg}) \tag{3.63}$$

($\sigma_{ii} = -u$); this is the electromagnetic virial theorem.[9] Let us multiply equation (3.57) for the k-th component by $\varepsilon_{ijk}x_j$:

$$\begin{aligned} \varepsilon_{ijk}x_j f_k &= \varepsilon_{ijk}x_j\partial_l(\sigma_{kl}) - \varepsilon_{ijk}x_j\frac{\partial}{\partial t}g_k = \\ &= \partial_l(\varepsilon_{ijk}x_j\sigma_{kl}) - \varepsilon_{ijk}x_j\frac{\partial}{\partial t}g_k \quad , \end{aligned} \tag{3.64}$$

because σ_{kl} is a symmetric tensor. We recognize in equation (3.64) the torque of the force on the left, the rate of change in time of the angular momentum and the flux of the torque (flow of the angular momentum) on the right. This is the law of conservation of the angular momentum for the electromagnetic field.

3.8 Longitudinal field

Let us represent the scalar potential in the volume V as the Fourier expansion

$$\Phi(t, \boldsymbol{r}) = \frac{1}{\sqrt{V}}\sum_{\boldsymbol{k}}\Phi_{\boldsymbol{k}}(t)e^{i\boldsymbol{kr}} \quad , \tag{3.65}$$

where $\Phi^*_{-\boldsymbol{k}} = \Phi_{\boldsymbol{k}}$. Similarly, the vector potential is represented as

$$\boldsymbol{A}(t, \boldsymbol{r}) = \frac{1}{\sqrt{V}}\sum_{\boldsymbol{k}}\boldsymbol{A}_{\boldsymbol{k}}(t)e^{i\boldsymbol{kr}} + \frac{1}{\sqrt{V}}\sum_{\boldsymbol{k}}\frac{i\boldsymbol{k}}{ck^2}\dot{\Phi}_{\boldsymbol{k}}(t)e^{i\boldsymbol{kr}} \quad , \tag{3.66}$$

[8] W. Heitler, *The Quantum Theory of Radiation*, Oxford University Press, London (1954).

[9] For a mechanical particle $\mathbf{rf} = d(\mathbf{rp})/dt - \mathbf{vp} = d(\mathbf{rp})/dt - 2T$, where T is the kinetic energy; for $\mathbf{f} = const \cdot \mathbf{r}/r^3$, as for Coulomb force, $\mathbf{rf} = V$, where V is the potential energy. We get $d(\mathbf{rp})/dt = 2T + V$, whose average over a finite (bound) motion leads to $2T + V = 0$.

where $\boldsymbol{k}\boldsymbol{A}_{\boldsymbol{k}} = 0$. The first term in equation (3.66) is the transverse part of the vector potential, while the second term is its longitudinal part. The Lorenz gauge is satisfied by this representation. The electric field is given by

$$\boldsymbol{E} = \frac{1}{\sqrt{V}} \sum_{\boldsymbol{k}} \left(-\frac{1}{c} \dot{\boldsymbol{A}}_{\boldsymbol{k}} \right) e^{i\boldsymbol{k}\boldsymbol{r}} +$$

$$+\frac{1}{\sqrt{V}} \sum_{\boldsymbol{k}} \frac{i\boldsymbol{k}}{k^2} \left(-\frac{1}{c^2} \ddot{\Phi}_{\boldsymbol{k}} - k^2 \Phi_{\boldsymbol{k}} \right) e^{i\boldsymbol{k}\boldsymbol{r}} = \tag{3.67}$$

$$= \frac{1}{\sqrt{V}} \sum_{\boldsymbol{k}} \left(-\frac{1}{c} \dot{\boldsymbol{A}}_{\boldsymbol{k}} \right) e^{i\boldsymbol{k}\boldsymbol{r}} - \frac{1}{\sqrt{V}} \sum_{\boldsymbol{k}} \frac{4\pi i\boldsymbol{k}}{k^2} \rho_{\boldsymbol{k}} e^{i\boldsymbol{k}\boldsymbol{r}} \ ,$$

where $\rho_{\boldsymbol{k}}$ is the Fourier component of the charge density; it has both a transverse part and a longitudinal part. The magnetic field is

$$\boldsymbol{H} = \frac{1}{\sqrt{V}} \sum_{\boldsymbol{k}} i\boldsymbol{k} \times \boldsymbol{A}_{\boldsymbol{k}} e^{i\boldsymbol{k}\boldsymbol{r}} \ ; \tag{3.68}$$

it is a transverse field. The energy of the electromagnetic field can now be written as

$$U = \frac{1}{8\pi} \int d\boldsymbol{r} \left(E^2 + H^2 \right) =$$

$$= \frac{1}{8\pi} \sum_{\boldsymbol{k}} \left(\frac{1}{c^2} \dot{\boldsymbol{A}}_{\boldsymbol{k}} \dot{\boldsymbol{A}}_{-\boldsymbol{k}} + k^2 \boldsymbol{A}_{\boldsymbol{k}} \boldsymbol{A}_{-\boldsymbol{k}} \right) + \tag{3.69}$$

$$+ \sum_{\boldsymbol{k}} \frac{2\pi}{k^2} \rho_{\boldsymbol{k}} \rho_{-\boldsymbol{k}} \ .$$

We recognize in the last row of equation (3.69) the Coulomb energy (with $4\pi/\sqrt{V}k^2$ the Fourier transform of the Coulomb potential). We have replaced $\frac{1}{c^2}\ddot{\Phi}_{\boldsymbol{k}} + k^2\Phi_{\boldsymbol{k}}$ in equation (3.67) by $4\pi\rho_{\boldsymbol{k}}$, as if the potential Φ is generated by a charge distribution. Consequently, the Coulomb energy in equation (3.69) arises from a self-interaction.

Let us now compute the interaction energy $V = \int dt d\boldsymbol{r} \boldsymbol{j}\boldsymbol{E}$, where $\boldsymbol{j}$ is the current density, by using the electric field given by equation (3.67). We get

$$V = \sum_{\boldsymbol{k}} \int dt \left(-\frac{1}{c} \boldsymbol{j}_{\boldsymbol{k}} \dot{\boldsymbol{A}}_{-\boldsymbol{k}} \right) + \sum_{\boldsymbol{k}} \int dt \frac{4\pi i \boldsymbol{k} \boldsymbol{j}_{\boldsymbol{k}}}{k^2} \rho_{-\boldsymbol{k}} \ . \tag{3.70}$$

Since $\sum_{\boldsymbol{k}}\left(-\frac{1}{c}\boldsymbol{j}_{\boldsymbol{k}}\dot{\boldsymbol{A}}_{-\boldsymbol{k}}\right)\Delta t=\sum_{\boldsymbol{k}}\left(-\frac{1}{c}\boldsymbol{j}_{\boldsymbol{k}}\Delta\boldsymbol{A}_{-\boldsymbol{k}}\right)$ is a mechanical work done in time Δt, it follows that $\sum_{\boldsymbol{k}}\left(-\frac{1}{c}\boldsymbol{j}_{\boldsymbol{k}}\boldsymbol{A}_{-\boldsymbol{k}}\right)$ is an interaction energy; therefore,

$$\begin{aligned}V&=\sum_{\boldsymbol{k}}\left(-\tfrac{1}{c}\boldsymbol{j}_{\boldsymbol{k}}\boldsymbol{A}_{-\boldsymbol{k}}\right)+\sum_{\boldsymbol{k}}\int dt\tfrac{4\pi i\boldsymbol{k}\boldsymbol{j}_{\boldsymbol{k}}}{k^2}\rho_{-\boldsymbol{k}}=\\&=\sum_{\boldsymbol{k}}\left(-\tfrac{1}{c}\boldsymbol{j}_{\boldsymbol{k}}\boldsymbol{A}_{-\boldsymbol{k}}\right)-\sum_{\boldsymbol{k}}\int dt\tfrac{4\pi}{k^2}\dot{\rho}_{\boldsymbol{k}}\rho_{-\boldsymbol{k}}\ .\end{aligned}\tag{3.71}$$

In the second summation in equation (3.71) we change $\boldsymbol{k}$ into $-\boldsymbol{k}$ and write $\frac{1}{2}\left(\dot{\rho}_{\boldsymbol{k}}\rho_{-\boldsymbol{k}}+\dot{\rho}_{-\boldsymbol{k}}\rho_{\boldsymbol{k}}\right)=\frac{1}{2}\frac{d}{dt}\left(\dot{\rho}_{\boldsymbol{k}}\rho_{-\boldsymbol{k}}\right)$. Finally, we get the interaction

$$V=\sum_{\boldsymbol{k}}\left(-\frac{1}{c}\boldsymbol{j}_{\boldsymbol{k}}\boldsymbol{A}_{-\boldsymbol{k}}\right)-\sum_{\boldsymbol{k}}\frac{2\pi}{k^2}\rho_{\boldsymbol{k}}\rho_{-\boldsymbol{k}}\ .\tag{3.72}$$

The Coulomb interaction in equation (3.72) is a self-interaction. We can see that this interaction cancels out the Coulomb energy in equation (3.69). The longitudinal degrees of freedom do not participate in the dynamics of the charges, the currents and the electromagnetic field. This amounts to view the field as a free radiation field, with a vanishing scalar potential, interacting with currents; although the longitudinal part of the field remains in the electric field (equation (3.67)). We are left with the transverse field energy

$$U=\frac{1}{8\pi}\sum_{\boldsymbol{k}}\left(\frac{1}{c^2}\dot{\boldsymbol{A}}_{\boldsymbol{k}}\dot{\boldsymbol{A}}_{-\boldsymbol{k}}+k^2\boldsymbol{A}_{\boldsymbol{k}}\boldsymbol{A}_{-\boldsymbol{k}}\right)\tag{3.73}$$

and the transverse part of the interaction

$$V=\sum_{\boldsymbol{k}}\left(-\frac{1}{c}\boldsymbol{j}_{\boldsymbol{k}}\boldsymbol{A}_{-\boldsymbol{k}}\right)\ .\tag{3.74}$$

It is worth noting that this interaction is not a self-interaction. We can see that the field is a set of harmonic oscillators with eigenfrequencies $\pm\omega_k$, $\omega_k=ck$, with momenta $\boldsymbol{P}_{\boldsymbol{k}}=\frac{1}{4\pi c^2}\dot{\boldsymbol{A}}_{\boldsymbol{k}}$ and coordinates $\boldsymbol{A}_{\boldsymbol{k}}$, whose equation of motion is the Maxwell-Ampere equation

$$\frac{1}{c^2}\ddot{\boldsymbol{A}}_{\boldsymbol{k}}+k^2\boldsymbol{A}_{\boldsymbol{k}}=\frac{4\pi}{c}\boldsymbol{j}_{\boldsymbol{k}}\ .\tag{3.75}$$

Noteworthy, $\boldsymbol{j_k}$ in this equation is the transverse part of the current. Since $\boldsymbol{A}^*_{-\boldsymbol{k}} = \boldsymbol{A_k}$, we may write $\boldsymbol{A_k} = \alpha_k \boldsymbol{e_k} a_{\boldsymbol{k}}(t) + \alpha_k \boldsymbol{e}_{-\boldsymbol{k}} a^*_{-\boldsymbol{k}}(t)$, where α_k and $a_{\boldsymbol{k}}$ are undetermined coefficients and $\boldsymbol{e_k}$ is the polarization vector ($\boldsymbol{e_k e_k} = 1$, $\boldsymbol{e_k k} = 0$). For free fields $a_{\boldsymbol{k}}(t) = a_{\boldsymbol{k}} e^{-i\omega_k t}$, $\omega_k = ck$, because $\boldsymbol{A_k}$ should satisfy the free wave equation $\frac{1}{c^2}\ddot{\boldsymbol{A}}_{\boldsymbol{k}} + k^2 \boldsymbol{A_k} = 0$. It follows that we can write the vector potential of the transverse (radiation) field as

$$\boldsymbol{A} = \frac{1}{\sqrt{V}} \sum_{\boldsymbol{k}} \alpha_k \boldsymbol{e_k} \left(a_{\boldsymbol{k}} e^{-i\omega_k t + i\boldsymbol{kr} + i\gamma_{\boldsymbol{k}}} + a^*_{\boldsymbol{k}} e^{i\omega_k t - i\boldsymbol{kr} - i\gamma_{\boldsymbol{k}}} \right) \ , \tag{3.76}$$

where $\gamma_{\boldsymbol{k}}$ are undetermined phases. The free field energy becomes

$$U = \sum_{\boldsymbol{k}} \frac{\omega_k^2 \alpha_k^2}{4\pi c^2} \left(a^*_{\boldsymbol{k}} a_{\boldsymbol{k}} + a_{\boldsymbol{k}} a^*_{\boldsymbol{k}} \right) \ . \tag{3.77}$$

Similarly, the total momentum of the free field (equation (3.60)) is

$$\begin{aligned} \boldsymbol{G} &= \tfrac{1}{4\pi c} \int d\boldsymbol{r} \boldsymbol{E} \times \boldsymbol{H} = \tfrac{1}{c^2} \int d\boldsymbol{r} \boldsymbol{S} = \\ &= \sum_{\boldsymbol{k}} \tfrac{\omega_k \alpha_k^2}{4\pi c^2} \boldsymbol{k} \left(a^*_{\boldsymbol{k}} a_{\boldsymbol{k}} + a_{\boldsymbol{k}} a^*_{\boldsymbol{k}} \right) \end{aligned} \tag{3.78}$$

and the total angular momentum is vanishing. An additional two-valued polarization label is tacitly assumed in the summations over $\boldsymbol{k}$. Also, we note that the volume integral of the *div* of the Fourier transform of any vector is vanishing, due to the cyclic boundary conditions (*e.g.*, $div\boldsymbol{S}$).

3.9 Photons

We choose the coefficients

$$\alpha_k = c\sqrt{\frac{2\pi\hbar}{\omega_k}} \ , \tag{3.79}$$

in equation (3.76), where $\hbar$ is Planck's constant; the transverse vector potential reads

$$\begin{aligned} \boldsymbol{A} &= \tfrac{1}{\sqrt{V}} \sum_{\boldsymbol{k}} \alpha_k \left(\boldsymbol{e_k} a_{\boldsymbol{k}} + \boldsymbol{e}_{-\boldsymbol{k}} a^*_{-\boldsymbol{k}} \right) e^{i\boldsymbol{kr}} = \\ &= \tfrac{1}{\sqrt{V}} \sum_{\boldsymbol{k}} \alpha_k \boldsymbol{e_k} \left(a_{\boldsymbol{k}} e^{i\boldsymbol{kr}} + a^*_{\boldsymbol{k}} e^{-i\boldsymbol{kr}} \right) \end{aligned} \tag{3.80}$$

(leaving aside the phases $\gamma_{\boldsymbol{k}}$). According to equation (3.77), we can view the coefficients $a_{\mathbf{k}}$ and $a^*_{\mathbf{k}}$ as coordinates of a motion with energy U. We quantize this motion by imposing commutation relations

$$[a_{\mathbf{k}}, a^*_{\mathbf{k}'}] = \delta_{\mathbf{k}\mathbf{k}'} \ , \ [a_{\mathbf{k}}, a_{\mathbf{k}'}] = 0 \ ; \tag{3.81}$$

the energy and the momentum (equations (3.77) and (3.78)) become

$$U = \sum_{\boldsymbol{k}} \frac{1}{2}\hbar\omega_k \left(a^*_{\boldsymbol{k}} a_{\boldsymbol{k}} + a_{\boldsymbol{k}} a^*_{\boldsymbol{k}}\right) = \sum_{\boldsymbol{k}} \hbar\omega_k \left(a^*_{\boldsymbol{k}} a_{\boldsymbol{k}} + 1/2\right) \tag{3.82}$$

and

$$\boldsymbol{G} = \sum_{\boldsymbol{k}} \frac{1}{2}\hbar\boldsymbol{k} \left(a^*_{\boldsymbol{k}} a_{\boldsymbol{k}} + a_{\boldsymbol{k}} a^*_{\boldsymbol{k}}\right) = \sum_{\boldsymbol{k}} \hbar\boldsymbol{k} \left(a^*_{\boldsymbol{k}} a_{\boldsymbol{k}} + 1/2\right) \ . \tag{3.83}$$

The time dependence of the coefficients $a_{\mathbf{k}}$ and $a^*_{\mathbf{k}}$ is obtained by using the free hamiltonian of the electromagnetic field

$$H_{em} = U = \sum_{\mathbf{k}} \hbar\omega_k \left(a^*_{\mathbf{k}} a_{\mathbf{k}} + \frac{1}{2}\right) \tag{3.84}$$

in the quantum-mechanical equations of motion

$$i\hbar\frac{da_{\mathbf{k}}}{dt} = [a_{\mathbf{k}}, H_{em}] \ . \tag{3.85}$$

We can see that the motion of the electromagnetic field is realized by quantum-mechanical harmonic oscillators, labelled by $\boldsymbol{k}$. The corresponding quanta of the electromagnetic field are the photons. The eigenstates $\mid n_{\mathbf{k}} >$ of H_{em} are defined by non-negative integers $n_{\mathbf{k}} \geq 0$, such that the action of the occupation number $n_{\mathbf{k}} = a^*_{\mathbf{k}} a_{\mathbf{k}}$ is

$$n_{\mathbf{k}} \mid n_{\mathbf{k}} >= a^*_{\mathbf{k}} a_{\mathbf{k}} \mid n_{\mathbf{k}} >= n_{\mathbf{k}} \mid n_{\mathbf{k}} > \ ; \tag{3.86}$$

it follows immediately

$$\begin{aligned} a_{\mathbf{k}} \mid n_{\mathbf{k}} > &= \sqrt{n_{\mathbf{k}}} \mid n_{\mathbf{k}} - 1 > \ , \\ a^*_{\mathbf{k}} \mid n_{\mathbf{k}} > &= \sqrt{n_{\mathbf{k}} + 1} \mid n_{\mathbf{k}} + 1 > \end{aligned} \tag{3.87}$$

and the existence of a vacuum state $\mid 0_{\mathbf{k}}>$, such that $a_{\mathbf{k}} \mid 0_{\mathbf{k}} >= 0$. $a_{\mathbf{k}}$ is a photon destruction (annihilation) operator and $a^*_{\mathbf{k}}$ is a photon creation operator. The states $\mid n_{\mathbf{k}} >$ are normalized to unity. The polarization label should be added to these states.

According to the Maxwell equations the vector potential is a polar vector which changes sign under time reversal. This symmetry is preserved for photons, providing the destruction of a $\boldsymbol{k}$-photon amounts to the creation of a $-\boldsymbol{k}$-photon.

3.10 Charge in electromagnetic field

Let us assume a pointlike particle with charge q and mass m, placed in an electromagnetic field with the potentials Φ and $\boldsymbol{A}$. The interaction given by equation (3.54) ($\int d\boldsymbol{r}\left(\rho\Phi - \frac{1}{c}\boldsymbol{j}\boldsymbol{A}\right)$) reads $q\Phi - \frac{q}{c}\boldsymbol{v}\boldsymbol{A}$, which leads to $-q\Phi + \frac{q}{c}\boldsymbol{v}\boldsymbol{A}$ in the lagrangian, where $\boldsymbol{v}$ is the velocity of the particle. Hence, we get immediately the Hamilton-Jacobi equation for a relativistic particle[10]

$$\left(\frac{H - q\Phi}{c}\right)^2 = m^2c^2 + \left(\boldsymbol{p} - \frac{q}{c}\boldsymbol{A}\right)^2 \ , \tag{3.88}$$

where H is the hamiltonian and $\boldsymbol{p}$ is the momentum of the particle. Since we need the non-relativistic limit we write $H = mc^2 + \mathcal{E}$, where $\mathcal{E}$ is the energy, and equation (3.88) as

$$-\frac{1}{2m}\left[\left(mc + \frac{\mathcal{E} - q\Phi}{c}\right)^2 - \left(\boldsymbol{p} - \frac{q}{c}\boldsymbol{A}\right)^2\right] + \frac{1}{2}mc^2 = 0 \ ; \tag{3.89}$$

up to terms of the order $1/c$ the energy is

$$\mathcal{E} = \frac{1}{2m}p^2 - \frac{q}{mc}\boldsymbol{p}\boldsymbol{A} + q\Phi \ . \tag{3.90}$$

The equations of motion of the particle are

$$\boldsymbol{v} = \frac{\boldsymbol{p}}{m} - \frac{q}{mc}\boldsymbol{A} \tag{3.91}$$

[10] L. Landau and E. Lifshitz, *Course of Theoretical Physics*, vol. 2, *Classical Theory of Fields*, Butterworth-Heinemann, Oxford (1987).

and

$$\dot{p}_i = \frac{q}{mc} p_j \partial_i A_j - q\partial_i \Phi \ , \tag{3.92}$$

or

$$m\dot{v}_i + \frac{q}{c}\left(\frac{\partial A_i}{\partial t} + v_j \partial_j A_i\right) = \frac{q}{mc} p_j \partial_i A_j - q\partial_i \Phi \ , \tag{3.93}$$

which is the non-relativistic equation of motion

$$m\dot{\boldsymbol{v}} = q\boldsymbol{E} + \frac{q}{c}\boldsymbol{v} \times \boldsymbol{H} \tag{3.94}$$

with the Lorentz force, $\boldsymbol{E} = -\frac{1}{c}\frac{\partial \boldsymbol{A}}{\partial t} - grad\Phi$ being the electric field and $\boldsymbol{H} = curl\boldsymbol{A}$ being the magnetic field.

We note that

$$V = -\frac{q}{mc}\boldsymbol{pA} + q\Phi \tag{3.95}$$

is the interaction hamiltonian. From equation (3.91) we get the current

$$\boldsymbol{J} = q\boldsymbol{v} = \frac{q}{m}\boldsymbol{p} - \frac{q^2}{mc}\boldsymbol{A} \ ; \tag{3.96}$$

we can see that besides the particle current

$$\boldsymbol{J}_p = \frac{q}{m}\boldsymbol{p} \ , \tag{3.97}$$

there appears also a current

$$J_m = -\frac{q^2}{mc}\boldsymbol{A} \ , \tag{3.98}$$

which we call the magnetic current. For a uniform magnetic field $\boldsymbol{H}$ the potential is $\boldsymbol{A} = \frac{1}{2}\boldsymbol{H} \times \boldsymbol{r}$ and the interaction term $-\frac{q}{mc}\boldsymbol{pA}$ reads

$$-\frac{q}{mc}\boldsymbol{pA} = -\frac{q}{2mc}\boldsymbol{p}(\boldsymbol{H} \times \boldsymbol{r}) = -\frac{q}{2mc}\boldsymbol{HL} \ , \tag{3.99}$$

where $\boldsymbol{L} = \boldsymbol{r} \times \boldsymbol{p}$ is the orbital angular momentum of the charge; but $\boldsymbol{m} = \frac{q}{2mc}\boldsymbol{L}$ is the magnetic moment of the charge, such that

$$-\frac{q}{mc}\boldsymbol{pA} = -\boldsymbol{mH} \tag{3.100}$$

is the energy of the charge in the magnetic field. Therefore, $-\frac{q}{mc}\boldsymbol{pA}$ is a paramagnetic interaction and $\boldsymbol{J}_p$ is a paramagnetic current, while the induced current $\boldsymbol{J}_m$ is a diamagnetic current.

The current densities are

$$\boldsymbol{j}_p = \frac{q}{m}n(\boldsymbol{r})\boldsymbol{p} \; , \; \boldsymbol{j}_m = -\frac{q^2}{mc}n(\boldsymbol{r})\boldsymbol{A} \; , \tag{3.101}$$

where $n(\boldsymbol{r})$ is the particle density.

Within the non-relativistic limit we may write

$$-\frac{q}{mc}\boldsymbol{pA} = -\frac{q}{c}\boldsymbol{vA} = -\frac{1}{c}\boldsymbol{J}_p\boldsymbol{A} \; ; \tag{3.102}$$

if we include the diamagnetic current, we need to keep the second-order term $\frac{q^2}{2mc^2}A^2$ in the hamiltonian, such that

$$-\frac{q}{mc}\boldsymbol{pA} + \frac{q^2}{2mc^2}A^2 = -\frac{q}{c}\boldsymbol{JA} - \frac{q^2}{2mc^2}A^2 \tag{3.103}$$

(and an additional term $-\frac{1}{2mc^2}(p^2/2m)^2$ appears in the hamiltonian); the term $\frac{q^2}{2mc^2}A^2$ is a diamagnetic interaction.

For a radiation field $\Phi = 0$ and $\boldsymbol{A}$ is transverse. The interaction term in the mechanical action

$$\begin{aligned} \int dt \left(-\tfrac{q}{c}\boldsymbol{vA}\right) &= \int dt \cdot \tfrac{q}{c}\boldsymbol{r}\left(\tfrac{\partial \boldsymbol{A}}{\partial t} + (\boldsymbol{v}grad)\boldsymbol{A}\right) \simeq \\ &\simeq -\int \boldsymbol{dt} \cdot \boldsymbol{dE} \; , \end{aligned} \tag{3.104}$$

where $\boldsymbol{d}$ is the dipole moment, providing $\boldsymbol{A}$ has a slight spatial variation. The same result is obtained for an extended particle, providing the field varies slowly over the dimension of the particle (dipole approximation).

3.11 Scattering off a charge distribution

Let us assume a radiation field with the vector potential $\boldsymbol{A}_0 e^{-i\omega t + i\boldsymbol{kr}}$, the electric field $\frac{i\omega}{c}\boldsymbol{A}_0 e^{-i\omega t + i\boldsymbol{kr}}$ (and the magnetic field $i\boldsymbol{k}\times\boldsymbol{A}_0 e^{-i\omega t + i\boldsymbol{kr}}$).

Let us assume that this field acts upon a pointlike particle with charge q and mass m, placed at $\boldsymbol{r}=0$; its equation of motion is

$$m\ddot{\boldsymbol{u}}+m\omega_0^2\boldsymbol{u}+m\gamma\dot{\boldsymbol{u}}=\frac{iq\omega}{c}\boldsymbol{A}_0e^{-i\omega t} \ , \tag{3.105}$$

where ω_0 is the eigenfrequency and γ is the damping coefficient. The displacement $\boldsymbol{u}$ is

$$\boldsymbol{u}=-\frac{iq\omega}{mc}\boldsymbol{A}_0\frac{1}{\omega^2-\omega_0^2+i\omega\gamma}e^{-i\omega t} \ . \tag{3.106}$$

this displacement generates a (dia)magnetic current

$$\boldsymbol{J}_m=q\dot{\boldsymbol{u}}=-\frac{q^2\omega^2}{mc}\boldsymbol{A}_0\frac{1}{\omega^2-\omega_0^2+i\omega\gamma}e^{-i\omega t} \ . \tag{3.107}$$

For a free particle $J_m=-\frac{q^2}{mc}\boldsymbol{A}_0e^{-i\omega t}$, which is precisely equation (3.98). The magnetic current is the current induced by the external field (superposed over the particle current $\boldsymbol{J}_p$), which is relevant for the scattering.

The magnetic current density is

$$\boldsymbol{j}_m=q\dot{\boldsymbol{u}}\delta(\boldsymbol{r})=-\frac{q^2\omega^2}{mc}\boldsymbol{A}_0\frac{1}{\omega^2-\omega_0^2+i\omega\gamma}e^{-i\omega t+i\boldsymbol{k}\boldsymbol{r}}\delta(\boldsymbol{r}) \ ; \tag{3.108}$$

it generates a scattered wave with the vector potential

$$\begin{aligned}\boldsymbol{A}(\boldsymbol{r},t)=&\\ =-\frac{q^2\omega^2}{mc^2}\boldsymbol{A}_0\frac{1}{\omega^2-\omega_0^2+i\omega\gamma}\int d\boldsymbol{r}'\frac{e^{-i\omega(t-|\boldsymbol{r}-\boldsymbol{r}'|/c)+i\boldsymbol{k}\boldsymbol{r}'}}{|\boldsymbol{r}-\boldsymbol{r}'|}\delta(\boldsymbol{r}')=&\\ =-\frac{q^2\omega^2}{mc^2}\boldsymbol{A}_0\frac{1}{\omega^2-\omega_0^2+i\omega\gamma}\frac{e^{-i\omega t+ikr}}{r} \ ;&\end{aligned} \tag{3.109}$$

we can see in equation (3.109) the outgoing spherical wave.

For a particle density $n(\boldsymbol{r})$ the vector potential is

$$\begin{aligned}\boldsymbol{A}(\boldsymbol{r},t)=&\\ =-\frac{q^2\omega^2}{mc^2}\boldsymbol{A}_0\frac{1}{\omega^2-\omega_0^2+i\omega\gamma}\int d\boldsymbol{r}'\frac{e^{-i\omega(t-|\boldsymbol{r}-\boldsymbol{r}'|/c)+i\boldsymbol{k}\boldsymbol{r}'}}{|\boldsymbol{r}-\boldsymbol{r}'|}n(\boldsymbol{r}')\simeq&\\ \simeq-\frac{q^2\omega^2}{mc^2}\boldsymbol{A}_0\frac{1}{\omega^2-\omega_0^2+i\omega\gamma}\frac{e^{-i\omega t+ikr}}{r}\int d\boldsymbol{r}'n(\boldsymbol{r}')e^{i(\boldsymbol{k}-\boldsymbol{k}')\boldsymbol{r}'}&\end{aligned} \tag{3.110}$$

in the far-field region ($r \gg r'$), where $\boldsymbol{k}' = k\boldsymbol{r}\boldsymbol{r}'/r$ is the scattered wavevector. On the other hand, the main contribution comes from $kl \ll 1$, *i.e.* $l \ll \lambda$, where l is the dimension of the charge distribution and λ is the wavelength of the field, which is the effective contribution for velocities of the charge much smaller than the velocity of light. We can see in equation (3.110) the Fourier transform of the charge density. The scattered electric fied is

$$\boldsymbol{E}(t,r) = \\ = -\frac{iq^2\omega^3}{mc^3}\boldsymbol{A}_0\frac{1}{\omega^2-\omega_0^2+i\omega\gamma}\frac{e^{-i\omega t+ikr}}{r}\int d\boldsymbol{r}'n(\boldsymbol{r}')e^{i(\boldsymbol{k}-\boldsymbol{k}')\boldsymbol{r}'} \ , \tag{3.111}$$

the scattered magnetic field is

$$\boldsymbol{H}(t,r) = \\ = -\frac{iq^2\omega^2 k}{mc^2}\frac{\boldsymbol{r}\times\boldsymbol{A}_0}{r}\frac{1}{\omega^2-\omega_0^2+i\omega\gamma}\frac{e^{-i\omega t+ikr}}{r}\int d\boldsymbol{r}'n(\boldsymbol{r}')e^{i(\boldsymbol{k}-\boldsymbol{k}')\boldsymbol{r}'} \ . \tag{3.112}$$

The energy scattered per unit time and unit area is given by the Poynting vector

$$\boldsymbol{S} = \frac{c}{8\pi}\left(\boldsymbol{E}^*\times\boldsymbol{H}+\boldsymbol{E}\times\boldsymbol{H}^*\right) = \\ = \frac{q^4\omega^6}{4\pi m^2c^5}\frac{\boldsymbol{r}A_0^2-(\boldsymbol{r}\boldsymbol{A}_0)\boldsymbol{A}_0}{r^3}\frac{1}{\left(\omega^2-\omega_0^2\right)^2+\omega^2\gamma^2}\cdot \\ \cdot\left|\int d\boldsymbol{r}'n(\boldsymbol{r}')e^{i(\boldsymbol{k}-\boldsymbol{k}')\boldsymbol{r}'}\right|^2 \ ; \tag{3.113}$$

the energy scattered in the solid angle do is obtained by multiplying this expression by $r^2\boldsymbol{r}do/r$. The incident energy per unit time and unit area is $\frac{\omega}{4\pi}\boldsymbol{k}A_0^2$, such that the cross-section

$$d\sigma = \frac{q^4\omega^4}{m^2c^4}\frac{1-\cos^2\alpha}{\left(\omega^2-\omega_0^2\right)^2+\omega^2\gamma^2}\left|\int d\boldsymbol{r}'n(\boldsymbol{r}')e^{i(\boldsymbol{k}-\boldsymbol{k}')\boldsymbol{r}'}\right|^2 do \ , \tag{3.114}$$

where α is the angle between $\boldsymbol{r}$ and $\boldsymbol{A}_0$. For an unpolarized radiation we take $\boldsymbol{k} = k(0,0,1)$, $\boldsymbol{A}_0 = A_0(\cos\varphi', \sin\varphi', 0)$ and $\boldsymbol{r} = r(\sin\theta\cos\varphi, \sin\theta\sin\varphi, \cos\theta)$, where θ is the angle between $\boldsymbol{k}$ and $\boldsymbol{r}$

(scattering angle); by averaging over polarizations, we get

$$d\sigma = \frac{q^4\omega^4}{m^2c^4}\frac{1-\frac{1}{2}\sin^2\theta}{\left(\omega^2-\omega_0^2\right)^2+\omega^2\gamma^2}\left|\int d\boldsymbol{r}'n(\boldsymbol{r}')e^{i(\boldsymbol{k}-\boldsymbol{k}')\boldsymbol{r}'}\right|^2 do \ , \qquad (3.115)$$

which is the Rayleigh cross-section for a charge distribution (see equation (3.29)). q^2/mc^2 is the classical radius of the charge.

3.12 The atom

The electrons in the atom move in the Coulomb potential

$$-\sum_{i=1}^{Z}\frac{Ze^2}{r_i}+\frac{1}{2}\sum_{i\neq j}\frac{e^2}{\mid \boldsymbol{r}_i-\boldsymbol{r}_j\mid} \ , \qquad (3.116)$$

where $-e$ is the electron charge and $\boldsymbol{r}_{i,j}$, $i,j=1,2...Z$, are the positions of the Z electrons with respect to the nucleus placed at the origin; the first term is the electron-nucleus attraction, the second term is the electron-electron repulsion. The classical motion of this multi-particle assembly cannot be given in one-particle terms. For an assembly of N particles with mass m_i the motion of the centre of mass with position

$$\boldsymbol{R}=\frac{1}{M}\sum_i m_i\boldsymbol{r}_i \ , \ M=\sum_i m_i \qquad (3.117)$$

can be separated, by using either the baricentric coordinates

$$\boldsymbol{\rho}_i=\boldsymbol{r}_i-\boldsymbol{R} \ , \qquad (3.118)$$

or the relative coordinates

$$\boldsymbol{r}_{ij}=\boldsymbol{r}_i-\boldsymbol{r}_j \ , \ r_i=\boldsymbol{r}_i=\boldsymbol{R}+\frac{1}{M}\sum_j m_j\boldsymbol{r}_{ij} \ . \qquad (3.119)$$

Indeed, in the first case we get

$$\frac{1}{2}\sum_i m_i\dot{\boldsymbol{r}}_i^2=\frac{1}{2}M\dot{\boldsymbol{R}}^2+\frac{1}{2}\sum_i m_i\dot{\boldsymbol{\rho}}_i^2 \ , \qquad (3.120)$$

while in the second case we get

$$\frac{1}{2}\sum_i m_i \dot{\boldsymbol{r}}_i^2 = \frac{1}{2}M\dot{\boldsymbol{R}}^2 + \frac{1}{2M^2}\sum_{ijk} m_i m_j m_k \dot{\boldsymbol{r}}_{ij}\dot{\boldsymbol{r}}_{ik} \ , \tag{3.121}$$

since $\sum_i m_i \boldsymbol{\rho}_i = 0$ and $\boldsymbol{r}_{ij} = -\boldsymbol{r}_{ji}$. In the former case we remain with an interaction depending on $\boldsymbol{\rho}_i - \boldsymbol{\rho}_j$, while in the latter we get a non-diagonal kinetic energy, which amounts to an interaction, and an excessive number of non-independent coordinates. These variables are useful for two particles. For an atom the problem is even more complicated because there exist two kinds of particles, the nucleus and the electrons, and the motion is a quantum-mechanical motion.

An electron in the atom moves in the centrally-symmetric field of the nucleus and the field produced by all the other electrons; this is a self-consistent field. With properly symmetrized wavefunctions constructed by one-particle wavefunctions the self-consistent field leads to a closed system of coupled (and non-linear) Schroedinger equations, which are the Hartree-Fock equations. This method works for light atoms. For a stationary one-electron state the lowest kinetic energy of the electron $\hbar^2/ma^2$ should be approximately equal to its potential energy e^2/a, hence the dimension of the atom is of the order of the Bohr radius a_H, $a = a_H = \hbar^2/me^2$, where $\hbar$ is Planck's constant and m is the electron mass. The potential energy e^2/a leads to a second-order variation $(e^2/a_H^3)u^2 = m\omega_0^2 u^2$, where u is the small displacement of the electron from its equilibrium position and $\omega_0 = \sqrt{e^2/ma_H^3} \simeq 5\times10^{16}s^{-1}$ ($\simeq 30eV$; $-e = 4.8\times10^{-10}esu$, $m = 10^{-27}g$, $\hbar = 10^{-27}erg\cdot s$, $1eV = 1.6\times10^{-12}erg$). This estimate holds also for heavy atoms, where the outer electron sees a fully screened nuclear potential. However, in heavy atoms an additional particularity occurs. In order to get a minimum of the potential energy given by equation (3.116) the distances should be $r_i \simeq \mid \boldsymbol{r}_i - \boldsymbol{r}_j \mid \simeq R$; therefore, $ZR^3 \simeq a_H^3$, *i.e.* the electrons are quasi-uniformly distributed within a distance $R \simeq a_H/Z^{1/3}$ from the nucleus; therefore, they see a screened potential $\simeq -Ze\cdot e^{-qr}/r$, where $q \simeq 1/R$, and their density is proportional to $Ze^{-qr}/R^2 r$. This is the Thomas-Fermi model for heavy atoms.[11] This equilibrium state changes by small displacements of the electron cloud as a whole. The total potential

[11] M. Apostol, "Giant dipole oscillations and ionization of heavy atoms by intense

energy is $\simeq Z^2e^2/R$, which leads to a change $(Z^2e^2/R^3)u^2 = M\Omega^2u^2$, where $M = Zm$. Hence, we get the frequency $\Omega = Z\omega_0$.

Let us assume two particles with masses $m_{1,2}$ and charges $q_{1,2}$, placed at $\boldsymbol{r}_{1,2}$. Equations (3.119) give

$$\boldsymbol{r}_1 = \boldsymbol{R} + \frac{m_2}{M}\boldsymbol{r} \ , \ \boldsymbol{r}_2 = \boldsymbol{R} - \frac{m_1}{M}\boldsymbol{r} \ , \tag{3.122}$$

where $\boldsymbol{R}$ is the position of the centre of mass and $\boldsymbol{r} = \boldsymbol{r}_1 - \boldsymbol{r}_2$ is the relative coordinate. The kinetic energy is

$$\frac{1}{2}m_1\dot{\boldsymbol{r}}_1^2 + \frac{1}{2}m_2\dot{\boldsymbol{r}}_2^2 = \frac{1}{2}M\dot{\boldsymbol{R}}^2 + \frac{1}{2}\mu\dot{\boldsymbol{r}}^2 \ , \tag{3.123}$$

where $\mu = m_1m_2/M$ is the reduced mass. The interaction energy with an external electric field $\boldsymbol{E}$ is

$$-q_1\boldsymbol{E}\boldsymbol{r}_1 - q_2\boldsymbol{E}\boldsymbol{r}_2 = -(q_1+q_2)\boldsymbol{E}\boldsymbol{R} - \frac{1}{M}(q_1m_2 - q_2m_1)\boldsymbol{E}\boldsymbol{r} \ . \tag{3.124}$$

The equations of motion for the displacements $\boldsymbol{U}$ of the centre of mass and the displacement $\boldsymbol{u}$ of the relative coordinate are

$$\begin{aligned} M\ddot{\boldsymbol{U}} &= (q_1+q_2)\boldsymbol{E} \ , \\ \mu\ddot{\boldsymbol{u}} + \mu\omega_0^2\boldsymbol{u} &= \tfrac{1}{M}(q_1m_2 - q_2m_1)\boldsymbol{E} \ . \end{aligned} \tag{3.125}$$

We can see that the external field induces a magnetic current $(q_1 + q_2)\dot{\boldsymbol{U}}$ of the centre of mass and a magnetic current $\frac{1}{M}(q_1m_2 - q_2m_1)\dot{\boldsymbol{u}}$ of the relative motion. These currents generate the scattered field of an external radiation field. Let us consider an atom (or a ion) with the nucleus placed at $\boldsymbol{r}_1$ and a spherically symmetric electron cloud, with the centre of mass placed at $\boldsymbol{r}_2$. Equations (3.125) can be used, so we can compute the scattering cross-section of the radiation. For a neutral atom $q_1 + q_2 = 0$, and we are left with the second equation (3.125), where $\mu = M_nZm/(M_n + Zm) \simeq Zm$ and $\omega_0 \simeq \Omega$, M_n being the mass of the nucleus.[12] Also, $\frac{1}{M}(q_1m_2 - q_2m_1) = -Ze$. The cross-section is given by equation (3.115), where $m \to \mu = Zm$, $q \to -Ze$

electromagnetic fields", Roum. Reps. Phys. **67** 837 (2015); "Resonant coupling of the electron cloud with the nucleus in heavy atoms", Roum. Reps. Phys. **71** 210 (2019).

[12]For ions Ω is slightly modified, see M. Apostol, *loc cit.*

and $n(\boldsymbol{r}) \rightarrow \delta(\boldsymbol{r})$. We get

$$d\sigma = Z^2 \left(\frac{e^2\omega^2}{mc^2}\right)^2 \frac{1-\frac{1}{2}\sin^2\theta}{\left(\omega^2-\Omega^2\right)^2+\omega^2\gamma^2} do \; ; \tag{3.126}$$

since $\Omega \gg \omega$, we may omit ω in the denominator. The energy radiated per unit time in the solid angle do is the incident energy per unit time and unit area ($\frac{c}{4\pi}E^2$, Poynting vector) multiplied by the cross-section. From the equation of motion of the damped harmonic oscillator we can compute the energy absorbed continuously from the radiation and the energy dissipated per unit time.

The equation of motion of an oscillator with mass m, eigenfrequency ω_0 and damping coefficient γ ($\ll \omega_0$) is

$$m\ddot{u} + m\omega_0^2 u + m\gamma\dot{u} = f_0 \cos\omega t \;\; , \tag{3.127}$$

where $f_0 \cos\omega t$ is an external force (the electric field is invariant under time reversal). The solution vanishing at the initial moment of time is

$$u = A\left(\cos\omega t - e^{-\gamma t/2}\cos\omega_0 t\right) + B\sin\omega t \;\; , \tag{3.128}$$

where

$$A = -\frac{f_0}{m}\frac{\omega^2-\omega_0^2}{\left(\omega^2-\omega_0^2\right)^2+\omega^2\gamma^2} \;\; , \;\; B = \frac{f_0}{m}\frac{\omega\gamma}{\left(\omega^2-\omega_0^2\right)^2+\omega^2\gamma^2} \;\; . \tag{3.129}$$

The energy conservation from equation (3.127) leads to

$$\frac{d}{dt}\left(\frac{1}{2}m\dot{u}^2 + \frac{1}{2}m\omega^2 u^2\right) + m\gamma\dot{u}^2 = \dot{u}f_0\cos\omega t \; . \tag{3.130}$$

Hence, leaving aside the transient solution of the free oscillator, we get the energy

$$\begin{gathered}\mathcal{E} = \tfrac{1}{2}m\dot{u}^2 + \tfrac{1}{2}m\omega^2 u^2 = \\ = \tfrac{1}{2}m[\left(A^2\omega_0^2 + B^2\omega^2\right)\cos^2\omega t + \\ + \left(A^2\omega^2 + B^2\omega_0^2\right)2\sin^2\omega t + \\ +2AB\left(\omega_0^2-\omega^2\right)\sin\omega t\cos\omega t] \;\; ,\end{gathered} \tag{3.131}$$

the dissipated energy per unit time

$$m\gamma\dot{u}^2 = \\ = m\gamma\omega^2\left(A^2\sin^2\omega t + B^2\cos^2\omega t - 2AB\sin\omega t\cos\omega t\right) \tag{3.132}$$

and the work done by force per unit time

$$\dot{u}f_0\cos\omega t = f_0\omega\left(-A\sin\omega t\cos\omega t + B\cos^2\omega t\right) \ . \tag{3.133}$$

By averaging over a period, the energy is

$$\overline{\mathcal{E}} = \frac{f_0^2}{4m}\frac{\omega^2+\omega_0^2}{\left(\omega^2-\omega_0^2\right)^2+\omega^2\gamma^2} \ , \tag{3.134}$$

and the dissipated energy, equal to the mechanical work per unit time, is

$$\overline{m\gamma\dot{u}^2} = \overline{\dot{u}f_0\cos\omega t} = \frac{f_0^2}{2m}\frac{\gamma\omega^2}{\left(\omega^2-\omega_0^2\right)^2+\omega^2\gamma^2} \ . \tag{3.135}$$

During the transient regime the oscillator absorbs the energy $\overline{\mathcal{E}}$ on the average, which is preserved during the motion, while the force performs a mechanical work which goes into dissipated energy.

The amplitude of the displacement of the electron cloud in an atom, for $\omega_0 = \Omega = 5Z \times 10^{16}s^{-1} \gg \omega$, is

$$A \simeq \frac{eE_0}{m\Omega^2} = 2\times 10^{-16}E_0/Z^2cm \ , \tag{3.136}$$

where E_0 is the radiation electric field. A laser intensity $I = 10^{19}w/cm^2$ gives a field $E_0 = \sqrt{8\pi I/c} \simeq 3\times 10^8 esu$, such that the amplitude is $A \simeq 6\times 10^{-8}/Z^2cm$ (usual laboratory static fields are of the order $10^2 esu$, at most, solar radiation on Earth's surface gives a field of the order $10^{-2}esu$). For an external frequency $\omega = 10^{15}s^{-1}$ the velocity of the electron cloud is of the order $\omega A \simeq 6\times 10^7 cm/s \ll c$.

If the force in equation (3.127) is $f_0e^{-i\omega t}$, we have in fact two identical particles, with two independent displacements, acted by two distinct forces $f_0\cos\omega t$ and $f_0\sin\omega t$, correlated by their phases (two motions). Then, the energy has not oscillating factors anymore.

In general, the change in the particle density is $-ndiv\boldsymbol{u}$, such that it gives a change $-nqdiv\boldsymbol{u}$ in the charge density, a polarization field $\boldsymbol{P} = 4\pi nq\boldsymbol{u}$ and an internal field $\boldsymbol{E}_i = -\boldsymbol{P}$ (from the Gauss law). Since $\boldsymbol{u}$ depends on the external field $\boldsymbol{E}_0$, we get the (averaged) polarizability α from $\overline{\mathcal{E}} = -\frac{1}{2}\alpha E_0^2$ ($\boldsymbol{P} = \alpha\boldsymbol{E}_0$ in the static case) and the susceptibility χ from $\boldsymbol{P} = \chi\boldsymbol{E}$, where $\boldsymbol{E}$ is the total electric field.[13]

3.13 Quantum theory of radiation

The Thomas-Fermi model works well for the collective motion of the electron cloud of the heavy atoms. The ground state of the hydrogen atom has the energy $-13.6eV$; hydrogen-like atoms have energies $\varepsilon \simeq -13.6Z^2eV$. Single-electron energies in atoms vary in the range $100eV - 100keV$, while the rest energy of the electron is $mc^2 = 0.5MeV$. The non-relativistic approximation holds good in atoms, though relativistic corrections like spin-orbit or spin-spin coupling are important in heavy atoms. It follows that the ratio of the squared electron velocity to the squared velocity of light is $v^2/c^2 \simeq \varepsilon/mc^2$, in the range 10^{-4} to 10^{-1}. The interaction with the electromagnetic radiation affects mainly the periphereal electrons. In atomic nuclei the typical single-nucleon energy is $8MeV$, while the rest energy of a nucleon is of the order $1GeV$; the squared velocity ratio is of the order $v^2/c^2 \simeq 10^{-3}$.

In the non-relativistic limit the interaction of a pointlike charge q with mass m with the electromagnetic radiation is

$$H_1 = -\frac{q}{mc}\boldsymbol{pA} \ , \tag{3.137}$$

where $\boldsymbol{A}$ is the vector potential and the momentum of the particle is

$$\boldsymbol{p} = m\boldsymbol{v} \tag{3.138}$$

(equations (3.91) and (3.95)). In the dipole approximation this interaction can also be written as

$$H_1 = -\boldsymbol{dE} \ , \ H_1 = -\frac{1}{c}\boldsymbol{J}_p\boldsymbol{A} \ , \tag{3.139}$$

[13] M. Apostol, *Essays in Electromagnetism and Matter*, Lambert, Saarbruecken (2013); *Radiation an Matter*, Nova, NY (2022).

where $\boldsymbol{d}$ is the dipole moment of the charge (equation (3.104)), $\boldsymbol{E} = -\frac{1}{c}\frac{\partial \boldsymbol{A}}{\partial t}$ is the electric field and $\boldsymbol{J}_p = q\boldsymbol{v}$ is the (paramagnetic) current (particle current, equation (3.102)).

The next-order corrections imply

$$H_2 = \frac{q^2}{2mc^2}A^2 \tag{3.140}$$

and

$$\boldsymbol{p} = m\boldsymbol{v} + \frac{q}{c}\boldsymbol{A} \ , \tag{3.141}$$

which modifies the interaction H_1 into

$$H_1 = -\frac{1}{c}\left(\boldsymbol{J} - J_m\right)\boldsymbol{A} \ , \tag{3.142}$$

where $\boldsymbol{J} = q\boldsymbol{v}$ is the total current and $\boldsymbol{J}_m = -\frac{q^2}{mc}\boldsymbol{A}$ is the diamagnetic current; the interaction becomes

$$H_1 + H_2 = -\frac{1}{c}\boldsymbol{J}\boldsymbol{A} - \frac{q^2}{2mc^2}A^2 \ . \tag{3.143}$$

The spin current can be added, as well as the spin-orbit coupling.

For a quantum particle $\boldsymbol{p}$ is $-i\hbar grad$, such that the interaction reads

$$H_1 = -\frac{q}{mc}\boldsymbol{A}\boldsymbol{p} \ , \tag{3.144}$$

since $(\boldsymbol{p}\boldsymbol{A}) = -i\hbar div\boldsymbol{A} = 0$ for a radiation field.

Let us consider the typical problem of perturbation

$$i\hbar\frac{\partial\psi}{\partial t} = (H_0 + H_i)\psi \ , \tag{3.145}$$

with unperturbed states given by

$$H_0\varphi_n = E_n\varphi_n \ . \tag{3.146}$$

We look for a wavefunction

$$\psi = \sum_n c_n\varphi_n e^{-\frac{i}{\hbar}E_n t} \ ; \tag{3.147}$$

it leads immediately to the equations

$$i\hbar\dot{c}_n = \sum_m c_m (H_i)_{nm} e^{\frac{i}{\hbar}(E_n-E_m)t} \tag{3.148}$$

(with φ_n normalized to unity). These are the basic equations of the quantum theory of radiation.[14] This system of equations is solved by means of the perturbation theory, where $H_i \sim \lambda$ and $c_n = c_n^{(0)} + \lambda c_n^{(1)} + \lambda^2 c_n^{(2)} + ...$, λ being the perturbation parameter; we get

$$i\hbar\dot{c}_n^{(0)} = 0 \ ,$$

$$i\hbar\dot{c}_n^{(1)} = \sum\nolimits_m c_m^{(0)} (H_i)_{nm} e^{\frac{i}{\hbar}(E_n-E_m)t} \ , \tag{3.149}$$

$$i\hbar\dot{c}_n^{(2)} = \sum\nolimits_m c_m^{(1)} (H_i)_{nm} e^{\frac{i}{\hbar}(E_n-E_m)t} \ , ... \ .$$

Let us assume that we are in the unperturbed stationary state i at the initial moment of time, which we take it at $t \to -\infty$, by introducing an adiabatic factor $e^{\alpha t}$, $\alpha \to 0^+$, in interaction; we assume a time-independent interaction. Then, $c_i^{(0)} = 1$ and all the others $c_n^{(0)}$ are zero. From the second equation (3.149) we get

$$c_{fi}^{(1)} = -(H_i)_{fi} \frac{e^{\frac{i}{\hbar}(E_f-E_i)t+\alpha t}}{E_f - E_i - i\alpha\hbar} \ , \tag{3.150}$$

where

$$\left| c_{fi}^{(1)} \right|^2 = |(H_i)_{fi}|^2 \frac{e^{2\alpha t}}{(E_f - E_i)^2 + \alpha^2\hbar^2} \tag{3.151}$$

is the probability of the final state f after a long time; by taking the time derivative and letting α go zero, we obtain the number of transitions per unit time

$$w_{fi}^{(1)} = \frac{2\pi}{\hbar} |(H_i)_{fi}| \delta(E_f - E_i) \ , \tag{3.152}$$

[14] P. A. M. Dirac, "The quantum theory of the emission and absorption of radiation", Proc. Roy. Soc. **A114** 243 (1927); "The quantum theory of dispersion", Proc. Roy. Soc. **A114** 710 (1927); "On the annihilation of electrons and protons", Proc. Camb. Phil. Soc. **26** 361 (1930); E. Fermi, "Quantum theory of radiation", Revs. Mod. Phys. **4** 87 (1932); W. Heitler, *The Quantum Theory of Radiation*, Dover (1984).

which is Fermi's golden rule. In the next-order approximation we have

$$c_{fi}^{(2)} = \sum_m \frac{(H_i)_{fm}(H_i)_{mi}}{E_m - E_i - i\alpha} \frac{e^{\frac{i}{\hbar}(E_f - E_i)t + 2\alpha t}}{E_f - E_i - 2i\alpha\hbar} , \tag{3.153}$$

which leads to the number of transitions per unit time

$$w_{fi}^{(2)} = \frac{2\pi}{\hbar} \left| \sum_m \frac{(H_i)_{fm}(H_i)_{mi}}{E_m - E_i - i\alpha} \right|^2 \delta(E_f - E_i) ; \tag{3.154}$$

of course, $w_{fi}^{(2)}$ is used only when $w_{fi}^{(1)} = 0$.

For the interactions $H_{1,2}$ given above we have, besides a charge, or an assembly of charges, the photons entering the vector potential. We denote by $\varphi(\boldsymbol{r})$ the wavefunction of a charge. A $\boldsymbol{k}$-photon is a state $\mid \boldsymbol{k} >= a_{\boldsymbol{k}}^* \mid 0 >$, or $a_{\boldsymbol{k}}^* \mid n_{\boldsymbol{k}} > / \sqrt{n_{\boldsymbol{k}} + 1} = \mid n_{\boldsymbol{k}} + 1 >$ for $n_{\boldsymbol{k}}$ photons already present (normalized to unity). For the initial and final states $\mid i, f >$ of a charge and a photon we use the mixed notation $\varphi(\boldsymbol{r}) \mid \boldsymbol{k} >$. Of course, we can consider also multi-photon states, whose transition probability is very small.[15]

3.14 Emission and absorption: free charges

Let us consider a pointlike particle with mass m and charge q, an initial state

$$\mid i >= \frac{1}{\sqrt{V}} e^{i\boldsymbol{K}\boldsymbol{r}} \mid n_{\boldsymbol{k}} > , \; E_i = \frac{\hbar^2 K^2}{2m} \tag{3.155}$$

and a final state

$$\mid f >= \frac{1}{\sqrt{V}} e^{i\boldsymbol{K}'\boldsymbol{r}} \mid n_{\boldsymbol{k}} + 1 > , \; E_f = \frac{\hbar^2 K'^2}{2m} + \hbar\omega_k . \tag{3.156}$$

This process is an emission of a photon. Making use of equation (3.152), the interaction hamiltonian H_1 given by equation (3.137) and

[15]M. Goeppert-Mayer, "Ueber Elementarakte mit zwei Quantensprungen", Ann. Phys. **401** 273 (1931).

the electromagnetic potential given by equation (3.80), the number of transitions per unit time is

$$\begin{gathered} w_{fi}^{(1)} = \frac{4\pi^2\hbar^2 q^2}{m^2 ckV}(\boldsymbol{K}\boldsymbol{e_k})^2(n_{\boldsymbol{k}}+1)\cdot \\ \cdot\delta_{\boldsymbol{K}'+\boldsymbol{k},\boldsymbol{K}}\delta\left(\frac{\hbar^2 K'^2}{2m}+\hbar\omega_k-\frac{\hbar^2 K^2}{2m}\right) . \end{gathered} \tag{3.157}$$

The momentum and energy conservation

$$\begin{gathered} \boldsymbol{K}' = \boldsymbol{K}-\boldsymbol{k} , \\ \frac{\hbar^2 K'^2}{2m}+\hbar\omega_k = \frac{\hbar^2 K^2}{2m} \end{gathered} \tag{3.158}$$

leads to a scattering angle $\theta = (\boldsymbol{k}, \boldsymbol{K})$ given by

$$\cos\theta = \frac{k+2mc/\hbar}{2K} . \tag{3.159}$$

The Compton momentum mc is much greater than the momenta $\hbar k$, $\hbar K$; for an electron the Compton wavelength is $\lambda_c = \hbar/mc = 3.8\times 10^{-11}cm$. The momentum and energy conservation can only be satisfied for $k=0$, *i.e.* without any spontaneous photon emission. For absorption the initial and final states are inversed. The result is also valid for the interaction hamiltonian H_2, or second-order approximation, as well as in the relativistic case. Indeed, for a relativistic particle the interaction hamiltonian $-(q/mc)\boldsymbol{A}\boldsymbol{p}$ is valid, as well as the perturbation theory, and we find again that the momentum and energy conservation are not satisfied for $\boldsymbol{k}\neq 0$.

3.15 Emission and absorption: bound charges

We consider an assembly of bound charges in an excited state 2 with energy E_2 which decays to a state 1, with energy E_1, by emitting a photon $\boldsymbol{k}$, in the presence of a radiation field (*e.g.*, an excited atom). Equations (3.148) give

$$\begin{gathered} i\hbar\dot{c}_2 = \sum_{\boldsymbol{k}} c_{1\boldsymbol{k}}\left(H_1\right)_{2,1\boldsymbol{k}} e^{\frac{i}{\hbar}(E_2-E_1-\hbar\omega_k)t} , \\ i\hbar\dot{c}_{1\boldsymbol{k}} = c_2\left(H_1\right)_{1\boldsymbol{k},2} e^{-\frac{i}{\hbar}(E_2-E_1-\hbar\omega_k)t} , \end{gathered} \tag{3.160}$$

where $H_1 = -(q/mc)\boldsymbol{Ap}$ and summation over $\boldsymbol{k}$ implies also a summation over the two polarizations of the photon; q, m and $\boldsymbol{p}$ are generic notations for the charge, mass and momentum of the assembly. The vector potential $\boldsymbol{A}$ is represented as in equation (3.80). We assume the photon wavelengths much longer than the dimension of the assembly, so we neglect the exponentials $e^{i\boldsymbol{kr}}$ (dipole approximation). Equations (3.160) become

$$\begin{aligned} i\hbar\dot{c}_2 &= -\sum\nolimits_{\boldsymbol{k}} c_{1\boldsymbol{k}} \tfrac{q\alpha_k\sqrt{n_{\boldsymbol{k}}+1}}{mc\sqrt{V}}(\boldsymbol{e_k p}_{21})e^{i(\omega_0-\omega_k)t} \ , \\ i\hbar\dot{c}_{1\boldsymbol{k}} &= -c_2\tfrac{q\alpha_k\sqrt{n_{\boldsymbol{k}}+1}}{mc\sqrt{V}}(\boldsymbol{e_k p}_{12})e^{-i(\omega_0-\omega_k)t} \ , \end{aligned} \tag{3.161}$$

where $\alpha_k = c\sqrt{2\pi\hbar/\omega_k}$, $\omega_0 = \omega_2 - \omega_1$ and $n_{\boldsymbol{k}}$ is the occupation number of the $\boldsymbol{k}$-photon state. We assume a decay $\sim e^{-\gamma t}$ for the coefficient c_2, which is unity at the initial moment of time $t = 0$. Then we get

$$c_{1\boldsymbol{k}} = -\frac{q\alpha_k\sqrt{n_{\boldsymbol{k}}+1}}{\hbar mc\sqrt{V}}(\boldsymbol{e_k p}_{12})\frac{e^{-i(\omega_0-\omega_k)t-\gamma t}-1}{\omega_0-\omega_k-i\gamma} \tag{3.162}$$

and

$$\gamma = \sum_{\boldsymbol{k}} \frac{2\pi i q^2(n_{\boldsymbol{k}}+1)}{\hbar m^2\omega_k V}\left|\boldsymbol{e_k p}_{12}\right|^2 \frac{1-e^{i(\omega_0-\omega_k)t+\gamma t}}{\omega_0-\omega_k-i\gamma} \ . \tag{3.163}$$

In equation (3.163) we assume that the polarization is undetermined, so we average over polarizations; the result is $\overline{\left|\boldsymbol{e_k p}_{12}\right|^2} = \frac{1}{3}p_{12}^2$; further on, we assume that the transition corresponds to a linear harmonic oscillator with coordinate u, such that $p_{12} = m\dot{u}_{12} = im\omega_0 u_{12}$; also, we assume that the occupation number does not depend on the direction of the wavevector $\boldsymbol{k}$; we pass from summation over $\boldsymbol{k}$ to integration over ω, by multiplying by the number of states $2V\omega^2 do d\omega/(2\pi)^3c^3$, such that

$$\gamma = \frac{2iq^2}{3\pi\hbar c^3}\omega_0^2\left|u_{12}\right|^2\int d\omega(n_\omega+1)\omega\frac{1-e^{i(\omega_0-\omega)t+\gamma t}}{\omega_0-\omega-i\gamma} \ . \tag{3.164}$$

For long times only the exponential remains in the integral; then, for small γ, $(\omega_0-\omega-i\gamma)^{-1} = P(\omega_0-\omega)^{-1}+i\pi\delta(\omega_0-\omega)$, and the principal value gives zero; we get an estimate of the damping coefficient

$$\gamma = \frac{2q^2}{3\hbar c^3}\omega_0^3\left|u_{12}\right|^2(n_0+1) \ , \tag{3.165}$$

where $n_0 = n_{\omega_0}$. $1/2\gamma$ is the lifetime of the state s . We compare this damping coefficient, for $n_0 = 0$, with the classical damping coefficient $\gamma_c/2$ given by equation (3.14),

$$2\gamma/\gamma_c \simeq \frac{2\omega_0 ma^2}{\hbar} \simeq \frac{a^2}{\lambda_0 \lambda_c} \ , \tag{3.166}$$

where u_{12} is of the order of the dimension a of the assembly, λ_0 is the photon wavelength and λ_c is the Compton wavelength. For electrons $2\gamma/\gamma_c \simeq 10^{-1}$ ($a = 10^{-8}cm$, $\lambda_0 = 10^{-4}cm$, $\lambda_c = 3.8 \times 10^{-11}cm$). In the presence of the radiation field the damping coefficient is much higher (stimulated emission). For a harmonic oscillator $u_{n,n-1} = \sqrt{\hbar n/2m\omega_0}$, which is of the order $\sqrt{na^2/2}$ ($\hbar^2/ma^2 = \hbar\omega_0$).

In the classical limit the occupation number n is proportional to the squared electric field. Indeed, from equations (3.76) and (3.80) we get

$$\begin{aligned} \boldsymbol{E} &= \tfrac{1}{\sqrt{V}} \textstyle\sum_{\boldsymbol{k}} \tfrac{i\alpha_k \omega_k}{c} \boldsymbol{e}_{\boldsymbol{k}} \left(a_{\boldsymbol{k}} + a^*_{-\boldsymbol{k}} \right) e^{i\boldsymbol{kr}} = \\ &= \tfrac{1}{\sqrt{V}} \textstyle\sum_{\boldsymbol{k}} \boldsymbol{E}_{\boldsymbol{k}} e^{i\boldsymbol{kr}} \ , \\ \boldsymbol{H} &= \tfrac{1}{\sqrt{V}} \textstyle\sum_{\boldsymbol{k}} i\alpha_k \boldsymbol{k} \times \boldsymbol{e}_{\boldsymbol{k}} \left(a_{\boldsymbol{k}} - a^*_{-\boldsymbol{k}} \right) e^{i\boldsymbol{kr}} = \\ &= \tfrac{1}{\sqrt{V}} \textstyle\sum_{\boldsymbol{k}} \boldsymbol{H}_{\boldsymbol{k}} e^{i\boldsymbol{kr}} \end{aligned} \tag{3.167}$$

and

$$\begin{aligned} &\tfrac{1}{8\pi} \textstyle\int d\boldsymbol{r} E^2 = \\ &= \tfrac{1}{4} \textstyle\sum_{\boldsymbol{k}} \hbar\omega_k \left(a_{\boldsymbol{k}} a^*_{\boldsymbol{k}} + a_{\boldsymbol{k}} a_{-\boldsymbol{k}} + a^*_{-\boldsymbol{k}} a^*_{\boldsymbol{k}} + a^*_{\boldsymbol{k}} a_{\boldsymbol{k}} \right) \ , \\ &\tfrac{1}{8\pi} \textstyle\int d\boldsymbol{r} H^2 = \\ &= \tfrac{1}{4} \textstyle\sum_{\boldsymbol{k}} \hbar\omega_k \left(a_{\boldsymbol{k}} a^*_{\boldsymbol{k}} - a_{\boldsymbol{k}} a_{-\boldsymbol{k}} - a^*_{-\boldsymbol{k}} a^*_{\boldsymbol{k}} + a^*_{\boldsymbol{k}} a_{\boldsymbol{k}} \right) \ . \end{aligned} \tag{3.168}$$

The average over a quantum state with a determined number of photons $n_{\boldsymbol{k}}$ leads to

$$\int d\boldsymbol{r} \overline{E^2} = \sum_{\boldsymbol{k}} \overline{|\boldsymbol{E}_{\boldsymbol{k}}|^2} = 4\pi \sum_{\boldsymbol{k}} \hbar\omega_k (n_{\boldsymbol{k}} + 1/2) \ . \tag{3.169}$$

In the classical limit $|\boldsymbol{E_k}|^2 = 4\pi\hbar\omega_k n_{\boldsymbol{k}}$. For a field $\boldsymbol{E} = \boldsymbol{E}_0 e^{i\boldsymbol{k}_0\boldsymbol{r}}$ we get $\boldsymbol{E_k} = \sqrt{V}\boldsymbol{E}_0\delta_{\boldsymbol{k},\boldsymbol{k}_0}$ and $VE_0^2 = 4\pi\hbar\omega_0 n_0$ (the energy in the mode ω_0).

The squared coefficient $c_{1\boldsymbol{k}}$ gives the number of transitions; its time derivative gives the number of transitions per unit time; multiplying by the number of states $2V\omega^2 do d\omega/(2\pi)^3c^3$, we get the number of transitions per unit time in the frequency interval $d\omega$

$$dw = \frac{4q^2(n_0+1)}{3\hbar c^3}\omega_0^3 \left|u_{12}\right|^2 \delta(\omega_0 - \omega)d\omega \ . \tag{3.170}$$

The total number of transitions is obtained from $\left|c_{1\boldsymbol{k}}\right|^2$ for $t \to \infty$. By using the same conditions as above, noting that the result is proportional to $\delta(\omega_0 - \omega)$ and multiplying by $\hbar\omega_0$, we get the total emitted energy

$$d\mathcal{E} = \frac{2q^2(n_0+1)}{3\pi c^3}\omega_0^4 \mid u_{12} \mid^2 \frac{1}{(\omega_0 - \omega)^2 + \gamma^2} d\omega \ , \tag{3.171}$$

where $d\omega$ extends mainly over the range $\Delta\omega = \pi\gamma$, since

$$\int d\omega\delta(\omega) = \frac{1}{\pi}\int d\omega \frac{\gamma}{\omega^2+\gamma^2} = \frac{1}{\pi\gamma}\Delta\omega = 1 \ . \tag{3.172}$$

We get the frequency distribution of the energy

$$\begin{aligned} d\mathcal{E}/d\omega &= \tfrac{2q^2(n_0+1)}{3\pi c^3}\omega_0^4 \mid u_{12} \mid^2 \tfrac{1}{\gamma}\tfrac{\gamma}{(\omega_0-\omega)^2+\gamma^2} = \\ &= \mathcal{E}_0 \tfrac{1}{\pi}\tfrac{\gamma}{(\omega_0-\omega)^2+\gamma^2} \ , \end{aligned} \tag{3.173}$$

where $\mathcal{E}_0$ is the total energy. This distribution has the same form as the distribution of the energy emitted by a classical charge which oscillates with frequency ω, as given by equation (3.17).

For absorption of a photon from the state 1 to the state 2 equations (3.161) are valid with the change $1 \longleftrightarrow 2$ and $\omega_0 \longleftrightarrow -\omega_0$, $\omega_{\boldsymbol{k}} \longleftrightarrow -\omega_{\boldsymbol{k}}$ in the exponentials. The absorption rate and the absorption distribution are the same as those of emission, and have the same form as the classical ones.

3.16 Propagation of light, interference

Let us assume that we have two atoms, separated by distance r. We denote their states by n, n', such that the fundamental equations (3.148) read

$$i\hbar\dot{c}_{nn'} = \sum_{mm'} c_{mm'}(H_i)_{nn',mm'} e^{\frac{i}{\hbar}(E_n+E_{n'}-E_m-E_{m'})t} \ . \quad (3.174)$$

One of the atoms decays quickly with a broad spectrum of photons, to its ground-state $m = 1$, while the second atom remains in its ground-state $m' = 1$. This $mm' = 11$ state is the initial state in equations (3.174). The coefficient c_{11} is given by equation (3.162) at $t = 0$, including the spatial factor of the emitting atom placed at $\boldsymbol{r}_1$,

$$c_{11\boldsymbol{k}} = \frac{q\alpha_k\sqrt{n_{\boldsymbol{k}}+1}}{\hbar mc\sqrt{V}}(\boldsymbol{e_k p}_{12})\frac{e^{-i\boldsymbol{kr}_1}}{\omega_0-\omega_k-i\gamma} \ ; \quad (3.175)$$

of course, p_{12} is the matrix element for the decaying atom. We are interested in the final state of the excited second atom, which absorbed the photon emitted by the former atom, *i.e.* $n = 1$ and $n' = 2$. From equations (3.174) we get

$$i\hbar\dot{c}_{12} = \sum_{\boldsymbol{k}} c_{11\boldsymbol{k}}(H_1)_{12,11} e^{\frac{i}{\hbar}(E_2^{'}-E_1^{'}-\hbar\omega_k)t} \ , \quad (3.176)$$

or

$$i\hbar\dot{c}_{12} = -\sum_{\boldsymbol{k}} c_{11\boldsymbol{k}}\frac{q\alpha_k\sqrt{n_{\boldsymbol{k}}+1}}{mc\sqrt{V}}(\boldsymbol{e}_k\boldsymbol{p}_{21}^{'})e^{i\boldsymbol{kr}_2}e^{i(\omega_0^{'}-\omega_k)t} \ , \quad (3.177)$$

$$\dot{c}_{12} = \sum_{\boldsymbol{k}} \frac{2\pi i q^2(n_{\boldsymbol{k}}+1)}{\hbar m^2\omega_k V}(\boldsymbol{e_k p}_{12})(\boldsymbol{e}_k\boldsymbol{p}_{21}^{'})\cdot$$
$$\cdot e^{i\boldsymbol{k}(\boldsymbol{r}_2-\boldsymbol{r}_1)}\frac{e^{i(\omega_0^{'}-\omega_k)t}}{\omega_0-\omega_k-i\gamma} \ , \quad (3.178)$$

and

$$c_{12} = \sum_{\boldsymbol{k}} \frac{2\pi q^2(n_{\boldsymbol{k}}+1)}{\hbar m^2\omega_k V}(\boldsymbol{e_k p}_{12})(\boldsymbol{e}_k\boldsymbol{p}_{21}^{'})e^{i\boldsymbol{k}(\boldsymbol{r}_2-\boldsymbol{r}_1)}\cdot$$
$$\cdot\frac{e^{i(\omega_0^{'}-\omega_k)t}-1}{(\omega_0-\omega_k-i\gamma)(\omega_0^{'}-\omega_k)} \ . \quad (3.179)$$

We average this expression over the directions of $\boldsymbol{k}$ and the polarizations. The average over directions can be done by fixing the direction of $\boldsymbol{k}$ and averaging over the directions of $\boldsymbol{r} = \boldsymbol{r}_2 - \boldsymbol{r}_1$. We use the parametrization

$$\begin{gathered} \boldsymbol{r} : \, (0,0,1) \; , \\ \boldsymbol{k} : \, (\sin\theta\cos\varphi, \sin\theta\sin\varphi, \cos\theta) \; , \\ \boldsymbol{e}_{\boldsymbol{k}} = (-\sin\varphi\cos\Phi - \cos\theta\cos\varphi\sin\Phi, \\ \cos\varphi\cos\Phi - \cos\theta\sin\varphi\sin\Phi, \sin\theta\sin\Phi) \end{gathered} \tag{3.180}$$

and get immediately

$$\frac{1}{4}(p_1 p_1^{'} + p_2 p_2^{'})(1+\cos^2\theta) + \frac{1}{2} p_3 p_3^{'} \sin^2\theta \; ; \tag{3.181}$$

the average over θ leads to

$$\overline{(\boldsymbol{e}_{\boldsymbol{k}} \boldsymbol{p}_{12})(\boldsymbol{e}_k \boldsymbol{p}_{21}^{'}) e^{i\boldsymbol{k}\boldsymbol{r}}} = \frac{1}{2}(p_1 p_1^{'} + p_2 p_2^{'}) \frac{\sin kr}{kr} \tag{3.182}$$

for $kr \gg 1$. By using the notation $\frac{1}{2}(p_1 p_1^{'} + p_2 p_2^{'}) = m^2 \omega_0 \omega^{'} (uu')$ we get

$$\begin{gathered} c_{12} = \frac{1}{r} \frac{q^2}{\pi \hbar c^2} \omega_0 \omega_0^{'} (uu') \int d\omega (n_\omega + 1) \cdot \\ \cdot \frac{\sin \frac{\omega r}{c} \left[^{i(\omega_0^{'} - \omega)t} - 1 \right]}{(\omega_0 - \omega - i\gamma)(\omega_0^{'} - \omega)} \; . \end{gathered} \tag{3.183}$$

The main contribution to the integral arises from $\omega \simeq \omega_0^{'}$, so we may take the other, regular, factors out of the integral; similarly, we may extend the integral to $-\infty$. Then, we use the new variable $\omega - \omega_0^{'} = x$ and the integrals

$$\begin{gathered} \int_{-\infty}^{+\infty} dx \frac{\sin px}{x} = \pi sgn(p) \; , \\ \int_{-\infty}^{+\infty} dx \frac{\sin px \cos qx}{x} = \pi\theta(\mid p \mid > q \; , \; (q>0) \; ; \end{gathered} \tag{3.184}$$

finally, we get

$$c_{12} = \frac{q^2 (n_0^{'} + 1)}{\hbar c^2} \frac{\omega_0 \omega_0^{'} (uu')}{\omega_0 - \omega_0^{'} - i\gamma} \begin{cases} \frac{e^{i\omega_0^{'} r/c}}{r} \; , \; r < ct \\ 0 \; , \; r > ct \; ; \end{cases} \; , \tag{3.185}$$

as expected.

The probability to find the second atom in an excited state is

$$| c_{12} |^2 \sim \frac{1}{(\omega_0 - \omega_0^{'})^2 + \gamma^2} , \tag{3.186}$$

which has the form of the classical absorption.

Let us assume now a perfectly reflecting mirror placed at the origin and two atoms placed to the left at distances x_1 and x_2 on a line perpendicular to the mirror. The atom 1 emits radiation with a short lifetime and a broad spectrum. The distance x_1 is sufficiently large for viewing the emitted radiation as consisting of plane waves. The atom 2 absorbs radiation, with the transition coefficient given by equation (3.179). The spatial factors $e^{i\boldsymbol{kr}}$ in this equation are replaced by $\sin kx_{1,2}$, since the mirror is perfectly reflecting the radiation. The product $\sin kx_1 \sin kx_2$ is $\frac{1}{2}[\cos k(x_1 - x_2) - \cos k(x_1 + x_2)]$. The average over directions implies

$$\frac{1}{2} \int d\theta \sin\theta \cos(kx\cos\theta) = \frac{1}{kx} \sin kx , \tag{3.187}$$

such that we get the factor

$$\frac{1}{x_-} \sin kx_- - \frac{1}{x_+} \sin kx_+ , \tag{3.188}$$

where $k = \omega/c$ and $x_\pm = x_1 \pm x_2$. Now, multiplying by the factor $\left[e^{i(\omega_0^{'} - \omega)t} - 1\right] / (\omega_0^{'} - \omega)$ in equation (3.179) and performing the integration over ω we get the factor

$$\frac{1}{x_-} e^{i\omega_0^{'} x_- / c} - \frac{1}{x_+} e^{i\omega_0^{'} x_+ / c} \tag{3.189}$$

in the transition coefficient c_{12} for sufficiently long times. This is an interference factor, and it has a maximum value for

$$\frac{\omega_0^{'}}{c}(x_1 - x_2) - \frac{\omega_0^{'}}{c}(x_1 + x_2) = 2\pi n , \tag{3.190}$$

where n is an integer; we get $x_2 = \frac{\lambda'}{2} \times \textit{integer}$, which is the condition of occurrence of Lippman's fringes. We can see that the quantum theory of radiation recovers the interference condition of the classical radiation.[16]

3.17 Doppler effect

Let us consider two pointlike charges $\pm q$ with masses $m_{1,2}$ placed at $\boldsymbol{r}_{1,2}$ with the hamiltonian

$$\begin{aligned} H = \tfrac{1}{2m_1}p_1^2 + \tfrac{1}{2m_2}p_2^2 + V(\boldsymbol{r}_1 - \boldsymbol{r}_2) + H_{0rad} - \\ -\tfrac{q}{m_1 c}(\boldsymbol{A}_1\boldsymbol{p}_1) + \tfrac{q}{m_2 c}(\boldsymbol{A}_2\boldsymbol{p}_2) + \tfrac{q^2}{2m_1c^2}A_1^2 + \tfrac{q^2}{2m_2c^2}A_2^2 \ , \end{aligned} \tag{3.191}$$

where $\boldsymbol{p}_{1,2}$ are the momenta of the two particles, V is an interaction, H_{0rad} is the free hamiltonian of the radiation field and the rest of terms are the charges-radiation interaction in the non-relativistic approximation (including the $1/c^2$-correction). We introduce the centre-of-mass coordinate

$$\boldsymbol{R} = \frac{m_1\boldsymbol{r}_1 + m_2\boldsymbol{r}_2}{m_1 + m_2} \tag{3.192}$$

and the relative coordinate $\boldsymbol{r} = \boldsymbol{r}_1 - \boldsymbol{r}_2$, such that

$$\begin{aligned} \boldsymbol{r}_1 = \boldsymbol{R} + \tfrac{m_2}{M}\boldsymbol{r} \ , \ \boldsymbol{r}_2 = \boldsymbol{R} - \tfrac{m_1}{M}\boldsymbol{r} \\ \boldsymbol{p}_1 = \tfrac{m_1}{M}\boldsymbol{P} + \boldsymbol{p} \ , \ \boldsymbol{p}_2 = \tfrac{m_2}{M}\boldsymbol{P} - \boldsymbol{p} \ , \end{aligned} \tag{3.193}$$

$$\boldsymbol{P} = M\dot{\boldsymbol{R}} = \boldsymbol{p}_1 + \boldsymbol{p}_2 \ , \ \boldsymbol{p} = \mu\dot{\boldsymbol{r}} = \mu\left(\frac{\boldsymbol{p}_1}{m_1} - \frac{\boldsymbol{p}_2}{m_2}\right) \ , \tag{3.194}$$

where $M = m_1 + m_2$ is the total mass and $\mu = m_1 m_2/M$ is the reduced mass. We assume that the dimension of the assembly of the two particles is much smaller than the radiation wavelength (dipole approximation), such that the hamiltonian becomes

$$\begin{aligned} H = \tfrac{1}{2M}P^2 + \tfrac{1}{2\mu}p^2 + V(\boldsymbol{r}) + H_{0rad} - \\ -\tfrac{q}{\mu c}\boldsymbol{A}(\boldsymbol{R})\boldsymbol{p} + \tfrac{q^2}{2\mu c^2}A^2(\boldsymbol{R}) \ . \end{aligned} \tag{3.195}$$

[16]E. Fermi, "Quantum theory of radiation", Revs. Mod. Phys. **4** 87 (1932).

We note the coupling of the centre of mass ($\boldsymbol{R}$) to the internal motion ($\boldsymbol{p}$). In the matrix elements of the transition coefficients (equations (3.149)) the free wavefunctions of the centre of mass $\frac{1}{\sqrt{V}}e^{i\boldsymbol{K}\boldsymbol{R}}$ and the wavefunctions of the relative coordinate appear; in addition, $\boldsymbol{A}(\boldsymbol{R})$ includes factors $e^{\pm i\boldsymbol{k}\boldsymbol{R}}$ (or products of such factors), such that these matrix elements are proportional to

$$\frac{1}{V}\int d\boldsymbol{R}e^{-i\boldsymbol{K}'\boldsymbol{R}+i\boldsymbol{K}\boldsymbol{R}\pm i\boldsymbol{k}\boldsymbol{R}} = \delta_{\boldsymbol{K}',\boldsymbol{K}\pm\boldsymbol{k}} \; ; \tag{3.196}$$

this is the momentum conservation $\boldsymbol{K}^{'} = \boldsymbol{K}\pm\boldsymbol{k}$, where $\hbar\boldsymbol{K}'$ is the final momentum of the centre of mass, $\hbar\boldsymbol{K}$ is the initial momentum of the centre of mass and $\hbar\boldsymbol{k}$ represents the photon momenta. In addition, from equations (3.149) we have the energy conservation $E_f = E_i$ in the final and initial states.

For instance, let we assume the emission of a photon with momentum $\hbar\boldsymbol{k}$ and energy $\hbar\omega_k$ by the internal transition between two states with energy difference $\hbar\omega_0$ (these states are generated by the interaction $V(\boldsymbol{r})$). The momentum and energy conservation reads

$$\begin{gathered}\boldsymbol{K}' = \boldsymbol{K} - \boldsymbol{k} \; , \\ \frac{\hbar^2K'^2}{2M} + \hbar\omega_k = \frac{\hbar^2K^2}{2M} + \hbar\omega_0 \; ;\end{gathered} \tag{3.197}$$

from these two equations we get the photon frequency

$$\omega_k = \omega_0\left(1 + \frac{V}{c}\cos\theta\right) \; , \tag{3.198}$$

where $\boldsymbol{V} = \hbar\boldsymbol{K}/M$ is the initial velocity of the centre of mass and θ is the angle between $\boldsymbol{K}$ and $\boldsymbol{k}$ (emission angle); this is the Doppler effect (in the non-relativistic approximation).[17] It is worth noting that the Doppler effect is zero if the centre of mass is initially at rest.

[17]The frequency in the moving frame is $\omega' = \omega_0(1 - \frac{V}{c}\cos\alpha)/\sqrt{1 - V^2/c^2}$, where α is the angle between $\boldsymbol{V}$ and the x-axis of the rest frame. In our case the rest frame is the $\boldsymbol{K}'$-particle and the moving frame is the frame where the momentum of the particle is $\boldsymbol{K} - \boldsymbol{K}' = \boldsymbol{k}$. Therefore, the angle α is equal to θ, but the velocity is $-\boldsymbol{V}$; the Doppler effect is $\omega = \omega_0(1 + \frac{V}{c}\cos\alpha)/\sqrt{1 - V^2/c^2}$, which, in the non-relativistic limit, is equation (3.198).

For the emission of a photon by such an assembly of two bound charges equations (3.165) and (3.170) for the damping coefficient and the number of transitions per unit time apply, providing the factor ω_0^3 is replaced by $\omega_0^2\omega$ and n_0 is replaced by n_ω, where $\omega = \omega_k$ (and the mass is the reduced mass). For instance, the number of transitions per unit time in the interval $d\omega$ and the solid angle do is

$$d^2w = \frac{q^2(n_0+1)}{3\pi\hbar c^3}\omega_0^2\omega\left|u_{12}\right|^2\cdot \\ \cdot\delta\left(\frac{\hbar K'^2}{2M}+\omega-\frac{\hbar K^2}{2M}-\omega_0\right)d\omega do\ , \tag{3.199}$$

or

$$d^2w = \frac{q^2(n_0+1)}{3\pi\hbar c^3}\omega_0^2\omega\left|u_{12}\right|^2\delta\left(\omega-\omega_0(1+\frac{V}{c}\cos\theta\right)d\omega do \tag{3.200}$$

(where n_0 is an average over directions). By using this equation we can compute the momentum and the energy transferred to the centre of mass, as well as the radiated energy, per unit time. The transferred energy and the radiated energy are not affected by the Doppler, within this approximation. The transferred momentum is $\hbar\boldsymbol{k}$. The total transferred momentum per unit time is the force; it is non-zero only along the forward direction (the z-direction); we get

$$f_z = \frac{4q^2(n_0+1)}{9c^4}\omega_0^4\left|u_{12}\right|^2\cdot\frac{V}{c}\ . \tag{3.201}$$

3.18 Scattering: first-order perturbation

We consider an initial state $\mid i >=\mid n_{\boldsymbol{k}}+1>$ and a final state $\mid f >=\mid n_{\boldsymbol{k}}, 1_{\boldsymbol{k}'}>$, where $n_{\boldsymbol{k},\boldsymbol{k}'}$ are the occupation numbers of the photon states: a $\boldsymbol{k}$-photon is scattered into a $\boldsymbol{k}'$-photon. Obviously, the interaction hamiltonian $-\frac{q}{\mu c}\boldsymbol{A}(\boldsymbol{R})\boldsymbol{p}$ brings a zero contribution to the first-order of the perturbation theory. To this order of the perturbation theory we use the interaction hamiltonian

$$H_2 = \frac{q^2}{2\mu c^2}A^2(\boldsymbol{R}) \tag{3.202}$$

from equation (3.195), where

$$A(R) = \frac{1}{\sqrt{V}} \sum_{k} \alpha_k e_k \left(a_k e^{ikR} + a_k^* e^{-ikR} \right) \tag{3.203}$$

(equation (3.80)). The number of transitions per unit time is $w_{fi} = \frac{2\pi}{\hbar} |(H_2)_{fi}|^2 \delta(E_f - E_i)$ (equation (3.152)). It is convenient to give up the notations $\boldsymbol{k}$ and $\boldsymbol{k}'$, and use $'$ for the latter. The matrix elements arise from products aa'^* and a'^*a; the spatial factors give the momentum conservation $\boldsymbol{K}' + \boldsymbol{k}' = \boldsymbol{K} + \boldsymbol{k}$, where $\hbar\boldsymbol{K}$, $\hbar\boldsymbol{K}'$ are the momenta of the centre of mass; we get

$$\begin{aligned} w_{fi} = (2\pi)^3 \left(\frac{q^2}{2\mu c^2} \right)^2 \frac{\hbar c^4}{\omega\omega' V^2} (n+1) \mid (\boldsymbol{e}\boldsymbol{e}' \mid^2 \cdot \\ \cdot \delta \left(\frac{\hbar^2 K'^2}{2M} + \hbar\omega' - \frac{\hbar^2 K^2}{2M} - \hbar\omega \right) . \end{aligned} \tag{3.204}$$

The momentum and energy conservation leads to $k \simeq k'$, since the Compton momentum Mc is much larger than the photon momentum $\hbar k$. The scattering is elastic. Therefore, we may write

$$w_{fi} = (2\pi)^3 \left(\frac{q^2}{2\mu c^2} \right)^2 \frac{c^4}{\omega^2 V^2} (n+1) \mid (\boldsymbol{e}\boldsymbol{e}' \mid^2 \delta(\omega' - \omega) . \tag{3.205}$$

In order to average over polarizations we use equation (3.180) with $\varphi = 0$ for $\boldsymbol{e}'$ and $\boldsymbol{e} = (\cos\psi, \sin\psi, 0)$; we get $\overline{\mid (\boldsymbol{e}\boldsymbol{e}' \mid^2} = \frac{1}{4}(1 + \cos^2\theta)$, where θ is the scattering angle (the angle between $\boldsymbol{k}$ and $\boldsymbol{k}'$). Now, we multiply by the number of states $2V\omega'^2 d\omega' do/(2\pi)^3 c^3$ and integrate over ω', with the result

$$dw_{fi} = \frac{1}{2} \left(\frac{q^2}{2\mu c^2} \right)^2 \frac{c}{V} (n+1)(1 + \cos^2\theta) do . \tag{3.206}$$

In the classical limit for the incident radiation $n \gg 1$ and the energy flux density of the incident photons is $cE_0^2/4\pi$, where E_0 is the amplitude of the electric field. Therefore, the flux density of the incident photons is $cE_0^2/4\pi\hbar\omega$, such that we get the differential cross-section

$$d\sigma = \frac{1}{2} \left(\frac{q^2}{2\mu c^2} \right)^2 \frac{4\pi\hbar\omega}{VE_0^2} n(1 + \cos^2\theta) do ; \tag{3.207}$$

according to equation (3.169) $VE_0^2 = 4\pi\hbar\omega n$, such that we get the differential cross-section

$$d\sigma = \frac{1}{2}\left(\frac{q^2}{2\mu c^2}\right)^2 (1+\cos^2\theta)do \ , \qquad (3.208)$$

which is the classical Thomson cross-section (equation (3.22)).

3.19 Scattering: second-order perturbation

Now we use the interaction hamiltonian $H_1 = -\frac{q}{\mu c}\boldsymbol{A}(\boldsymbol{R})\boldsymbol{p}$ given by equation (3.195) in the second-order of the perturbation theory. The number of transitions per unit time is

$$w_{fi} = \frac{2\pi}{\hbar}\left|\sum_n \frac{(H_1)_{fn}(H_1)_{ni}}{E_n - E_i - i\alpha}\right|^2 \delta(E_f - E_i) \qquad (3.209)$$

(equation (3.154)). Similarly, the matrix elements arise from products aa'^* and a'^*a, where the intermediate states n give denominators $E_n - E_i - \hbar\omega - i\alpha$, with $E_n - E_i > 0$ in the first case (absorption and emission) and $E_n - E_i + \hbar\omega' - i\alpha$, with $E_n - E_i < 0$ in the second case (emission and absorption), where $E_{n,i}$ denote the energies of the atom states. The spatial factors give the momentum conservation $\boldsymbol{K}' + \boldsymbol{k}' = \boldsymbol{K} + \boldsymbol{k}$ of the centre of mass. We get

$$w_{fi} = \frac{(2\pi)^3\hbar q^4}{\mu^4 V^2}\frac{n+1}{\omega\omega'}\cdot$$

$$\cdot\left|\sum\nolimits_n \frac{(\boldsymbol{e}'\boldsymbol{p}_{fn})(\boldsymbol{e}\boldsymbol{p}_{ni})}{E_n - E_i - \hbar\omega - i\alpha} + \sum\nolimits_n \frac{(\boldsymbol{e}\boldsymbol{p}_{fn})(\boldsymbol{e}'\boldsymbol{p}_{ni})}{E_n - E_i + \hbar\omega' - i\alpha}\right|^2 \cdot \qquad (3.210)$$

$$\cdot\delta(\Delta E_c + E_f + \hbar\omega' - E_i - \hbar\omega) \ ,$$

where $\Delta E_c = \hbar^2 K'^2/2\mu - \hbar^2 K^2/2\mu$ is the variation of the energy of the centre of mass. The first term in equation (3.210) corresponds to the absorption of the ω-photon and the transition from the initial state i to an upper intermediate state n, followed by the emission of the ω'-photon and the decaying to the lower final state f. The second term in equation (3.210) corresponds to the emission of the ω'-photon from the initial state i and the transition from the initial state i to a lower intermediate state n (which presuposes that i is an excited state), followed by the absorption of the ω-photon and the excitation to the upper final state f. If $i = f$ and $\omega = \omega'$ the scattering is elastic; this corresponds to the classical Rayleigh scattering. If $i \neq f$, $\omega \neq \omega'$ the scattering is the Raman scattering: for $\omega' < \omega$ the scattering is the Stokes scattering, for $\omega' > \omega$ the scattering is the anti-Stokes scattering.[18] For free charges $E_n = E_i$, such that we get the Compton effect (in the non-relativistic approximation).[19] Equation (3.210) was derived in the dispersion theory of light[20] before the advent of the Quantum Mechanics.[21]

In equation (3.210) we may neglect ΔE_c because the Compton momentum μc is very large in comparison with the photon momentum. We are left with $\delta(E_f + \hbar\omega' - E_i - \hbar\omega)$, where $\omega' > 0$ is equal to $\omega - \omega_{fi} > 0$, because $\omega_{fi} = (E_f - E_i)/c$ is always smaller than ω. The energy conservation leads to $\omega' = \omega - \omega_{fi}$, *i.e.* $k' = k - \omega_{fi}/c$ (> 0). From $\boldsymbol{K}' = \boldsymbol{k} - \boldsymbol{k}'$ we may get the momentum and the energy transferred to the centre of mass (assuming the particle initially at rest). Since

$$K'^2 = -4k\frac{\omega_{fi}}{c}\sin^2\frac{\theta}{2} + \frac{\omega_{fi}^2}{c^2} > 0 \ , \tag{3.211}$$

where θ is the angle between $\boldsymbol{k}$ and $\boldsymbol{k}'$ (the scattering angle), we get the conditions $\omega_{fi} < 0$ ($\omega' > \omega$, anti-Stokes scattering), or $\omega_{fi} > 4\omega\sin^2\frac{\theta}{2}$, *i.e.* $\sin\frac{\theta}{2} < 1/2$, $\omega' < \omega(1 - 4\sin^2\frac{\theta}{2}) < \omega$ (Stokes scattering).

---

[18] A. Smekal, "Zur Quantentheorie der Dispersion", Naturwiss. **11** 873 (1923); G. Landsberg and L. Mandelstam, "Eine neue Erscheinung bei der Lichtzerstreuung in Kristallen, Naturwiss. **16** 557, 772 (1928); C. V. Raman and K. S. Krishnan, "A new type of secondary radiation", Nature **121** 501 (1928); G. Placzek, *Rayleigh-Streuung und Raman Effekt*, in *Handbuch der Radiologie*, ed. E. Marx, Akademische Verlag, Leipzig (1934), vol. 6, p 205.

[19] The interaction hamiltonian H_1 remains valid for a relativistic charge, but the matrix elements are different; moreover, we can get the elastic scattering in the relativistic case, equivalent to the Thomson scattering, through particle-antiparticle excitations.

[20] H. A. Kramers and W. Heisenberg, "Ueber die Streuung von Strahlen durch Atome", Z. Phys. **31** 681 (1925).

[21] W. Heisenberg, "Ueber quantentheoretische Umdeutung kinematischer und mechanischer Beziehungen", Z. Phys. **33** 879 (1925).

Let us derive now the classical Rayleigh scattering from equation (3.210). In order to do this we assume $E_i = E_f$, multiply by the number of states $2V\omega'^2 d\omega' do/(2\pi)^3 c^3$ and integrate over ω'. Also, we assume a linear harmonic oscillator with frequency $\omega_0 = \omega$. From equation (3.210) we get

$$dw = \frac{2q^4(n+1)}{\mu^4 V\hbar^2 c^3} \mid (\boldsymbol{ep})(\boldsymbol{e'p}) \mid^2 \frac{1}{(\alpha/2)^2} do \rightarrow$$
$$\rightarrow \frac{2q^4(n+1)}{\mu^4 V\hbar^2 c^3} \mid (\boldsymbol{ep})(\boldsymbol{e'p}) \mid^2 \frac{1}{(\omega-\omega_0)^2+(\alpha/2)^2} do \ , \tag{3.212}$$

where $\boldsymbol{p}$ is the matrix element $i\mu\omega_0\boldsymbol{u} = i\mu\sqrt{n\hbar\omega_0/2\mu}\boldsymbol{s}$, with $n = 1$ (ground-state) and $\boldsymbol{s}$ a unit vector; equation (3.212) becomes

$$dw = \frac{q^4(n+1)}{2\mu^2 V c^3} \mid (\boldsymbol{es})(\boldsymbol{e's}) \mid^2 \frac{\omega_0^2}{(\omega-\omega_0)^2+(\alpha/2)^2} do \ . \tag{3.213}$$

The average over directions and polarizations is made by $e_i e_j^{'} s_i s_j \rightarrow (e_i e_i^{'}) s_i^2$, where $\boldsymbol{e} = (\cos\psi, \sin\psi, 0)$ and for $\boldsymbol{e}'$ we use equation (3.180); we get $\mid (\boldsymbol{es})(\boldsymbol{e's}) \mid^2 = \frac{1}{4}(1+\cos^2\theta)$ and

$$dw = \frac{q^4(n+1)}{8\mu^2 V c^3}(1+\cos^2\theta)\frac{\omega_0^2}{(\omega-\omega_0)^2+(\alpha/2)^2} do \ . \tag{3.214}$$

In the classical limit $n \gg 1$ and we use $VE_0^2 = 4\pi\hbar\omega_0 n$, where E_0 is the amplitude of the electric field, to get

$$dw = \frac{q^4 E_0^2}{32\pi\hbar\mu^2 c^3}(1+\cos^2\theta)\frac{\omega_0}{(\omega-\omega_0)^2+(\alpha/2)^2} do \ . \tag{3.215}$$

We can use this equation for computing the force f along the forward direction (momentum $\hbar k(1-\cos\theta)$ transferred per unit time)

$$f = \frac{q^4 E_0^2}{6\mu^2 c^4} \frac{\omega_0^2}{(\omega-\omega_0)^2+(\alpha/2)^2} =$$
$$= \frac{1}{6} a^2 E_0^2 \frac{\omega_0^2}{(\omega-\omega_0)^2+(\alpha/2)^2} \ , \tag{3.216}$$

where $a = q^2/\mu c^2$ is the classical radius of the charge (the energy transfer is zero).

By dividing the number of transitions per unit time given by equation (3.215) to the incident photon flux density $cE_0^2/4\pi\hbar\omega$, we get the differential cross-section

$$d\sigma = \frac{1}{8}a^2(1+\cos^2\theta)\frac{\omega_0^2}{(\omega-\omega_0)^2+(\alpha/2)^2}do \ , \qquad (3.217)$$

which is the Rayleigh cross-section given by equations (3.23) and (3.29).

It is worth noting that the derivation given here is for resonance fluorescence, when ω is close to ω_0; the exact formula is

$$d\sigma = \frac{1}{2}a^2(1+\cos^2\theta)\frac{\omega^4}{(\omega-\omega_0)^4+\alpha^2}do \ ; \qquad (3.218)$$

for $\omega_0 = 0$ we get the Thomson formula. Indeed, under the classical conditions used here the interaction term $\frac{q^2}{2\mu c^2}A^2$ should be neglected; it is remarkable that its contribution is given by the term $-\frac{q}{\mu c}\boldsymbol{Ap}$ in the second order of the perturbation theory.

It may happen that an intermediate state in equation (3.210) be at a frequency close to ω with respect to the initial state; in addition, it is quasi-discrete, *i.e.* it has a certain width Γ_n; then, it gives the main contribution to the summation over n, such that we may limit ourselves only to this term, where α is replaced by $\Gamma_n/2$.[22] Similarly, an excited state which emits a photon may have a width Γ, so we should be interested in the total number of transitions

$$w_t = w\Delta t = w\frac{1}{\Gamma} \ , \qquad (3.219)$$

where w is the number of transitions per unit time; on the other hand, we can write

$$\frac{2\pi}{\Gamma} = \int d\omega \frac{1}{\omega_0^2+\frac{1}{4}\Gamma^2} \ , \qquad (3.220)$$

[22] V. F. Weisskopf, "Zur Theorie der Resonanzfluoreszenz", Ann. Phys. **9** 23 (1931).

such that the spectral distribution of the emitted line ω_0 is[23]

$$dw_t = w \cdot \frac{1}{2\pi} \frac{d\omega}{\omega_0^2 + \frac{1}{4}\Gamma^2} \ . \tag{3.221}$$

Finally we note that the vector potential can be removed from the hamiltonian by the Goeppert-Mayer canonical transformation[24] and the Kramers-Henneberger canonical transformation,[25] which have the effect of shifting the position coordinate. This transformation was used for scattering of charges by laser fields[26] and for fast atom ionization.[27]

[23] V. F. Weisskopf and E. Wigner, "Berechnung der natuerlichen Linienbreite auf Grund der Diracschen Lichttheorie", Z. Phys. **63** 54 (1930); "Ueber die naturliche Linienbreite in der Strahlung des harmonischen Oszillators", Z. Phys. **65** 18 (1930).

[24] M. Goeppert-Mayer, "Ueber Elementarakte mit zwei Quantensprungen", Ann. Phys. **9** 273 (1931).

[25] W. C. Henneberger, "Perturbation method for atoms in intense light beams", Phys. Rev. Lett. **21** 838 (1968); H. A. Kramers, "Non-relativistic quantum-electrodynamics and the correspondence principle", Solvay Congress 1948, in *Collected Scientific Papers*, North-Holland, Amsterdam (1956), p. 845; N. G. van Kampen, "Contribution to the quantum theory of light scattering", Kgl. Danske Videnskab. Selskab, Mat.-fys. Medd. **26** (15) (1951); W. Pauli and M. Fierz, "Zur Theorie der Emission langwelliger Lichtquanten", Nuovo Cimento **15** 167 (1938).

[26] M. Apostol, "Scattering of non-relativistic charged particles by electromagnetic radiation", Z. Naturforsch. **72a** 1173 (2017); M. Apostol amd L. C. Cune, "Analysis of the cross-section of charge scattering by electromagnetic radiation", Roum J. Phys. **63** 110 (2018).

[27] M. Apostol, "Fast atom ionization in strong electromagnetic radiation", Z. Naturforsch. **73a** 461 (2018).

4 X-Rays Diffraction

4.1 X-rays

Typical energies of the X-rays are of the order $10keV$ (frequency $\omega \simeq 10^{19}s^{-1}$). The corresponding wavelength is $\lambda \sim 1\text{Å}$, which is the order of magnitude of the inter-atomic distance in condensed matter (solids, liquids). Typical flux densities of the X-ray sources are $\Phi = 10^{12} photons/cm^2 \cdot s$.[1] The X-rays act upon the atomic charges, mainly by the electric field. The electric field in the X-rays is macroscopic (classical, with low intensity). The coupling is mainly with the outer electrons of the atoms (molecules), since the atomic nuclei have a much higher inertia and they are screened by inner electrons. The X-rays energy is much higher than the energy of the (active) electrons in condensed matter ($\omega \simeq 10^{16}s^{-1}$), so the electrons are forced to oscillate with the frequency of the incoming X-rays and generate outgoing X-rays. Therefore, the X-rays scattering is mainly an elastic scattering.

Let us assume an external electric field $\boldsymbol{E}_0 \sim e^{i\boldsymbol{kr}-i\omega t}$ with frequency $\omega = ck$ and wavelength $\lambda = 2\pi/k = 2\pi c/\omega$ (where $c = 3 \times 10^{10}cm/s$ is the speed of light in vacuum). We assume also that this is a macroscopic (classical) field, *i.e.* the number of photons (occupation number of the $\boldsymbol{k}$-state) is much larger than unity. Then, the force $q\boldsymbol{E}_0$ acting on a pointlike charge q is a classical quantity. The force defines a displacement $\boldsymbol{u}$ of the charge, by the interaction energy $-q\boldsymbol{E}_0\boldsymbol{u}$. The displacement field $\boldsymbol{u}$ can be either classical or quantum-mechanical (the interaction can be so). The characteristic length of variation of the displacement $\boldsymbol{u}$ is λ. Since the external field is weak, the displacement $\boldsymbol{u}$ is small; usually, it is much smaller than λ. If λ is larger than

[1] A. H. Compton and S. K. Allison, *X-rays in Theory and Experiment*, Van Nostrand, NY (1935).

the spatial range of variation of the wavefunctions of the charge (*e.g.*, the atomic dimension, or the inter-atomic distance), the displacement $\boldsymbol{u}$ has a slow variation over this spatial range of the wavefunctions of the charge, so the matrix elements of the displacement are reduced, approximately, to the mean value of the displacement. In this case, the displacement is, practically, a classical variable. However, undetermined phase factors of the wavefunctions in the matrix elements may exist. If the wavelength λ is much shorter than the spatial range of variation of the wavefunctions, the displacement may vary appreciably over this range, so its matrix elements between distinct states are not vanishing. Then, the displacement is a quantum-mechanical variable. Even if a classical treatment is valid, the process of interaction is absorption and emission of photons, so it is a quantum-mechanical process and, therefore, a statistical process; undetermined phase factors are preserved, attached to the "classical" variable u. The final state of scattering corresponds to a distinct state, even it has the same energy, the same quantum numbers and is non-degenerate.

For a flux density Φ of the radiation the energy density is $w = E_0^2/4\pi = \Phi\hbar\omega/c$, such that the amplitude of the electric field is $E_0 = \sqrt{4\pi\Phi\hbar\omega/c}$. For $\Phi = 10^{12} photons/cm^2 \cdot s$ and $\omega = 10^{19}s^{-1}$ the electric field is $E_0 \simeq 10^{-3}esu$. For an electron with mass m and charge e ($m = 10^{-27}g$, $e = -4.8 \times 10^{-10}esu$) the amplitude of the displacement $u = eE_0/m\omega^2$ is $u \simeq 10^{-24}cm$ (and the electron velocity is $10^{-5}cm/s$). We can see that $u \ll \lambda$, and $\lambda = 10^{-8}cm$ is of the order of the atomic dimensions. Therefore, the dynamics of the displacement $\boldsymbol{u}$ may be viewed as a classical dynamics. Usually, the condition $u \ll \lambda$ is fulfilled, but the existence of $\boldsymbol{u}$ (the existence of a force), or the condition λ larger than the spatial range of the wavefunctions is not always met. For instance, for more energetic X-rays we need a quantum-mechanical treatment. In this case the X-rays can produce atomic excitations, photoelectric effect, Auger effect, or, for higher energies (gamma rays), Compton effect, which are quantum-mechanical processes.

4.2 Extinction

Let us assume a homogeneous continuous medium with pointlike charges q, with mass m and uniform density n. We assume that these charges can be displaced from their equilibrium positions by an electric field. Also, we assume that these charges obey a classical equation of motion under the action of the electric field. A small, local displacement $\boldsymbol{u}$ generates a local charge density imbalance $\delta\rho = -qn div\boldsymbol{u}$ and a current density $\delta\boldsymbol{j} = nq\dot{\boldsymbol{u}}$. Therefore, internal electric and magnetic fields appear. The equation of motion for the displacement $\boldsymbol{u}$ is

$$m\ddot{\boldsymbol{u}} + m\omega_c^2\boldsymbol{u} + m\gamma\dot{\boldsymbol{u}} = q(\boldsymbol{E}_0 + \boldsymbol{E}) \ , \tag{4.1}$$

where ω_c is a characteristic frequency of the equilibrium state of the charges, γ is a (small) damping coefficient which governs the dissipation, $\boldsymbol{E}_0$ is the external electric field and $\boldsymbol{E}$ is the internal electric field. We may neglect the spatial dispersion in the equation of motion (arising from inhomogeneities and the interaction of the charges); the displacement is a smooth function. The characteristic frequency ω_c is of the order of low atomic frequencies in dielectrics, while $\omega_c = 0$ in metals. It is convenient to introduce the polarization $\boldsymbol{P} = nq\boldsymbol{u}$, such that the equation of motion can be written as

$$\ddot{\boldsymbol{P}} + \omega_c^2\boldsymbol{P} + \gamma\dot{\boldsymbol{P}} = \frac{\omega_p^2}{4\pi}(\boldsymbol{E}_0 + \boldsymbol{E}) \ , \tag{4.2}$$

where $\omega_p = (4\pi nq^2/m)^{1/2}$ is the plasma frequency. For electrons ($n = 10^{22}cm^{-3}$) the plasma frequency is $\omega_p \simeq 10^{16}s^{-1}$. The internal fields $\boldsymbol{E}$ and $\boldsymbol{H}$ are given by Maxwell equations

$$\begin{gathered} div\boldsymbol{E} = -4\pi div\boldsymbol{P} \ , \ div\boldsymbol{H} = 0 \ , \\ curl\boldsymbol{E} = -\frac{1}{c}\frac{\partial\boldsymbol{H}}{\partial t} \ , \ curl\boldsymbol{H} = \frac{1}{c}\frac{\partial\boldsymbol{E}}{\partial t} + \frac{4\pi}{c}\frac{\partial\boldsymbol{P}}{\partial t} \ . \end{gathered} \tag{4.3}$$

This is the Drude-Lorentz model of polarization.[2]

Let us assume an external radiation field $\boldsymbol{E}_0 \sim e^{i\boldsymbol{kr}-i\omega t}$ (and a similar external magnetic field), with $\omega = ck$ and $\boldsymbol{kE}_0 = 0$ (transverse field, as

[2]See M. Apostol, *Essays in Electromagnetism and Matter*, Lambert, Saarbruecken (2013) and References therein.

for an X-ray radiation). Since $\boldsymbol{E}$ is given by $\boldsymbol{P}$ from the Maxwell equations, it is easy to see, from equation (4.2), that $\boldsymbol{P}, \boldsymbol{E}, \boldsymbol{H} \sim e^{i\boldsymbol{kr}-i\omega t}$, which means that $\boldsymbol{E}$ and $\boldsymbol{H}$ are free fields, and, consequently$\boldsymbol{P} = 0$. Indeed, from the Maxwell equations (4.3) we get

$$curl\, curl\boldsymbol{E} = -\frac{1}{c^2}\frac{\partial^2 \boldsymbol{E}}{\partial t^2} - \frac{4\pi}{c^2}\frac{\partial^2 \boldsymbol{P}}{\partial t^2} \ , \tag{4.4}$$

or

$$\frac{1}{c^2}\frac{\partial^2 \boldsymbol{E}}{\partial t^2} - \Delta \boldsymbol{E} = -grad\, div\boldsymbol{E} - \frac{4\pi}{c^2}\frac{\partial^2 \boldsymbol{P}}{\partial t^2} \ . \tag{4.5}$$

Since $\omega = ck$, the *lhs* of this equation is zero, which means that $\boldsymbol{P}$ is longitudinal and equal to the longitudinal component $\boldsymbol{E}_{\parallel}$ of $\boldsymbol{E}$. From the Gauss equation $div\boldsymbol{E} = -4\pi div\boldsymbol{P}$ we get $\boldsymbol{P} = -\boldsymbol{E}_{\parallel}/4\pi$ and, from equation (4.2), $\boldsymbol{P} = 0$ and $\boldsymbol{E}_{\parallel} = 0$. Also, from the same equation (4.2), we get $\boldsymbol{E}_{\perp} = -\boldsymbol{E}_0$, *i.e.* the total electric field $\boldsymbol{E}_0 + \boldsymbol{E}$ is zero (as in a metal). Similarly, from the Maxwell equations we get that the total magnetic field is vanishing. We arrive at the conclusion that a radiation field cannot be propagated in a polarizable medium. This is known as the Ewald-Oseen extinction theorem.[3]

This is valid for an infinite (homogeneous and continuous) medium. A radiation ray incident on a surface is refracted in the medium: its original incidence path is extincted, but the radiation emitted by the polarization beyond the boundaries of this path interferes constructively in the refracted path.[4]

4.3 Polarization

The situation is very different for finite bodies. Let us assume an external radiation field incident on a finite body, such that the whole

[3]see M. Apostol, *loc cit* and References therein.

[4]M. Apostol and G. Vaman, "Electromagnetic field interacting with a semi-infinite plasma", J. Opt. Soc. Am. **A26** 1747 (2009); "Reflection and refraction of the electromagnetic field in a semi-infinite plasma", Opt. Commun. **282** 4329 (2009); "Plasmons and polaritons in a semi-infinite plasma and a plasma slab", Physica **B404** 3775 (2009); "Reflected and refracted electromagnetic fields in a semi-infinite body", Solid State Commun. **149** 1936 (2009); M. Apostol, "Propagation of electromagnetic pulses through the surface of dispersive bodies", Roum. J. Phys. **58** 1298 (2013).

body is covered by the incident radiation. We leave aside the absorption as well as surface effects. Also, we may assume a spatial variation of the density n. The charges are displaced by $\boldsymbol{u}$ along the direction of the (total) electric field. Also, they oscillate as a whole. Therefore, a polarization $\boldsymbol{P} = nq\boldsymbol{u}$ and a current $\dot{\boldsymbol{P}}$ appear inside the body. The induced charge density is $\delta\rho = -q div(n\boldsymbol{u}) = -div\boldsymbol{P}$. The equation of motion (4.2) and the Maxwell equations (4.3) are valid, though the spatial variation includes the spatial variation of n (spatial dispersion) as well as the spatial variation at the surface. If we leave aside for the moment these effects, the Gauss equation $div\boldsymbol{E} = -4\pi div\boldsymbol{P}$ has the solution $\boldsymbol{E} = -4\pi\boldsymbol{P}$.[5] Indeed, on the surface $\boldsymbol{P} = nq\boldsymbol{u}$ is a superficial charge density, which generates an electric field $\boldsymbol{E} = -4\pi\boldsymbol{P}$. We can see that the presence of the surface produces a non-vanishing polarization. From the equation of motion (4.2) we get

$$\boldsymbol{P} = -\frac{\omega_p^2}{4\pi}\boldsymbol{E}_0\frac{1}{\omega^2 - \omega_c^2 - \omega_p^2 + i\omega\gamma} \tag{4.6}$$

and an internal field

$$\boldsymbol{E} = -4\pi\boldsymbol{P} = \boldsymbol{E}_0\frac{\omega_p^2}{\omega^2 - \omega_c^2 - \omega_p^2 + i\omega\gamma} \ , \tag{4.7}$$

which is different from $\boldsymbol{E} = -\boldsymbol{E}_0$ for an infinite, homogeneous, continuous medium. The total field

$$\boldsymbol{E}_t = \boldsymbol{E}_0\frac{\omega^2 - \omega_c^2}{\omega^2 - \omega_c^2 - \omega_p^2 + i\omega\gamma} \tag{4.8}$$

is vanishing for metals ($\omega_c = 0$) in the static limit $\omega \to 0$.

Equation (4.6) defines the polarizability

$$\alpha = -\frac{\omega_p^2}{4\pi}\frac{1}{\omega^2 - \omega_c^2 - \omega_p^2 + i\omega\gamma} \tag{4.9}$$

by $\boldsymbol{P} = \alpha\boldsymbol{E}_0$. Also, the total field $\boldsymbol{E}_t = \boldsymbol{E}_0 + \boldsymbol{E} = \boldsymbol{E}_0 - 4\pi\boldsymbol{P}$ given by equation (4.8) defines the electric permittivity

$$\varepsilon = 1 - \frac{\omega_p^2}{\omega^2 - \omega_c^2 + i\omega\gamma} \tag{4.10}$$

[5] An additional $curl\boldsymbol{G}$ in $\boldsymbol{E}$ leads to the Maxwell equation $curl\boldsymbol{H} = \frac{1}{c}\frac{\partial\boldsymbol{E}}{\partial t} + \frac{4\pi}{c}\frac{\partial\boldsymbol{P}}{\partial t}$, by $\boldsymbol{H} = \frac{1}{c}\frac{\partial\boldsymbol{G}}{\partial t}$.

by $\boldsymbol{E}_0 = \varepsilon \boldsymbol{E}_t$ and the electric susceptibility

$$\chi = -\frac{\omega_p^2}{4\pi} \frac{1}{\omega^2 - \omega_c^2 + i\omega\gamma} \tag{4.11}$$

by $\boldsymbol{P} = \chi \boldsymbol{E}_t$ ($\varepsilon = 1 + 4\pi\chi = 1/(1 - 4\pi\alpha)$, $\alpha = \chi/\varepsilon$). The induced current $\delta\boldsymbol{j} = \dot{\boldsymbol{P}} = \sigma \boldsymbol{E}_t$ defines the electric conductivity

$$\sigma = \frac{\omega_p^2}{4\pi} \frac{i\omega}{\omega^2 - \omega_c^2 + i\omega\gamma} \ . \tag{4.12}$$

Since $\boldsymbol{E} + 4\pi\boldsymbol{P} = 0$, the internal magnetic field satisfies the Maxwell equations $div\boldsymbol{H} = 0$ and $curl\boldsymbol{H} = 0$ (equations (4.3)), which means $\boldsymbol{H} = 0$ and $curl\boldsymbol{E} = 0$. The latter equation can be written as $curl\boldsymbol{P} = curl(\alpha\boldsymbol{E}_0) = 0$, which means that $\alpha\boldsymbol{E}_0$ derives from a gradient.

The results presented above are only partly correct. They neglect the effect of the surface discontinuities and the spatial dispersion. The surface discontinuities produce non-local effects. In addition, an internal electric field exists due to the inhomogeneities. It follows that the external field $\boldsymbol{E}_0$ is changed inside the body into a displacement field $\boldsymbol{D}$ (electric induction), such that the total field is $\boldsymbol{E}_t = \boldsymbol{E} + \boldsymbol{D}$. The equation of motion (4.2) becomes

$$\ddot{\boldsymbol{P}} + \omega_c^2 \boldsymbol{P} + \gamma \dot{\boldsymbol{P}} = \frac{\omega_p^2}{4\pi} \boldsymbol{E}_t \ , \tag{4.13}$$

such that $\boldsymbol{P} = \chi \boldsymbol{E}_t$ and the susceptibility χ is given by equation (4.11). Moreover, since $\boldsymbol{E} = -4\pi\boldsymbol{P}$, we get $\boldsymbol{E}_t = -4\pi\chi\boldsymbol{E}_t + \boldsymbol{D}$ and $\boldsymbol{D} = (1 + 4\pi\chi)\boldsymbol{E}_t = \varepsilon\boldsymbol{E}_t$, where the electric permittivity ε is given by equation (4.10). Similarly, the electric conductivity σ preserves its form given by equation (4.12). On the other hand, since $\boldsymbol{D}$ replaces $\boldsymbol{E}_0$, the Gauss equation becomes $div\boldsymbol{D} = 4\pi\rho_e$, where ρ_e is the external charge density (in our case $\rho_e = 0$). The Maxwell equations $div\boldsymbol{H} = 0$ and $curl\boldsymbol{H} = \frac{1}{c}\frac{\partial \boldsymbol{E}}{\partial t} + \frac{4\pi}{c}\frac{\partial \boldsymbol{P}}{\partial t} = 0$ (which in our case gives a zero internal magnetic field) are written usually for the total magnetic field as $div\boldsymbol{H}_t = 0$ and $curl\boldsymbol{H}_t = \frac{1}{c}\frac{\partial \boldsymbol{D}}{\partial t} + \frac{4\pi}{c}\boldsymbol{j}_e$, where $\boldsymbol{j}_e$ is the external current density. In the absence of the magnetization (as in our case) $\boldsymbol{H}_t$ is denoted by $\boldsymbol{B}$ (magnetic induction); when magnetization is present $\boldsymbol{H}_t$ is denoted simply by $\boldsymbol{H}$. The Faraday equation becomes $curl\boldsymbol{E}_t = -\frac{1}{c}\frac{\partial \boldsymbol{H}_t}{\partial t}$.

Let us focus on the Maxwell equations

$$curl\boldsymbol{E}_t = -\frac{1}{c}\frac{\partial \boldsymbol{H}_t}{\partial t} \ , \ curl\boldsymbol{H}_t = \frac{1}{c}\frac{\partial \boldsymbol{D}}{\partial t} \tag{4.14}$$

(where in our case $\boldsymbol{H}_t$ is the external magnetic field). We eliminate $\boldsymbol{H}_t$ from these equations and get

$$curl\, curl\boldsymbol{E}_t = -\frac{1}{c^2}\frac{\partial^2 \boldsymbol{D}}{\partial t^2} \ . \tag{4.15}$$

This equation can also be written as

$$curl\, curl\left(\frac{1}{\varepsilon}\boldsymbol{D}\right) = \frac{\omega^2}{c^2}\boldsymbol{D} \ , \tag{4.16}$$

where ε is given by equation (4.10). This ε exhibits a spatial dispersion through n in ω_p^2; also, it should be solved with appropriate boundary conditions. Consequently, it is an eigenvalue equation for $\boldsymbol{D}$. For a given ω the solutions involve several allowed wavevectors, which indicate that the external field suffers multiple reflections (and diffractions) in the medium, as if it would be partially trapped inside the medium. The solution $\boldsymbol{D}$ is determined by the external field $\boldsymbol{E}_0$ through boundary conditions. Since $\boldsymbol{P} = \frac{\chi}{\varepsilon}\boldsymbol{D}$, we get relations between various components of the polarization $\boldsymbol{P}$ and the incident field $\boldsymbol{E}_0$. We have various polarization coefficients α, generated by the spatial dispersion.

Since $\omega \gg \omega_{c,p}$ for X-rays, the electric permittivity is practically equal to unity ($\varepsilon \simeq 1$, transparent medium) and the solution of equation (4.16) is $\boldsymbol{D} \simeq \boldsymbol{E}_0$ ($\omega = ck$). Similarly, $\boldsymbol{P} = \frac{\chi}{\varepsilon}\boldsymbol{D} \simeq \chi\boldsymbol{E}_0$ provides the polarization $\alpha \simeq \chi$, according to equations (4.9) and (4.11) for $\omega \gg \omega_{c,p}$. Spatial-dispersion corrections to this solution are known as the "dyamical theory of X-ray diffraction".[6]

For an infinite homogeneous medium (without spatial dispersion) equation (4.16) has the solution $\varepsilon\omega^2 = c^2k^2$,where ε is given by equation (4.10). This equation defines the polaritons, which are propagating

[6]W. H. Zachariasen, *Theory of X-Ray Diffraction in Crystals*, Dover, NY (1945); Z. G. Pinsker, *Dynamical Scattering of X-Rays in Crystals*, Springer, Berlin (1978).

electromagnetic eigenmodes in matter. The zeros of the electric permittivity ($\omega^2 = \omega_c^2 + \omega_p^2$) define another type of (non-propagating) electromagnetic eigenmodes ($\boldsymbol{D} = \varepsilon \boldsymbol{E}_t$), called plasmons. In the static limit $\omega \to 0$ for finite bodies the equation $(1 + 4\pi\chi)\boldsymbol{E}_t = \boldsymbol{D}$ is modified into $(1 + 4\pi f\chi)\boldsymbol{E}_t = \boldsymbol{E}_0$, where f is called the depolarizing factor. This equation leads to $(1 + 4\pi f\chi)\boldsymbol{E}_t = \boldsymbol{P}/\alpha = \chi\boldsymbol{E}_t/\alpha$, or $4\pi\alpha = (\varepsilon - 1)/[1 + f(\varepsilon - 1)]$. f depends on the shape of the body. For a sphere $f = 1/3$ and we get the Clausius-Mossotti (Lorentz-Lorenz) relation $\frac{4\pi\alpha}{3} = (\varepsilon - 1)/(\varepsilon + 2)$.[7]

For an electron with mass $m = 10^{-27}g$ and charge $q = e = -4.8 \times 10^{-10}esu$, and for $\omega \gg \omega_{p,c}$, the energy variation is $\Delta E = m\omega^2 u^2 = e^2 E_0^2/m\omega^2$, *i.e.* $\Delta E = a_0\lambda^2 w/\pi$, where $a_0 = e^2/mc^2 = 2.8\times 10^{-13}cm^2$ is the classical radius of the electron, λ is the radiation wavelength and $w = E_0^2/4\pi$ is the energy density of the radiation. On the other hand the energy density is $w = \hbar\omega\Phi/c = \hbar\omega n_{ph}$, where $n_{ph} = \Phi/c$ is the density of photons. For $\Phi = 10^{12}photons/cm^2 \cdot s$ the photon density is $n_{ph} \simeq 30photons/cm^3$. Therefore, $\Delta E = \hbar\omega a_0\lambda^2 n_{ph}/\pi = \hbar\omega N_{ph}/\pi$, where N_{ph} is the number of photons in the volume $a_0\lambda^2$. We can see that ΔE is a small amount of energy. The condition $\omega^2 \gg \omega_p^2$ reads $a_0\lambda^2 \ll \pi a^3$, where a is the mean inter-electron separation. We always have $a^2 \ll a^3/a_0$, such that the conditions $a^2 \ll \lambda^2 \ll \pi a^3/a_0$ can be fulfilled. In this case the motion is classical.

The condition $u = eE_0/m\omega^2 \ll \lambda$ amounts to $eE_0\lambda \ll 4\pi^2 mc^2$, which means that the mechanical work done by the field should be much less than the relativistic energy of the particle. Also, the energy $\Delta E = m\omega^2 u^2$ is then $\Delta E \ll m\omega^2\lambda^2$, where $m\omega^2\lambda^2$ is of the order mc^2. Therefore, the condition $u \ll \lambda$ amounts to a non-relativistic interaction process. This condition is often fulfilled. It is the condition $\lambda \gg a$, where a is the range of the spatial variation of the wavefunctions, which may not be fulfilled, a case where a quantum-mechanical treatment is necessary. For a relativistic dynamics, where the equation of motion is $\frac{d}{dt}\left(m\dot{u}/\sqrt{1 - \dot{u}^2/c^2}\right) = eE_0$, the displacement is $u = \frac{\lambda}{2\pi}(eE_0/mc\omega)\left(1 + e^2E_0^2/m^2c^2\omega^2\right)^{1/2}$ which, in the ultra-relativistic limit tends to $u = \lambda/2\pi$.

[7] M. Apostol, *Essays in Electromagnetism and Matter*, Lambert, Saarbruecken (2013) and References therein.

4.4 Elastic scattering of X-rays

Equation (4.15) can also be written as

$$curl\, curl(\boldsymbol{D} - 4\pi \boldsymbol{P}) = -\frac{1}{c^2}\frac{\partial^2 \boldsymbol{D}}{\partial t^2} \ , \tag{4.17}$$

or

$$\frac{1}{c^2}\frac{\partial^2 \boldsymbol{D}}{\partial t^2} - \Delta\, \boldsymbol{D} = 4\pi curl\, curl \boldsymbol{P} \ . \tag{4.18}$$

This equation can be viewed as a wave equation for an electromagnetic field outside the body, generated by the polarization of the body. The source $4\pi curl\, curl\boldsymbol{P}$ can be viewed as a boundary condition for free electromagnetic waves. For $\omega \gg \omega_{p,c}$ the polarization is $\boldsymbol{P} \simeq \alpha \boldsymbol{E}_0$, where $\alpha \simeq -\frac{\omega_p^2}{4\pi\omega^2}$ (equation (4.9)). We denote the field outside the body by $\boldsymbol{D} = \boldsymbol{E}$. This is the scattered field. Equation (4.18) becomes

$$\Delta \mathbf{E} + k^2 \mathbf{E} = curl\, curl \left(\omega_p^2 \boldsymbol{E}_0/\omega^2\right) \ . \tag{4.19}$$

At large distances from the body and for large distances in comparison to the wavelength (far field, wave region) the solution is given by the well-known Kirchhoff potential

$$\mathbf{E} = \frac{e^2}{m\omega^2} \cdot \frac{e^{ikr}}{r} \cdot \mathbf{k}' \times (\mathbf{k}' \times \mathbf{E}_0) \int d\mathbf{r}' \cdot n(\boldsymbol{r}') e^{i\mathbf{K}\mathbf{r}'} \tag{4.20}$$

where $\mathbf{k}$ is the incident wavevector, $\mathbf{k}'$ is the scattered wavevector and $\mathbf{K} = \mathbf{k} - \mathbf{k}'$, $\boldsymbol{k}' = k\boldsymbol{r}/r$ and $\omega = ck = ck'$. It is worth noting that for each $\boldsymbol{r}$ there occurs a scattered wave with wavevector $\boldsymbol{k}' = k\boldsymbol{r}/r$. The integration is performed over the volume of the sample. We neglect the surface effects, as for a macroscopic sample. Also, we neglect the absorption, which is small at high frequencies. In addition, in equation (4.20) we limit ourselves to single scatterings. It is worth noting that the above approach to this problem is a perturbation-theory approach, where the change in electron density is small and the internal field is a local field. All these conditions are fairly well fulfilled for X-rays, such that multiple scatterings bring not a large contribution.

The cross-section is obtained as the ratio of the energy flux through the surface $R^2 do$ to the incident energy flux density, where do is the solid angle. We get the cross-section

$$d\sigma = a_0^2 \sin^2\Theta \left| \int d\mathbf{r} \cdot n(\boldsymbol{r}) e^{i\mathbf{Kr}} \right|^2 do \tag{4.21}$$

where $a_0 = e^2/mc^2 = 2.8 \times 10^{-13} cm$ is the classical radius of the electron and Θ is the angle between $\mathbf{E}_0$ and $\mathbf{k}'$. We use the parametrization $\mathbf{E}_0 = (E_0, 0, 0)$, $\mathbf{k} = (0, 0, k)$ and $\mathbf{k}' = k(\sin\theta\cos\varphi, \sin\theta\sin\varphi, \cos\theta)$, so $\cos\Theta = \sin\theta\cos\varphi$. By averaging over φ, as for natural polarization, we get

$$d\sigma = \frac{1}{2} a_0^2 (1 + \cos^2\theta) \left| \int d\mathbf{r} \cdot n(\boldsymbol{r}) e^{i\mathbf{Kr}} \right|^2 do \tag{4.22}$$

where θ is the scattering angle (the angle between $\mathbf{k}$ and $\mathbf{k}'$). This is the Thomson scattering. The cross-section does not depend on the X-rays frequency. The intensity of the diffracted X-rays is proportional to the square of the Fourier transform of the electron density. For $\mathbf{K} \to 0$ this is the average electron density $\overline{n}$ and, according to Maxwell's equations for continuous media, the X-rays are refracted with the refractive index $\sqrt{\varepsilon} = 1 - 2\pi\overline{n}e^2/m\omega^2 \simeq 1$. Therefore, the scattering at small angles is a refraction.

4.5 Diffraction peaks in crystals

The electron density $n(\mathbf{r})$ in a crystal is a periodic function with crystal periodicity. Consequently, it can be expanded in a Fourier series

$$n(\mathbf{r}) = \sum_{\mathbf{g}} n_{\mathbf{g}} e^{i\mathbf{gr}} \tag{4.23}$$

where $\mathbf{g}$ are the reciprocal vectors of the lattice.[8] They are generated by $\mathbf{a}^* = 2\pi(\mathbf{b} \times \mathbf{c})/v$, etc, where $\mathbf{a}$, $\mathbf{b}$, $\mathbf{c}$ are the basic vectors of the unit cell and $v = \mathbf{a}(\mathbf{b} \times \mathbf{c})$ is the volume of the unit cell. They have

[8] P. P. Ewald, Phys. Z. **14** 465 (1913); Z. Krist. **56** 129 (1921); M. Laue, Jahrb. Radioakt. Elektr. **11** 308 (1914).

the property $\mathbf{a}^*\mathbf{a} = 2\pi$, $\mathbf{a}^*\mathbf{b} = \mathbf{a}^*\mathbf{c} = 0$, etc, and the inverse formula is

$$n_{\mathbf{g}} = \frac{1}{v}\int_v d\mathbf{r} \cdot n(\mathbf{r})e^{-i\mathbf{gr}} \ . \tag{4.24}$$

For $n(\boldsymbol{r}) = \sum_{i=1}^{N} \delta(\boldsymbol{r} - \boldsymbol{r}_i)$ in a crystal $n_{\boldsymbol{g}} = 1/v$. For a small extension of $n(\boldsymbol{r})$ around $\boldsymbol{r}_i$ the coefficients $n_{\boldsymbol{g}}$ vary slowly over a large range of reciprocal vectors, such that $n(\boldsymbol{r}) = \sum_{i=1}^{N} \delta(\boldsymbol{r} - \boldsymbol{r}_i)$ and $n_{\boldsymbol{g}} = 1/v$ are fair representations of the density. The cross-section given by (4.22) becomes

$$d\sigma = \frac{1}{2}a_0^2(1 + \cos^2\theta)\left|\sum_{\mathbf{g}} n_{\mathbf{g}} \int d\mathbf{r} \cdot e^{i(\mathbf{K}+\mathbf{g})\mathbf{r}}\right|^2 do \tag{4.25}$$

whence one can see that the cross-section is maximum for peaks defined by Laue's condition[9] $\mathbf{K} = -\mathbf{g}$, *i.e.* $\boldsymbol{k}' = \boldsymbol{k} + \boldsymbol{g}$, or, equivalently, by Bragg's condition[10] $2\mathbf{kg} = g^2$, *i.e.* $2k\sin\theta/2 = g$. It is worth noting that θ depends on $\boldsymbol{g}$ (we can write $\theta_{\boldsymbol{g}}$). The vectors $\mathbf{k}$ and $\mathbf{k}'$ lie on the Ewald sphere; they are connected by reciprocal vectors $\mathbf{g}$.

A reciprocal vector $\mathbf{g} = h\mathbf{a}^* + k\mathbf{b}^* + l\mathbf{c}^*$, where hkl are integers, defines a set of parallel crystalline planes perpendicular to $\mathbf{g}$ according to $\mathbf{gr} = 2\pi \times integer$ (Miller indices). The wavevector $\mathbf{k}$ is reflected into $\mathbf{k}'$ at the angle $\theta/2$. We can say that the $\mathbf{g}$-peak comes from the reflection on the hkl-planes.

In equation (4.25) we have

$$\int d\mathbf{r} \cdot e^{i(\mathbf{K}+\mathbf{g})\mathbf{r}} = (2\pi)^3\delta(\mathbf{K} + \mathbf{g}) \tag{4.26}$$

and

$$\left|\int d\mathbf{r} \cdot e^{i(\mathbf{K}+\mathbf{g})\mathbf{r}}\right|^2 = (2\pi)^3 V\delta(\mathbf{K} + \mathbf{g}) \ , \tag{4.27}$$

[9]M. Laue, Munch. Sitzungsber. 363 (1912); Ann. Phys. **41** 989 (1913). The X-rays diffraction in crystals has been observed for the first time by W. Friedrich, P. Knipping and M. von Laue, Sitzungsber. der Bayerischen Akad. Wissenschaft., Math-Phys. Klasse, p. 303 (1912) (see also W. Barlow, Nature **29** 186, 205, 404 ((1883)).

[10]W. H. Bragg and W. L. Bragg, Proc. Roy. Soc. **88** 428 (1913); *ibid*, **89** 246 (1913).

where V is the volume of the (macroscopic) sample. The number of unit cells (*e.g.*, atoms, molecules) is N, such that $V = Nv$ and a macroscopic sample is defined by $N \gg 1$. We can see that the peak maximum is proportional to the square volume V^2 of the sample ($\delta(\boldsymbol{K}) = V/(2\pi)^3$ for $\Delta\boldsymbol{K} = (2\pi)^3/V$ around $\boldsymbol{K} = 0$, for large V, see Appendix). The cross-section reads

$$d\sigma = \frac{1}{2}a_0^2(1+\cos^2\theta)\cdot(2\pi)^3 V \sum_{\mathbf{g}} |n_{\mathbf{g}}|^2\,\delta(\mathbf{K}+\mathbf{g})\cdot do\ . \qquad (4.28)$$

Such large and sharp δ-diffraction peaks, whose maximum is proportional to V^2, define a coherent scattering.

The total intensity of a $\mathbf{k}' = \mathbf{k}+\mathbf{g}$-peak is obtained by integrating (4.25) over the solid angle about $\mathbf{k}'$. We denote by $\boldsymbol{q}$ this integration vector, and see that it is perpendicular to $\mathbf{k}'$ since $k' = k$. We get

$$\sigma_{\boldsymbol{g}} = \frac{1}{2k^2}a_0^2(1+\cos^2\theta_{\boldsymbol{g}})\,|n_{\mathbf{g}}|^2 \int dq_x dq_y \left|\int d\mathbf{r}\cdot e^{-i\mathbf{qr}}\right|^2 \qquad (4.29)$$

where the z-axis is taken along $\mathbf{k}'$. The integral in (4.29) is performed straightforwardly; we get

$$\begin{aligned}\sigma_{\boldsymbol{g}} &= \tfrac{2\pi^2}{k^2}a_0^2(1+\cos^2\theta_{\boldsymbol{g}})\,|n_{\mathbf{g}}|^2 Al^2 = \\ &= \tfrac{8\pi^2}{g^2}a_0^2\sin^2\theta_{\boldsymbol{g}}/2\cdot(1+\cos^2\theta_{\boldsymbol{g}})\,|n_{\mathbf{g}}|^2 Al^2\end{aligned} \qquad (4.30)$$

where A is the area of the cross-section of the sample perpendicular to $\mathbf{k}'$ and l is the length of the sample along this direction. For macroscopic samples, whose shape is unimportant, we may write $l = V^{1/3}$and $A = V^{2/3}$. It follows that the total intensity of a spot is proportional to $V^{4/3}$; by comparing it to the maximum intensity $\sim V^2$, we get the width of the spot $\sim V^{-2/3}$. In the limit $V \to \infty$ it gives an idea of the sharpness of the diffraction peaks. The dependence $V^{4/3}$ of the cross-section defines a coherent diffraction.

The same result can be obtained straightforwardly from equation (4.25) by

$$\begin{aligned}d\sigma &= \tfrac{1}{2}a_0^2(1+\cos^2\theta)\left|\textstyle\sum_{\mathbf{g}} n_{\mathbf{g}}\int d\mathbf{r}\cdot e^{i(\mathbf{K}+\mathbf{g})\mathbf{r}}\right|^2 do = \\ &= \tfrac{1}{2}a_0^2V^2\textstyle\sum_{\mathbf{g}}|n_{\mathbf{g}}|^2(1+\cos^2\theta_{\boldsymbol{g}})\delta_{\boldsymbol{K},-\boldsymbol{g}}do\end{aligned} \qquad (4.31)$$

and

$$\sigma = \frac{1}{2}a_0^2V^2\sum_{\mathbf{g}}|n_{\mathbf{g}}|^2(1+\cos^2\theta_{\boldsymbol{g}})\Delta o_{\boldsymbol{g}} = \\ = \frac{1}{2k^2}a_0^2V^2\sum_{\mathbf{g}}|n_{\mathbf{g}}|^2(1+\cos^2\theta_{\boldsymbol{g}})\frac{(2\pi)^2}{V^{2/3}} \ . \tag{4.32}$$

Of course $\sigma_{\boldsymbol{g}} \ll A$ or

$$a_0 \cdot \frac{V^{1/3}}{k}|n_{\mathbf{g}}| \ll 1 \ ; \tag{4.33}$$

indeed, the scattered energy flux cannot exceed the incident energy flux. Equation (4.33) can also be written approximately as $a_0\lambda l/v \ll 1$, where λ is the wavelength of the X-rays. This is a special, strong condition upon the size (volume) of the sample. If this condition is not satisfied, we are in an unphysical situation, which arises from neglecting of the statistical character of the scattering process.

4.6 Form-factor

Let $\boldsymbol{E}_i = \boldsymbol{E}_0 e^{i\boldsymbol{kr}-i\omega t}$ and $\boldsymbol{H}_i = \frac{\boldsymbol{k}}{k}\times\boldsymbol{E}_i$ be the incident electric field and magnetic field (the latter from $curl\boldsymbol{E}_i = -\frac{1}{c}\frac{\partial \boldsymbol{H}_i}{\partial t}$), where $\omega = ck$. We assume that this is a radiation field, *i.e.* $\boldsymbol{E}_0\boldsymbol{k} = 0$ ($\boldsymbol{E}_i\boldsymbol{k} = \boldsymbol{H}_i\boldsymbol{k} = \boldsymbol{E}_i\boldsymbol{H}_i = 0$). The incident energy density is $w_i = \frac{1}{4\pi}E_0^2$, the incident Poynting vector is $\boldsymbol{S}_i = \frac{c\boldsymbol{k}}{4\pi k}E_0^2$ (energy flux density) and the incident momentum density is $\boldsymbol{g}_i = \boldsymbol{S}_i/c^2$ (the bilinear forms like $\boldsymbol{EH}$ are written as $\frac{1}{2}(\boldsymbol{EH}^* + \boldsymbol{E}^*\boldsymbol{H})$ for complex quantities). Equation (4.19) reads

$$\Delta\boldsymbol{E} + k^2\boldsymbol{E} = curl\,curl(4\pi bn\boldsymbol{E}_i) \ , \tag{4.34}$$

where $b = e^2/m\omega^2$ and $n\boldsymbol{E}_i$ is restricted to the sample (like a boundary condition). We neglect the surface effects (large sample), the absorption (high frequencies) and the multiple scatterings. The solution of this equation reads

$$\boldsymbol{E}(\boldsymbol{r},t) = \\ = -b\int d\boldsymbol{r}'\frac{curl'\,curl'\left[n(\boldsymbol{r}')\boldsymbol{E}_0 e^{i\boldsymbol{kr}'-i\omega[t-|\boldsymbol{r}-\boldsymbol{r}'|/c]}\right]}{|\boldsymbol{r}-\boldsymbol{r}'|} \ . \tag{4.35}$$

At large distances from the sample $\mid \boldsymbol{r}-\boldsymbol{r}' \mid = r-\boldsymbol{r}\boldsymbol{r}'/r...$, so we have

$$\begin{gathered}\boldsymbol{E}(\boldsymbol{r},t)=-b\tfrac{e^{ikr}}{r}e^{-i\omega t}\cdot \\ \cdot\int d\boldsymbol{r}' curl' \, curl' \left[n(\boldsymbol{r}')e^{i(\boldsymbol{k}-\boldsymbol{k}')\boldsymbol{r}'}\boldsymbol{E}_0\right] \ ,\end{gathered} \tag{4.36}$$

in the far-field region, or

$$\begin{gathered}\boldsymbol{E}(\boldsymbol{r},t)=b\tfrac{e^{ikr}}{r}e^{-i\omega t}\cdot \\ \cdot[\boldsymbol{k}'\times(\boldsymbol{k}'\times\boldsymbol{E}_0)]\int d\boldsymbol{r}'n(\boldsymbol{r}')e^{i(\boldsymbol{k}-\boldsymbol{k}')\boldsymbol{r}'} \ ,\end{gathered} \tag{4.37}$$

where $\boldsymbol{k}'=k\boldsymbol{r}/r$; we introduce the form-factor

$$n_{\boldsymbol{K}}=\int d\boldsymbol{r}n(\boldsymbol{r})e^{i\boldsymbol{K}\boldsymbol{r}} \ , \ \ \boldsymbol{K}=\boldsymbol{k}-\boldsymbol{k}' \ . \tag{4.38}$$

it is worth noting that for each $\boldsymbol{r}$ there appears a scattered wave with wavevector $\boldsymbol{k}'=k\boldsymbol{r}/r$. We can see that the radiated field has the same frequency as the incident radiation. The magnetic field is given by $curl\boldsymbol{E}=-\frac{1}{c}\frac{\partial \boldsymbol{H}}{\partial t}$. We assume $r\gg\lambda$, where $\lambda=2\pi/k$ is the wavelength of the radiation. Consequently, $curl\boldsymbol{E}=ik\boldsymbol{r}\times\boldsymbol{E}/r=i\boldsymbol{k}'\times\boldsymbol{E}$ and $\boldsymbol{H}=\boldsymbol{k}'\times\boldsymbol{E}/k$; the derivatives appear only from e^{ikr} (not from $1/r$), such that $k\gg 1/r$, *i.e.* $r\gg\lambda$. λ is always greater than the dimension of the sample, because the motion inside the sample is finite. The condition $r\gg\lambda>l$, where l is the dimension of the sample, is the radiation condition, or the far-field (wave-zone) condition (if $r\simeq\lambda>l$ we are in the near-field region). Actually, $\lambda\gg l$, since the velocity of the charges is much smaller than the speed of light.

From equation (4.36) we get

$$\begin{gathered}\boldsymbol{E}=b[\boldsymbol{k}'(\boldsymbol{k}'\boldsymbol{E}_0)-\boldsymbol{E}_0k^2]n_{\boldsymbol{K}}\tfrac{e^{ikr}}{r}e^{-i\omega t} \ , \\ \boldsymbol{H}=-b\boldsymbol{k}'\times\boldsymbol{E}_0kn_{\boldsymbol{K}}\tfrac{e^{ikr}}{r}e^{-i\omega t} \ .\end{gathered} \tag{4.39}$$

We can see that this field is transverse, *i.e.* $\boldsymbol{E}\boldsymbol{k}'=\boldsymbol{H}\boldsymbol{k}'=\boldsymbol{E}\boldsymbol{H}=0$. The density of radiated energy is

$$w=\frac{a_0^2}{4\pi r^2}E_0^2\sin^2\Theta \mid n_{\boldsymbol{K}}\mid^2 \tag{4.40}$$

and the Poynting vector is

$$\boldsymbol{S} = \frac{ca_0^2}{4\pi r^2} E_0^2 \sin^2\Theta \mid n_{\boldsymbol{K}} \mid^2 \frac{\boldsymbol{k}'}{k} \ , \tag{4.41}$$

where $a_0 = e^2/mc^2 = 2.8 \times 10^{-13} cm$ is the classical radius of the electron and Θ is the angle between $\boldsymbol{k}'$ and $\boldsymbol{E}_0$. We use the parametrization $\boldsymbol{k} = k(0,0,1)$, $\boldsymbol{E}_0 = E_0(\cos\varphi, \sin\varphi, 0)$ and $\boldsymbol{k}' = k(\sin\theta\cos\varphi', \sin\theta\sin\varphi', \cos\theta)$, where θ is the angle between $\boldsymbol{k}$ and $\boldsymbol{k}'$ (scattering angle). For a natural (unpolarized) incident radiation we average over φ, such that $\sin^2\Theta$ becomes $\frac{1}{2}(1+\cos^2\theta)$. We get immediately the differential cross-section

$$d\sigma = \frac{1}{2} a_0^2 \left(1+\cos^2\theta\right) \mid n_{\boldsymbol{K}} \mid^2 do \tag{4.42}$$

through the solid angle do, as the ratio of the energy per unit time across the area $r^2 do$ in the direction of $\boldsymbol{k}'$ ($\boldsymbol{S}\boldsymbol{k}' r^2 do/k$) to the energy per unit time and unit area of the incident radiation ($\boldsymbol{S}_i\boldsymbol{k}/k = cE_0^2/4\pi$). The radiated intensity (energy per unit time, along a given direction) is the (differential) cross-section multiplied by the incident energy flux density. We note the occurrence in the cross-section of the form-factor $n_{\boldsymbol{K}}$ given by equation (4.38).

Let us compute the form-factor given by equation (4.38). For a homogeneous, continuous medium the density $n(\boldsymbol{r})$ is a constant, say $n(\boldsymbol{r}) = 1/v$, such that $Nv = V$, where V is the volume of the (macroscopic) medium and N ($N \gg 1$) is the number of atoms in the medium. Then

$$\begin{gathered} n_{\boldsymbol{K}} = \tfrac{1}{v} \int d\boldsymbol{r} e^{i\boldsymbol{K}\boldsymbol{r}} = \\ = \tfrac{1}{v} \frac{\sin(K_x l_x/2)}{K_x/2} \frac{\sin(K_y l_y/2)}{K_y/2} \frac{\sin(K_z l_z/2)}{K_z/2} \ , \end{gathered} \tag{4.43}$$

where $l_{x,y,z}$ are the dimensions of the medium along the x, y, z-directions. This is an oscillating, decreasing function of $K_{x,y,z}$, with a very high peak $n_0 = V/v = N$ at $\boldsymbol{K} = 0$. According to equation (4.43), this peak has a very small extension, of the order $K_x l_x = K_y l_y = K_z l_z = 2\pi$ in all directions. Moreover,

$$\int dK_x \frac{\sin(K_x l_x/2)}{K_x/2} = 2\pi \ , \tag{4.44}$$

a result which can also be obtained by $l_x \cdot (2\pi/l_x)$, *i.e.* the function at $K_x = 0$ multiplied by its extension; so, we can write

$$\int d\boldsymbol{r} e^{i\boldsymbol{K}\boldsymbol{r}} = (2\pi)^3 \delta(\boldsymbol{K}) \ . \tag{4.45}$$

The same results can be obtained by using discrete values of $K_{x,y,z}$, of the form $K_x = \frac{2\pi}{l_x} n_x$, etc, where n_x are integers, and, similarly, discrete values for x, y, z (see Appendix); then

$$\int d\boldsymbol{r} e^{i\boldsymbol{K}\boldsymbol{r}} = (2\pi)^3 \delta(\boldsymbol{K}) = V \delta_{\boldsymbol{K},0} \ . \tag{4.46}$$

The form-factor is

$$n_{\boldsymbol{K}} = \frac{(2\pi)^3}{v} \delta(\boldsymbol{K}) = \frac{V}{v} \delta_{\boldsymbol{K},0} = N \delta_{\boldsymbol{K},0} \ . \tag{4.47}$$

For a discrete medium the density is $n(\boldsymbol{r}) = \sum_i \delta(\boldsymbol{r} - \boldsymbol{r}_i)$, where i is the label of the atom placed at position $\boldsymbol{r}_i$. The form-factor becomes

$$n_{\boldsymbol{K}} = \sum_i e^{i\boldsymbol{K}\boldsymbol{r}_i} \ . \tag{4.48}$$

For random positions $\boldsymbol{r}_i$ (as in an amorphous solid, or a liquid) the phase factors $e^{i\boldsymbol{K}\boldsymbol{r}_i}$ are randomly distributed, such that we get $n_{\boldsymbol{K}} = N\delta_{\boldsymbol{K},0}$, *i.e.* the same result as in equation (4.47). If the density $n(\boldsymbol{r}) = \sum_i \chi(\boldsymbol{r} - \boldsymbol{r}_i)$ is given by a superposition of functions $\chi(\boldsymbol{r} - \boldsymbol{r}_i)$, localized over a small volume Δv, the form-factor is given by

$$n_{\boldsymbol{K}} = \chi(\boldsymbol{K}) \sum_i e^{i\boldsymbol{K}\boldsymbol{r}_i} \ , \tag{4.49}$$

where $\chi(\boldsymbol{K}) = \int_{\Delta v} d\boldsymbol{r} \chi(\boldsymbol{r}) e^{i\boldsymbol{K}\boldsymbol{r}}$ is the Fourier transform of the function $\chi(\boldsymbol{r})$. For random positions we get again $n_{\boldsymbol{K}} = N\chi(\boldsymbol{K} = 0)\delta_{\boldsymbol{K},0} = N\delta_{\boldsymbol{K},0}$, as in equation (4.47). We conclude that for random atomic positions the cross-section is

$$d\sigma = a_0^2 N^2 do_{\boldsymbol{k}'=\boldsymbol{k}} \tag{4.50}$$

and the total cross-section is

$$\sigma = a_0^2 N^2 \Delta o_{\boldsymbol{k}'=\boldsymbol{k}} = \frac{1}{k^2} a_0^2 N^2 \frac{(2\pi)^2}{V^{2/3}} = 4\pi^2 \frac{a_0^2}{k^2 v^{2/3}} N^{4/3} \ . \tag{4.51}$$

For a crystal the form-factor has a certain particularity. The density $n(\boldsymbol{r})$ is a periodic function with period equal to the inter-atomic separation. Such a function can be expanded in a Fourier series of reciprocal vectors $\boldsymbol{g}$, with expansion coefficients $n_{\boldsymbol{g}}$, as in equation (4.23). It is easy to see that $\boldsymbol{K}$ in the function $\chi(\boldsymbol{K})$ above becomes $-\boldsymbol{g}$, the volume Δv becomes v and $n_{\boldsymbol{g}} = \frac{1}{v}\chi(-\boldsymbol{g})$. The form-factor given by equation (4.38) becomes

$$\begin{aligned} n_{\boldsymbol{K}} &= \sum_{\boldsymbol{g}} n_{\boldsymbol{g}}(2\pi)^3\delta(\boldsymbol{K}+\boldsymbol{g}) = \\ &= \sum_{\boldsymbol{g}} V n_{\boldsymbol{g}}\delta_{\boldsymbol{K},-\boldsymbol{g}} = N\chi(\boldsymbol{K})\sum_{\boldsymbol{g}}\delta_{\boldsymbol{K},-\boldsymbol{g}} \ , \end{aligned} \tag{4.52}$$

whence, by comparing with equation (4.49), we get

$$\sum_i e^{i\boldsymbol{K}\boldsymbol{r}_i} = N\sum_{\boldsymbol{g}}\delta_{\boldsymbol{K},-\boldsymbol{g}} \ , \tag{4.53}$$

as expected. We can see that all the relevant vectors $\boldsymbol{K}$ are restricted to the first Brillouin zone, defined by a unit cell in the reciprocal space.

Since $\boldsymbol{K}$ may be restricted to the first Brillouin zone we may transform the summation in equation (4.49) in an integration, with the result

$$\begin{aligned} n_{\boldsymbol{K}} &= \frac{\chi(\boldsymbol{K})}{v}\int d\boldsymbol{r}e^{i\boldsymbol{K}\boldsymbol{r}} = \\ &= \chi(\boldsymbol{K})\sum_{\boldsymbol{g}}\frac{1}{v}\frac{\sin(K_x l_x/2)}{K_x/2}\frac{\sin(K_y l_y/2)}{K_y/2}\frac{\sin(K_z l_z/2)}{K_z/2} = \\ &= N\chi(\boldsymbol{K})\sum_{\boldsymbol{g}}\delta_{\boldsymbol{K},-\boldsymbol{g}} = \frac{(2\pi)^3}{v}\chi(\boldsymbol{K})\sum_{\boldsymbol{g}}\delta(\boldsymbol{K}+\boldsymbol{g}) \ , \end{aligned} \tag{4.54}$$

which is again the result given by equation (4.52).

For a crystal the radiated energy is concentrated in the Bragg peaks defined by $\boldsymbol{k}' = \boldsymbol{k}+\boldsymbol{g}$, situated on the Ewald sphere $k = k'$ ($2k\sin\theta/2 = g$, where θ is the scattering angle). The linear extension of each peak is $\Delta k'_x = 2\pi/l_x, ...$ If the peaks are sufficiently dense on the Ewald sphere, we may associate the area $(\Delta g)^2 = (2\pi)^2/v^{2/3}$ to each peak, such that the number of peaks is $4\pi k^2 v^{2/3}/(2\pi)^2 \gg 1$. This condition is equivalent to $\lambda/a \ll 1$, where a is the mean separation distance between the atoms.

Let us compute the total cross-section for a crystal

$$\sigma = \frac{1}{2}a_0^2 \int (1+\cos^2\theta) \left| \int d\mathbf{r} \cdot n(\boldsymbol{r}) e^{i\mathbf{Kr}} \right|^2 do \tag{4.55}$$

given by equation (4.42), for $n(\boldsymbol{r}) = \sum_i \delta(\boldsymbol{r} - \boldsymbol{r}_i)$. First, we use equations (4.54) to write

$$\left| \int d\mathbf{r} \cdot n(\boldsymbol{r}) e^{-i\mathbf{Kr}} \right|^2 = N\frac{(2\pi)^3}{v} \sum_{\boldsymbol{g}} \delta(\boldsymbol{K} + g) \; ; \tag{4.56}$$

then the total cross-section becomes

$$\sigma = \frac{1}{2}a_0^2 N \frac{(2\pi)^3}{v} \sum_{\boldsymbol{g}} (1+\cos^2\theta_{\boldsymbol{g}}) \int \delta(\boldsymbol{K}+\boldsymbol{g}) do_{\boldsymbol{g}} \; , \tag{4.57}$$

where the integration over the solid angle $o_{\boldsymbol{g}}$ is performed about the direction of the vector $\boldsymbol{k}' = \boldsymbol{k} + \boldsymbol{g}$. We can write $\delta(\boldsymbol{K}+\boldsymbol{g}) = \delta(q_x)\delta(q_y)\frac{l_z}{2\pi} = \frac{l_z}{(2\pi)^3}\int dxdye^{i(q_x x + q_y y)}$, where the z-direction is the direction of the vector $\boldsymbol{k}'$, and get

$$\begin{aligned} &\int \delta(\boldsymbol{K}+\boldsymbol{g}) do_{\boldsymbol{g}} = \\ &= \tfrac{l_z}{(2\pi)^3 k^2} \int dq_x dq_y dx dy e^{i(q_x x + q_y y)} = \tfrac{l_z}{2\pi k^2} \; ; \end{aligned} \tag{4.58}$$

also, we can write

$$\int \delta(\boldsymbol{K}+\boldsymbol{g}) do_{\boldsymbol{g}} = \frac{1}{k^2}\frac{V}{(2\pi)^3}\frac{(2\pi)^2}{l_x l_y} = \frac{l_z}{2\pi k^2} \; , \tag{4.59}$$

with the same result. We can see, from equation (4.57), that the width of the peak is proportional to Nl_z, or $N^{4/3}$ (or Vl_z). Now, the total cross-section (equation (4.57)) becomes[11]

$$\sigma = \frac{1}{2}a_0^2 N \frac{(2\pi)^2}{vk^2} l_z \sum_{\boldsymbol{g}} (1+\cos^2\theta_{\boldsymbol{g}}) \; . \tag{4.60}$$

[11] If the density $n(\boldsymbol{r}) = \sum_i \chi(\boldsymbol{r} - \boldsymbol{r}_i)$ is a sum of localized functions, the square Fourier coefficient $|n_{\boldsymbol{g}}|^2$ appears in this summation, as in equation (4.32).

The summation over $\boldsymbol{g}$ in equation (4.60) is performed over the Ewald sphere, such that we may write

$$\sum_{\boldsymbol{g}}(1+\cos^2\theta_{\boldsymbol{g}}) = \frac{v^{2/3}}{(2\pi)^2}k^2\int do(1+\cos^2\theta) = \frac{8}{3}\frac{v^{2/3}}{2\pi}k^2 \qquad (4.61)$$

for sufficiently dense peaks; the total cross-section is

$$\sigma = \frac{8\pi}{3}a_0^2 N l_z/v^{1/3} \ . \qquad (4.62)$$

This cross-section is of the order $N^{4/3}$ (coherent diffraction); of course, the cross-section should satisfy the condition $\sigma \ll A$, or $a_0 N^{1/3}/a \ll 1$, where a is the mean separation distance between the atoms and A is of the order of the area of the sample. This condition is similar to the condition (4.33). The total cross-section is of the order $\sigma \simeq a_0^2 N^{4/3}$, the number of peaks is of the order $(a/\lambda)^2$ and the cross-section of a peak is of the order $a_0^2 N^{4/3}(\lambda/a)^2$. We can see that the cross-section of the peak for a medium with random atomic positions (equation (4.51)) is smaller than the cross-section of a crystal by a factor of the order of the number of the Bragg peaks $(k^2 v^{2/3}/\pi)$. It is a coherent diffraction with only one (forward) peak.

4.7 Energy and momentum

According to the definition of the cross-section the radiated intensity I (energy pe unit time) is related to the incident intensity I_i by $dI = d\sigma I_i/A$, where A is of the order of the area of the sample; therefore, $I = \sigma I_i/A \ll I_i$. The incident radiation sets the charges in motion, and the mobile charges emit radiation. The intensity I should always be smaller than I_i; the remaining energy is transmitted (it is not scattered). If this condition is not fulfilled, we shall see that coherence domains appear inside the sample, as a consequence of the statistical nature of the scattering. Obviously, the limiting case is $\sigma = A$.

The momentum conservation reads

$$\mathbf{f} = \mathbf{t} - \frac{\partial \mathbf{g}}{\partial t} \ , \qquad (4.63)$$

where $\boldsymbol{f}$ is the density of Lorentz force, $t_i = \partial_j \sigma_{ij}$,

$$\sigma_{ij} = \frac{1}{4\pi}\left[E_iE_j + H_iH_j - \frac{1}{2}\delta_{ij}(E^2 + H^2)\right] \tag{4.64}$$

is the stress tensor and $\boldsymbol{g} = \frac{1}{c^2}\boldsymbol{S}$ is the momentum density ($\boldsymbol{S}$ being the Poynting vector). For the radiated field $\boldsymbol{f} = 0$ (no charges and no currents in vacuum). We integrate equation (4.63) over a spherical volume with radius r enclosed by a surface $\boldsymbol{s}$, with the element $\boldsymbol{ds}$ directed along the vector $\boldsymbol{k}'$. We get

$$\int d\boldsymbol{r}t_i = \oint \sigma_{ij}ds_j = -\oint wds_i \ , \ \int d\boldsymbol{r}\boldsymbol{t} = -\oint w\boldsymbol{ds} \ , \tag{4.65}$$

because the fields $\boldsymbol{E}$ and $\boldsymbol{H}$ are transverse (equations (4.39)), such that only the term with the energy density w remains. On the other hand,

$$\int d\boldsymbol{r}\frac{\partial \boldsymbol{g}}{\partial t} = c\oint \boldsymbol{g}ds = c\oint g\boldsymbol{ds} \ ; \tag{4.66}$$

$c\boldsymbol{g}$ is the momentum flux density, *i.e.* the momentum per unit time and unit area (force per unit area). The element of area is $\boldsymbol{ds} = r^2 do\frac{\boldsymbol{k}'}{k}$, such that, by using equations (4.40) and (4.42), we get

$$\oint w\boldsymbol{ds} = w_i\int d\sigma\frac{\boldsymbol{k}'}{k} \ . \tag{4.67}$$

The integration is performed by summing over the reciprocal vectors and integrating over the solid angle about each reciprocal vector, as above. It is easy to see that this operation implies the factor

$$\begin{gathered}\int d\varphi d\theta \sin\theta(1+\cos^2\theta)\cdot \\ \cdot(\sin\theta\cos\varphi, \sin\theta\sin\varphi, \cos\theta) = 0 \ ;\end{gathered} \tag{4.68}$$

the stress force of the radiated field is zero. Similarly,

$$\oint g\boldsymbol{ds} = g_i\int d\sigma\frac{\boldsymbol{k}'}{k} = 0 \ , \tag{4.69}$$

i.e. the rate of the momentum per unit time (the force arising from the change of the momentum) is zero. The same result, which is momentum conservation, is valid for a surface enclosing the sample, though close to the sample the field is not a radiation field.

A similar computation can be done for the incident momentum flux. The incident field is transverse, such that we get $\int d\boldsymbol{r}\boldsymbol{t} = -c\oint g d\boldsymbol{s} = E_0^2 A/4\pi$ (the integration is performed over the surface that encloses the field, whence the change of the sign); it follows that a force $F = E_0^2 A/4\pi$ acts upon the incidence surface of the sample. In the complex incident field we have two waves, $\boldsymbol{E}_0 \cos(\boldsymbol{k}\boldsymbol{r} - \omega t)$ and $\boldsymbol{E}_0 \sin(\boldsymbol{k}\boldsymbol{r} - \omega t)$, such that, if we limit to one wave and average over a period, we get $F = E_0^2 A/8\pi$. This is the radiation pressure; it is the energy density of the field. A similar force acts, in the opposite direction, on the other surface, where the radiation is transmitted through. The difference is given by the density of the radiated field w_r, which is related to the density of the incident energy through $w_r = \sigma w_i/A$. Therefore, the net force acting upon the sample, in the forward direction, is $F = E_0^2\sigma/4\pi$. It is due to the pressure of the incident radiation.

The same result is valid for a medium with random atomic positions, where $\int d\boldsymbol{r}\boldsymbol{t} = -w_i\sigma\boldsymbol{k}/k$ and $c\oint g d\boldsymbol{s} = cg_i\sigma\boldsymbol{k}/k$, such that the force acting upon the sample in the forward direction is $w_i\sigma = E_0^2\sigma/4\pi$. It is smaller than the force acting upon a crystal by the factor equal to the number of peaks in the cross-section of the crystal.

For an incident flux $10^{12} photons/cm^2 \cdot s$ with energy $10keV$ we get an incident energy density $w_i = 10^{12} \times 10keV/c = 5 \times 10^{-7} erg/cm^3$ and an electric field $E_0 = (4\pi w_i)^{1/2} \simeq 10^{-3} esu$; the pressure on the incidence surface is $5 \times 10^{-7} dyn/cm^2$ and the net force on the sample is $F = 5 \times 10^{-7}\sigma\, dyn$; it is of the order $10^{-6} \times 10^{-26} N^{4/3} = 10^{-32} N^{4/3} dyn$ (for a crystalline sample), valid for $10^{-32} N^{4/3} \ll 5 \times 10^{-7} A(cm^2)$. In the limiting case where the whole incident radiation is scattered, the force is of the order $5 \times 10^{-7} dyn$ for an area $1cm^2$; this is comparable to the gravitational force between two usual crystals ($100g$) separated by $1cm$. However, the validity condition $\sigma < A$ is not satisfied by macroscopic samples. We shall see in the Neutrino Scattering chapter that coherence domains appear in the sample, which reduce the cross-section to $\sigma \leq A$.

If most of the radiation is trapped inside the sample, the main contribution to the cross-section may arise from the surface. If, formally, we replace N by $N^{2/3}$ in the validity condition $a_0^2 N^{4/3} \ll A$

($a_0N^{1/3}/a \ll 1$), the order of magnitude of the cross-section becomes $a_0^2N^{8/9}$ and the validity condition reads $a_0N^{1/9} \ll a$, which is fulfilled for any reasonable-size sample. For the numerical data given above the net force on the sample is of the order $5\times10^{-7}\times10^{-5} = 5\times10^{-12}dyn$ for $1mol$. This is an extremely small force.

4.8 Average total intensity

In the Laue method for X-rays diffraction in crystals the direction of the wavevector $\mathbf{k}$ of the incident X-ray is kept fixed but its magnitude is distributed by $\rho(k)$ such that $\int dk\rho(k) = 1$. The total cross-section of a $\boldsymbol{g}$-peak is given by

$$\begin{aligned}
\sigma &= \tfrac{1}{2}(e^2/mc^2)^2\left|n_{\mathbf{g}}\right|^2\cdot \\
&\cdot \int dodk\rho(k)(1+\cos^2\theta)\left|\int d\mathbf{r}\cdot e^{(\mathbf{k}-\mathbf{k}'+\mathbf{g})\mathbf{r}}\right|^2 = \\
&= 4\pi^3V(e^2/mc^2)^2\left|n_{\mathbf{g}}\right|^2\cdot \\
&\cdot \int dodk\rho(k)(1+\cos^2\theta)\delta(\mathbf{k}'-\mathbf{k}-\mathbf{g}) = \\
&= 4\pi^3V(e^2/mc^2)^2\left|n_{\mathbf{g}}\right|^2(1+\cos^2\theta_0)\cdot \\
&\cdot \int dodk\rho(k)\delta(\mathbf{k}'-\mathbf{k}-\mathbf{g}) \ ,
\end{aligned} \tag{4.70}$$

where θ_0 is the angle between $\mathbf{k}_0$ and $\mathbf{k}_0^{'}$ which satisfy the Laue condition $\mathbf{k}_0^{'} = \mathbf{k}_0 + \mathbf{g}$. We introduce $d(k'^2)\delta(k^2-k'^2)$ and take into account that $d\mathbf{k}' = k'^2dk'do = k'd(k'^2)do/2$. We get

$$\begin{aligned}
&\int dodk\rho(k)\delta(\mathbf{k}'-\mathbf{k}-\mathbf{g}) = \\
&= \int d\mathbf{k}'dk(2/k)\rho(k)\delta(\mathbf{k}'-\mathbf{k}-\mathbf{g})\delta(k^2-k'^2) = \\
&= \int dk(2/k)\rho(k)\delta(2\mathbf{k}\mathbf{g}+g^2) = 2\rho(k_0)/g^2 \ ,
\end{aligned} \tag{4.71}$$

such that

$$\sigma = (2\pi)^3V(e^2/mc^2)^2\rho(k_0)\left(\left|n_{\mathbf{g}}\right|^2/g^2\right)(1+\cos^2\theta_0) \ . \tag{4.72}$$

It is proportional to $1/g^2$.[12]

In the Bragg method of rotating crystal the magnitude of the wavevector $\mathbf{k}$ is kept fixed but its direction is distributed about the unit vector $\mathbf{e}$ by $\rho(\psi)$, such that $\int d\psi\rho(\psi) = 1$. From equation (4.71) we get

$$\begin{aligned}\sigma = (2\pi)^3 V(e^2/mc^2)^2 \left|n_{\mathbf{g}}\right|^2 (1+\cos^2\theta_0)\cdot \\ \cdot \int d\psi (1/k)\rho(\psi)\delta(2\mathbf{kg}+g^2) \ .\end{aligned} \tag{4.73}$$

For a small inclination $\mathbf{k} = \mathbf{k}_0 + (\mathbf{e}\times\mathbf{k}_0)\psi$ and

$$\begin{aligned}\int d\psi(1/k)\rho(\psi)\delta(2\mathbf{kg}+g^2) = \\ = \int d\psi(1/k)\rho(\psi)\delta\left[(2\mathbf{g}(\mathbf{e}\times\mathbf{k}_0)\psi\right] = \\ = \tfrac{\rho(0)}{k_0}\cdot\tfrac{1}{2|\mathbf{g}(\mathbf{e}\times\mathbf{k}_0)|} = \\ = 2\rho(0)\left(\sin^2\theta_0/2\right)/g^2\left|\mathbf{g}(\mathbf{e}\times\mathbf{n}_0)\right| \ ,\end{aligned} \tag{4.74}$$

where $n_0 = \mathbf{k}_0/k_0$. The cross-section is given by

$$\begin{aligned}\sigma = 2(2\pi)^3 V(e^2/mc^2)^2\rho(0)\sin^2\theta_0/2\cdot \\ \cdot\tfrac{|n_{\mathbf{g}}|^2}{g^2|\mathbf{g}(\mathbf{e}\times\mathbf{n}_0)|}(1+\cos^2\theta_0) \ .\end{aligned} \tag{4.75}$$

It is proportional to $1/g^3$.

In the Debye-Scherrer method of X-rays diffraction the wavevector $\mathbf{k}$ is kept fixed but the sample consists of a powder of crystallites. The Laue condition reads now $\mathbf{k}_0^{'} = \mathbf{k} + \mathbf{g}_0$ and it is satisfied for $\mathbf{k}_0^{'}$ lying on conical surfaces of angle $2\theta_0$ about $\mathbf{k}$. For randomly distributed crystallites we have from equation (4.71)

$$\begin{aligned}\sigma = (2\pi)^3 V(e^2/mc^2)^2 \left|n_{\mathbf{g}}\right|^2 (1+\cos^2\theta_0)\cdot \\ \cdot\int(do_{\mathbf{g}}/4\pi)(1/k)\delta(2\mathbf{kg}+g^2) \ .\end{aligned} \tag{4.76}$$

[12] L. Landau and E. Lifshitz, *Course of Theoretical Physics*, vol. 8, *Electrodynamics of Continuous Media* (L. Landau, E. Lifshitz and L. Pitaevskii), Elsevier, Oxford (1993).

The integral in equation (4.76) is

$$\begin{gathered}\int (do_{\mathbf{g}}/4\pi)(1/k)(2\mathbf{kg}+g^2) = \\ (1/2)\int d\alpha \sin\alpha (1/k)\delta(2kg\cos\alpha + g^2) = \\ = 1/4k^2 g = \left(\sin^2\theta_0/2\right)/g^3 \ .\end{gathered} \tag{4.77}$$

The cross-section is given by

$$\begin{gathered}\sigma = (2\pi)^3 V(e^2/mc^2)^2 \sin^2\theta_0/2 \cdot \\ \cdot \frac{|n_{\mathbf{g}}|^2}{g^3}\cdot(1+\cos^2\theta_0) \ .\end{gathered} \tag{4.78}$$

It goes like $1/g^3$.

4.9 Diffuse thermal scattering of X-rays

The thermal fluctuations $\delta n(\mathbf{r})$ in the electron density produce a diffusion of the diffraction peaks.[13] If $u(\mathbf{r})$ is the displacement of the atoms from their equilibrium positions then

$$\delta n(\mathbf{r}) = n(\mathbf{r}+\mathbf{u}) - n(\mathbf{r}) = \mathbf{u}\partial n/\partial\mathbf{r} \ , \tag{4.79}$$

or, near a diffraction peak $\mathbf{g}$,

$$\delta n(\mathbf{r}) = i\mathbf{ug}n_{\mathbf{g}}e^{i\mathbf{gr}} \ . \tag{4.80}$$

The corresponding cross-section is obtained from equation (4.22) as

$$\begin{gathered}d\sigma = \tfrac{1}{2}(e^2/mc^2)^2(1+\cos^2\theta)\left|n_{\mathbf{g}}\right|^2 \cdot \\ \left\langle \left|\int d\mathbf{r}\cdot \mathbf{ug}e^{i\mathbf{qr}}\right|^2\right\rangle do \ ,\end{gathered} \tag{4.81}$$

where the brackets mean the thermal average and $\mathbf{q} = \mathbf{k} - \mathbf{k}' + \mathbf{g}$. One can see that the main contributions come from long-wavelength phonons.

[13] W. H. Zachariasen, "A theoretical study of the diffuse scattering of X-rays by crystals", Phys. Rev. **57** 597 (1940).

We get from equation (4.81)

$$d\sigma = \frac{1}{2}(e^2/mc^2)^2(1+\cos^2\theta)V^2\left|n_{\mathbf{g}}\right|^2 g_i g_j \langle u_i u_j\rangle\, do\ . \tag{4.82}$$

The elastic energy is $E = (1/2)V\lambda_{ijkl}u_{ij}u_{kl}$ where λ_{ijkl} is the elastic modulus tensor and $u_{ij} = (1/2)(\partial u_i/\partial x_j + \partial u_j/\partial x_i)$ is the strain tensor. Therefore $\langle E\rangle = (1/2)V\lambda_{ijkl}q_j q_l \langle u_i u_k\rangle$ and $\langle u_i u_k\rangle = (T/V)g_{ik}^{-1}$, where $g_{ik} = \lambda_{ijkl}q_j q_l$ and T is the temperature. Therefore the cross-section is given by

$$d\sigma = \frac{1}{2}(e^2/mc^2)^2(1+\cos^2\theta)\left|n_{\mathbf{g}}\right|^2 TV g_{ij}^{-1} g_i g_j do\ . \tag{4.83}$$

One can see that the cross-section depends on direction and goes like $\sim 1/q^2$, *i.e.* it decreases more slowly than the sharp peak. However, $Tg_{ij}^{-1}g_i g_j$ in equation (4.83) is an appreciable reduction factor.

4.10 Lattice distortion and phase defects

Let us assume that the positions of the atoms are displaced according to

$$\mathbf{r}' = \mathbf{r} + \mathbf{u}\cos(\mathbf{Q}\mathbf{r}+\varphi) \tag{4.84}$$

as for a Peierls distortion,[14] where u is much smaller than the size of the unit cell. This distortion produces a change in the electron density given by

$$n' = n + \mathbf{u}(\partial n/\partial \mathbf{r})\cos(\mathbf{Q}\mathbf{r}+\varphi)\ , \tag{4.85}$$

or, for the **g**-component of the Fourier transform,

$$n_{\boldsymbol{g}}^{'} = n_{\mathbf{g}}\left[1 + i\mathbf{u}\mathbf{g}\cos(\mathbf{Q}\mathbf{r}+\varphi)\right] \tag{4.86}$$

for $|\mathbf{ug}| \ll 1$. The cross-section is obtained from equation (4.25) as

$$\begin{gathered} d\sigma = \tfrac{1}{2}(e^2/mc^2)^2(1+\cos^2\theta)\left|n_{\mathbf{g}}\right|^2(2\pi)^3V\cdot \\ \cdot\{\delta(\mathbf{K}-\mathbf{g}) + \tfrac{1}{4}(\mathbf{ug})^2\cdot \\ \cdot\left[\delta(\mathbf{K}-\mathbf{g}+\mathbf{Q}) + \delta(\mathbf{K}-\mathbf{g}-\mathbf{Q})\right]\}do\ . \end{gathered} \tag{4.87}$$

[14] R. Peierls, *Quantum Theory of Solids*, Clarendon, Oxford (1955).

We can see that satellites of diffraction appear around the main peak $\mathbf{g}$, whose relative intensity goes like $(\mathbf{ug})^2$.

Phase defects may appear in the modulated atomic displacement. Let us focus on a satellite peak given by

$$d\sigma/do \sim \frac{1}{4}(\mathbf{ug})^2 \left| \int d\mathbf{r} \cdot e^{-i\mathbf{Kr}} \cdot e^{i[(\mathbf{g}+\mathbf{Q})\mathbf{r}+\varphi]} \right|^2 . \tag{4.88}$$

We assume that φ depends only on z and along this direction it has a domain-like structure, *i.e.* within each domain (z_n, z_{n+1}) it has the same constant value which may be taken zero, but at the boundaries of the domains it acquires a change, the same for each boundary.[15] This corresponds to the natural idea of phase defects. If φ varies slowly over large distances one speaks of weakly-pinned phases, while abrupt variations in φ over short distances are referred to as a strong pinning. The phase slippage at the domain boundaries proceeds as if the coordinate z changes by a constant amount p. From equation (4.88) one gets

$$\begin{gathered} d\sigma/do \sim \tfrac{1}{4}(\mathbf{ug})^2 \left| \sum_n e^{in\chi} \int_{z_n}^{z_{n+1}} dz e^{-iq_z z} \right|^2 \cdot \\ \cdot (2\pi)^2 A\delta(q_x)\delta(q_y) \ . \end{gathered} \tag{4.89}$$

where $\chi = (g_z + Q_z)p$, A is the sample cross-section perpendicular to the z-direction and $\mathbf{q} = \mathbf{K} - \mathbf{g} - \mathbf{Q}$. We have further on the following

[15] L. Landau, "The scattering of X-rays by crystals with variable lamellar structure", ZhETF **7** 1227 (1937) (Phys. Z. Sowjet. **12** 579 (1937)). See also M. Apostol, "X-ray scattering by a superlattice", Superlatt.&Microstr. **17** 419 (1995).

transformations

$$\begin{gathered}F(q_z)=\left|\sum_n e^{in\chi}\int_{z_n}^{z_{n+1}}dze^{-iq_zz}\right|^2=\\=\tfrac{1}{q_z^2}\left|1-e^{-i\chi}\right|^2\left|\sum_n e^{in\chi}e^{-iq_zz_n}\right|^2=\\=\tfrac{1}{q_z^2}\left|1-e^{-i\chi}\right|^2\sum_{nl}e^{il\chi}e^{-iq_z(z_n-z_{n-l})}=\\=\tfrac{N_z}{q_z^2}\left|1-e^{-i\chi}\right|^2\sum_l e^{il\chi}\left\langle e^{-iq_z(z_n-z_{n-l})}\right\rangle=\\=\tfrac{N_z}{q_z^2}\left|1-e^{-i\chi}\right|^2[1+\sum_{l=1}e^{il\chi}\left\langle e^{-iq_z(z_n-z_{n-l})}\right\rangle+\\+\sum_{l=1}e^{-il\chi}\left\langle e^{-iq_z(z_n-z_{n+l})}\right\rangle]\end{gathered}\tag{4.90}$$

where N_z is the number of domains and these domains are assumed to be uncorrelated. We introduce the domain thickness $d_n=z_{n+1}-z_n$ such as

$$\begin{gathered}z_n-z_{n-l}=d_{n-1}+d_{n-2}+...d_{n-l}\ ,\\z_n-z_{n+l}=-(d_n+d_{n+1}+...d_{n+l-1})\ .\end{gathered}\tag{4.91}$$

Equation (4.90) becomes

$$\begin{gathered}F(q_z)=\tfrac{N_z}{q_z^2}\left|1-e^{-i\chi}\right|^2\cdot\\\cdot\left[1+\sum_{l=1}e^{il\chi}\xi^{*l}+\sum_{l=1}e^{-il\chi}\xi^l\right]=\\=\tfrac{N_z}{q_z^2}\cdot 4\sin^2\chi/2\cdot\tfrac{1-|\xi|^2}{|1-\xi\exp(-i\chi)|^2}\ ,\end{gathered}\tag{4.92}$$

where $\xi=\langle\exp(-iq_zd)\rangle$. We may assume that the thickness is distributed according to the probability $\overline{d}^{-1}\exp(-d/\overline{d})$ which gives $\xi=(1+iq_z\overline{d})^{-1}$ where $\overline{d}$ is the mean thickness of the domains. We get

$$F(q_z)=4N_z\left(\overline{d}\right)^2\cdot\frac{\sin^2\chi/2}{(q_z\overline{d}+\sin\chi)^2+(1-\cos\chi)^2}\ .\tag{4.93}$$

It follows that the relative cross-section with respect to that of the satellite peak is

$$d\sigma/do = \frac{2\overline{d}}{\pi} \cdot \frac{\sin^2 \chi/2}{(q_z\overline{d} + \sin\chi)^2 + (1 - \cos\chi)^2} \ . \qquad (4.94)$$

We can see that the statistical distribution of the phase defects produces a diffuse X-rays scattering described by the Lorentz curve given by equation (4.94). The maximum of diffraction is shifted from $q_z = 0$ to $\Delta q_z = -\sin\chi/\overline{d}$ and the width of the peak increases by $\delta q_z = (1 - \cos\chi)/\overline{d}$. Since the energy of the defects is proportional to their density it follows that $\overline{d}$ decreases when temperature increases, so the diffuse character is increased.

4.11 Temperature dependence: X-rays scattering and phonons

The electron density can be represented as

$$n(\mathbf{r}) = \sum_{ij} f_j(\mathbf{r} - \mathbf{r}_i - \rho_j - \mathbf{u}_{ij}) \ , \qquad (4.95)$$

where $\mathbf{r}_i$ denotes the position of the unit cell, ρ_j denotes the position of the j-th atom in the unit cell ($j = 1, 2, ...r$) and $\mathbf{u}_{ij}$ is the displacement corresponding to atomic vibrations. The function f_j does not change during atomic motion, according to the adiabatic approximation. We introduce the Fourier transformation

$$n(\mathbf{r}) = \sum_{\mathbf{k}} \sum_{ij} \left[f_j(\mathbf{k}) e^{-i\mathbf{k}\mathbf{r}_i} e^{-i\mathbf{k}\rho_j} e^{-i\mathbf{k}\mathbf{u}_{ij}} \right] e^{i\mathbf{k}\mathbf{r}} \qquad (4.96)$$

in the cross-section

$$d\sigma/do = \frac{1}{2}(e^2/mc^2)^2(1 + \cos^2\theta) \left| \int d\mathbf{r} \cdot n e^{i\mathbf{K}\mathbf{r}} \right|^2 \qquad (4.97)$$

given by equation (4.22) and get

$$d\sigma/do = \frac{1}{2}(e^2/mc^2)^2(1+\cos^2\theta)V^2\cdot$$
$$\cdot\sum_{ii'jj'} f_j(K)f_{j'}^*(K)e^{i\mathbf{K}(\mathbf{r}_i-\mathbf{r}_{i'})}\cdot \qquad (4.98)$$
$$\cdot e^{i\mathbf{K}(\rho_j-\rho_{j'})}e^{i\mathbf{K}(\mathbf{u}_{ij}-\mathbf{u}_{i'j'})} .$$

We want to compute the thermal averages

$$\left\langle e^{i\mathbf{K}(\mathbf{u}_{ij}-\mathbf{u}_{i'j'})}\right\rangle = e^{-\left\langle\left[\mathbf{K}(\mathbf{u}_{ij}-\mathbf{u}_{i'j'})\right]^2\right\rangle/2} \qquad (4.99)$$

where

$$\left\langle\left[\mathbf{K}(\mathbf{u}_{ij}-\mathbf{u}_{i'j'})\right]^2\right\rangle = K_\alpha K_\beta\left\langle u_{ij\alpha}u_{ij\beta}\right\rangle +$$
$$+K_\alpha K_\beta\left\langle u_{i'j'\alpha}u_{i'j'\beta}\right\rangle - 2K_\alpha K_\beta\left\langle u_{ij\alpha}u_{i'j'\beta}\right\rangle . \qquad (4.100)$$

We use the normal-vibrations expansion

$$u_{ij\alpha} = (Nm_j)^{-1/2}\sum_{\nu\mathbf{q}} e_{j\alpha}^\nu(\mathbf{q})Q_\nu(\mathbf{q})e^{i\mathbf{q}\mathbf{r}_i} \qquad (4.101)$$

with usual notations ($\mathbf{e}_j^\nu$ is the polarization vector of the j-th atom in the ν-th phonon branch, α denotes the polarization component); the normal coordinates are

$$Q_\nu(\mathbf{q}) = \sqrt{\frac{\hbar}{2\omega_\nu(\mathbf{q})}}\left[a_\nu(\mathbf{q})+a_\nu^+(-\mathbf{q})\right] , \qquad (4.102)$$

where a, a^+ are phonon destruction and creation operators. We get

$$\left\langle Q_\nu(\mathbf{q})Q_{\nu'}(\mathbf{q}')\right\rangle = \frac{\hbar}{2\omega_\nu(\mathbf{q})}\left[2n_\nu(\mathbf{q})+1\right]\delta_{\nu\nu'}\delta_{\mathbf{q},-\mathbf{q}'} , \qquad (4.103)$$

where

$$n_\nu(\mathbf{q}) = \frac{1}{\exp\left[\hbar\omega_\nu(\mathbf{q})/T\right]-1} . \qquad (4.104)$$

The thermal averages in equation (4.100) read now

$$2W_j = K_\alpha K_\beta \langle u_{ij\alpha} u_{ij\beta} \rangle =$$
$$= (\hbar/2m_j) K_\alpha K_\beta \sum_{\nu\mathbf{q}} e^\nu_{j\alpha}(\mathbf{q}) e^\nu_{j\beta}(-\mathbf{q}) \cdot \frac{\coth[\hbar\omega_\nu(\mathbf{q})/2T]}{\omega_\nu(\mathbf{q})} \tag{4.105}$$

and

$$2M_{iji'j'} = 2K_\alpha K_\beta \langle u_{ij\alpha} u_{i'j'\beta} \rangle =$$
$$= [\hbar/N(m_j m_{j'})^{1/2}] K_\alpha K_\beta \sum_{\nu\mathbf{q}} e^\nu_{j\alpha}(\mathbf{q}) e^\nu_{j'\beta}(-\mathbf{q}) \cdot \tag{4.106}$$
$$\cdot \frac{\coth[\hbar\omega_\nu(\mathbf{q})/2T]}{\omega_\nu(\mathbf{q})} e^{i\mathbf{q}(\mathbf{r}_i - \mathbf{r}_{i'})} .$$

The former quantity given by equation (4.105) is known as the Debye-Waller factor.[16] The thermal average of the cross-section given by equation (4.98) can now be written as

$$d\sigma/do = \frac{1}{2}(e^2/mc^2)^2(1+\cos^2\theta)V^2 \cdot$$
$$\cdot \sum_{ii'jj'} f_j(\mathbf{K}) f^*_{j'}(\mathbf{K}) e^{-i\mathbf{K}(\mathbf{r}_i - \mathbf{r}_{i'})} e^{-i\mathbf{K}(\rho_j - \rho_{j'})} \cdot \tag{4.107}$$
$$\cdot e^{-W_j(\mathbf{K})} e^{-W_{j'}(\mathbf{K})} (1 + M_{iji'j'}) ,$$

since $M_{iji'j'}$ is much smaller than W. If we neglect the small contribution of $M_{iji'j'}$ the cross-section can be written as

$$d\sigma/do = \frac{1}{2}(e^2/mc^2)^2(1+\cos^2\theta)(2\pi)^3 V \cdot$$
$$\sum_{\mathbf{g}} \left| N \sum_j f_j(\mathbf{g}) e^{-i\mathbf{g}\rho_j} e^{-W_j(\mathbf{g})} \right|^2 \delta(\mathbf{K} - \mathbf{g}) . \tag{4.108}$$

We can see that the diffraction peaks remain sharp at the same positions but their intensity is reduced by the Debye-Waller factor as a consequence of the thermal phonons. The quantity in the square modulus in equation (4.108) is the structure factor of the lattice.

[16] P. Debye, "Interferenz von Roentgenstrahlen und Waermebewegung", Ann. Phys. **43** 49 (1914); I. Waller, "Theoretische Studien zur Interferenz- und Dispersionstheorie der Roentgenstrahlen". Dissertation, Uppsala (1925).

From equation (4.105) a fair estimation of the Debye-Waller factor is $W \simeq (\hbar^2/ma^2)/\hbar\omega_D$, where m is the atomic mass, a is the mean inter-atomic distance and ω_D is the Debye frequency. Typical Debye temperatures $\hbar\omega_D$ are of the order $10^2 K$, such that the Debye-Waller factor does not affect too much the cross-section.

The factor $M_{iji'j'}$ in equation (4.107) brings its main contribution for long-wavelength phonons. Making use of equations (4.106) and (4.108) we get

$$\begin{gathered}
d\sigma/do = \tfrac{1}{2}(e^2/mc^2)^2(1+\cos^2\theta)V^2\cdot \\
\cdot\sum\nolimits_{ii'jj'} N^2 f_j(\mathbf{g}) f^*_{j'}(\mathbf{g})\cdot \\
\cdot e^{-i\mathbf{g}(\rho_j-\rho_{j'})} e^{-W_j(\mathbf{g})} e^{-W_{j'}(\mathbf{g})}\cdot \frac{\hbar}{2N(m_j m_{j'})^{1/2}} g_\alpha g_\beta\cdot \\
\cdot\sum\nolimits_\nu e^\nu_{j\alpha}(0) e^\nu_{j'\beta}(0)\cdot \frac{\coth[\hbar\omega_\nu(0)/2T]}{\omega_\nu(0)}
\end{gathered} \tag{4.109}$$

in the neighbourhood of a diffraction peak. It agrees with the diffuse thermal scattering given by equation (4.83) except that equation (4.109) takes into account the basis of the unit cell and the Debye-Waller factor.

Inelastic X-rays scattering is also contained in equation (4.98), due to phonon absorption and emission processes, providing the cross-section given by equation (4.98) is analyzed with respect to energy too. However, the X-rays energy is much higher than the phonon energies, so, practically, the inelastic scattering cannot be seen experimentally.

4.12 Higher-order effects

The average electron density of a perfect crystal can be represented as

$$n(\mathbf{r}) = \sum_{ij} f_j(\mathbf{r}-\mathbf{r}_i-\rho_j) \ , \tag{4.110}$$

where $\mathbf{r}_i$ denotes the position of the unit cell and ρ_j denotes the position of the j-th atom in the unit cell. Its Fourier transform reads

$$n(\mathbf{r}) = \sum_{\mathbf{k}} \left[\sum_j f_j(\mathbf{k}) e^{-i\mathbf{k}\rho_j} \right] \sum_i e^{-i\mathbf{k}\mathbf{r}_i} e^{i\mathbf{k}\mathbf{r}} \ , \tag{4.111}$$

or

$$\begin{aligned} n(\mathbf{r}) &= N \sum_{\mathbf{g}} \left[\sum_j f_j(\mathbf{g}) e^{-i\mathbf{g}\rho_j} \right] e^{i\mathbf{g}\mathbf{r}} = \\ &= \sum_{\mathbf{g}} n_{\mathbf{g}} e^{i\mathbf{g}\mathbf{r}} \end{aligned} \tag{4.112}$$

with usual notations. The quantity $f_j(\mathbf{g})$ is the atomic structure factor. The function f_j is often approximated by $f_j(\mathbf{r}) = z_j \delta(\mathbf{r})$ where z_j is the number of electrons on the j-th atom which are active in the X-rays diffraction, so $f_j(\mathbf{g}) = z_j / V$.

Let $\mathbf{r}' = \mathbf{r} + \mathbf{u} \sin \mathbf{Q}\mathbf{r}$ be a modulation in the electron density

$$n(\mathbf{r}) = \sum_{\mathbf{g}} n_{\mathbf{g}} e^{i(\mathbf{g}\mathbf{r} + \mathbf{g}\mathbf{u} \sin \mathbf{Q}\mathbf{r})} = \sum_{\mathbf{g}n} n_{\mathbf{g}} J_n(\mathbf{g}\mathbf{u}) e^{i(\mathbf{g}+n\mathbf{Q})\mathbf{r}} \ , \tag{4.113}$$

where J_n in

$$e^{i\mathbf{g}\mathbf{u} \sin \mathbf{Q}\mathbf{r}} = \sum_{n=-\infty}^{\infty} J_n(\mathbf{g}\mathbf{u}) e^{in\mathbf{Q}\mathbf{r}} \ , \tag{4.114}$$

are the Bessel functions. It follows that the intensity of the diffracted X-rays is proportional to

$$(2\pi)^3 V \sum_{\mathbf{g}n} |n_{\mathbf{g}}|^2 J_n^2(\mathbf{g}\mathbf{u}) \delta(\mathbf{K} - \mathbf{g} - n\mathbf{Q}) \ , \tag{4.115}$$

i.e. there appear also higher-order satellites near a main peak as a consequence of the distortion. Some can be extinct, due to the zeros of the Bessel functions. For $|\mathbf{g}\mathbf{u}| \ll 1$ we have $J_n(\mathbf{g}\mathbf{u}) \simeq (\mathbf{g}\mathbf{u}/2)^n / n!$ and we recover equation (4.87) for the first-order satellites $n = \pm 1$.

If we include the distortion given above in the positions of the unit cells we get

$$\begin{aligned} n(\mathbf{r}) &= \tfrac{1}{N} \sum_{\mathbf{k}} n_{\mathbf{k}} \sum_i e^{-i(\mathbf{k}\mathbf{r}_i + \mathbf{k}\mathbf{u} \sin \mathbf{Q}\mathbf{r}_i)} e^{i\mathbf{k}\mathbf{r}} = \\ &= \tfrac{1}{N} \sum_{\mathbf{k}n} n_{\mathbf{k}} J_n(\mathbf{k}\mathbf{u}) \sum_i e^{-i(\mathbf{k}+n\mathbf{Q})\mathbf{r}_i} e^{i\mathbf{k}\mathbf{r}} = \\ &= \sum_{\mathbf{g}n} n_{\mathbf{g}} J_n[(\mathbf{g} - \mathbf{n}\mathbf{Q})\mathbf{u}] e^{-i(\mathbf{g}-n\mathbf{Q})\mathbf{r}} \ , \end{aligned} \tag{4.116}$$

which makes the intensities of the first-order satellites proportional to $|(\mathbf{g} \pm \mathbf{Q})\mathbf{u}|^2$.

The uniform or oscillating displacement of the atoms near the surface of a semi-infinite solid is reflected, in principle, in the X-rays scattering, as a higher-order effect.[17] In general, a comparison with experiment shows that the experimental X-rays diffraction peaks are more intense and wider than those estimated herein.

Finally, we note that at higher energies of the photons, comparable to the rest energy of the electron $mc^2 = 0.5MeV$ the Thomson scattering becomes the inelastic Compton scattering, whose cross-section is given by the Klein-Nishina formula.[18] Also, the photons can couple to the electron spin,[19] and the small cross-section is enhanced by a resonance, through a Rayleigh-type scattering formula (equations (4.9) and (4.18)), in a resonant magnetic X-ray diffraction.[20]

[17] M. Apostol, "A semi-infinite solid", J. Theor. Phys. **24** (1997).

[18] O. Klein and Y. Nishina, "Ueber die Streuung von Strahlung durch freie Elektronen nach der neuen relativistischen Quantendynamik von Dirac", Z. Phys. **52** 853 (1929).

[19] M. Gell-Mann and M. L. Goldberger, "Scattering of low-energy photons by particles of spin 1/2", Phys. Rev. **96** 1433 (1954).

[20] S. W. Lovesey and S. P. Collins, *X-ray scattering and absorption by magnetic materials*, Clarendon Press, Oxford (1996).

5 Neutrino Scattering

5.1 Neutrinos

Neutrinos (antineutrinos) can be obtained, for instance, from a tritium source, with energy $\simeq 10keV$ and activity $3000Ci$ ($1Ci = 3.7\times10^{10}Bq$, $1Bq$ =one decay per second). Also, the nuclear reactors may generate antineutrinos with energies in the range of $1MeV$ and flux density (at close distance) of approximately $10^{12}/cm^2\cdot s$. Another source is provided by Sun which gives $\simeq 10^{11}/cm^2\cdot s$ neutrinos (in all flavours) with an average energy $\simeq 300keV$. We consider an incident monoenergetic beam of neutrinos with a fixed momentum and a uniform distribution across the transverse area, occupying distinct pointlike states across this area. The numerical estimations are performed herein for a neutrino energy $E = 1MeV$ ($1eV = 1.6\times10^{-12}erg$), corresponding to a wavevector $k = E/\hbar c = 5\times10^{10}cm^{-1}$ (Planck's constant $\hbar = 10^{-27}erg\cdot s$, the speed of light $c = 3\times10^{10}cm/s$) and wavelength $\lambda = 2\pi/k \simeq 10^{-10}cm$. Also, we assume a neutrino flux density $\Phi = 10^{12}/cm^2\cdot s$.

Notoriously, neutrinos interact extremely weak with matter. For instance, their cross-section with a nucleon is of the order $\sigma_{nucl} \simeq G^2E^2/\hbar^4c^4$, where $G \simeq 10^{-4}MeV\cdot fm^3$ is Fermi's weak-interaction coupling constant ($1fm = 10^{-13}cm$). Similar neutrino-electron interactions exist. For an energy $E = 1MeV$ the nucleon cross-section is $\simeq 10^{-44}cm^2$.[1] With the available fluxes, the neutrino detection is extremely difficult. Large amounts of suitable fluids are used in order to detect them.[2]

[1] For comparison, nuclear cross-sections are of the order $1barn = 10^{-24}cm^2$ and the electromagnetic radiation cross-sections are of the order $10^{-25}cm^2$.

[2] Neutrinos have been discovered by such methods, see F. Reines and C. L. Cowan, Jr., "The neutrino", Nature **178** 446 (1956); C. L. Cowan, Jr., F. Reines, F. B. Harrison, H. W. Kruse and A. D. McGuire, "Detection of the free neutrino:

The nucleons are held together by strong forces in a nucleus, so it is conceivable that if a certain amount of energy is transmitted by a neutrino to a nucleon, then it is in fact transferred to the whole nucleus which will suffer the collision as a whole. The cross-section is the square of the sum of the scattering amplitudes arising from all the nucleons, so they may add coherently to give a cross-section $\sigma_0 \simeq (G^2E^2/\hbar^4c^4)A^2$, where A is the mass number of the nucleus. This cross-section goes like the square of the number of scatterers, which enhances it appreciably in comparison with an incoherent scattering where the cross-section goes like the first power of the number of scatterers. Actually, the proton scattering amplitude is of opposite sign with respect to the neutron scattering amplitude and it is also diminished by the (Weinberg) mixing angle θ ($\simeq 0.2$). The nuclear cross-section reads then $\sigma_0 \simeq (G^2E^2/\hbar^4c^4)[N-Z(1-4\sin^2\theta)]^2$, where N stands for the number of neutrons and Z for the number of protons. The possibility of a nuclear coherent scattering is supported by the neutrinos' wavelength, which is much larger than the nuclear dimensions (nuclear dimension is of the order of a few fm's); but it could be much shorter than inter-atomic distance, which is of the order of a few Å's, $1\text{Å} = 10^{-8}cm$. This observation opens up the way of investigating the coherent scattering of neutrinos on a larger scale (like by a stiff sapphire crystal[3]). We note that for the solar neutrinos the decrease in energy may be compensated by the factor $[N - Z(1 - 4\sin^2\theta)]^2$, while the decrease in the flux density may bring a factor 10^{-1}, such that the force ($\sim \Phi\sigma_0 E$, derived below) may decrease by an order of magnitude. For tritium neutrinos the cross-section may decrease by a factor $\simeq 10^{-3}$, but the flux density may increase by a factor 10^2, such that an overall factor 10^{-1} is obtained. Also, we note that a large unit cell, as for a sapphire crystal, may increase the cross-section σ_0. There is a large uncertainty in the input numerical data (*e.g.*, the single-particle cross-section may be larger, or the density of atomic nuclei in the target may be larger).

a confirmation", Science **124** 103 (1956).

[3]J. Weber, "Method for observation of neutrinos and antineutrinos", Phys. Rev. **C31** 1468 (1985); J. Weber, "Apparent observation of abnormally large coherent scattering cross sections using keV and MeV range antineutrinos, and solar neutrinos", Phys. Rev. **D38** 32 (1988).

5.2 Scattering parameters

Let us consider the scattering of a neutrino with momentum $\boldsymbol{p}$ with a nucleus with mass m and momentum $\boldsymbol{P}$. After collision the neutrino has a momentum $\boldsymbol{p}'$ and the nucleus has a momentum $\boldsymbol{P}'$. The momentum and energy conservation reads

$$\boldsymbol{P}' + \boldsymbol{p}' = \boldsymbol{P} + \boldsymbol{p} \ ,$$
$$\frac{P'^2}{2m} + cp' = \frac{P^2}{2m} + cp \ . \tag{5.1}$$

These are four equations with six unknowns ($\boldsymbol{P}'$ and $\boldsymbol{p}'$). Besides the azimuthal angles we introduce two free parameters, namely the angle θ between $\boldsymbol{p}$ and $\boldsymbol{p}'$ and the angle ψ between $\boldsymbol{p}$ and $\boldsymbol{P}$; the angle between $\boldsymbol{p}'$ and $\boldsymbol{P}$ is $\psi - \theta$. From the first equation (5.1) we get

$$P'^2 = P^2 + p^2 + p'^2 + 2Pp\cos\psi - 2pp'\cos\theta - 2Pp'\cos(\psi - \theta) \ , \tag{5.2}$$

which, introduced in the second equation (5.1), leads to

$$p'^2 + 2\left[mc - p\cos\theta - P\cos(\psi - \theta)\right]p' -$$
$$- (2mc - p - 2P\cos\psi)\, p = 0 \ . \tag{5.3}$$

Since P/mc, $p/mc \ll 1$, we expand the solution in powers of p/mc and P/mc, such that

$$p' = p - (1 - \cos\theta)\frac{p^2}{mc} + \left[\cos(\psi - \theta) - \cos\psi\right]\frac{pP}{mc} + \ldots \ . \tag{5.4}$$

We can see that $p' \simeq p$ and, from equation (5.3), $P' \simeq 2p\sin\frac{\theta}{2}$ for $p \gg P$ and $P' \simeq P$ for $p \ll P$. The momentum transfer to the nucleus along the direction of the projectile is $\Delta p = (1 - \cos\theta)p$, such that the force is $f = \Phi\sigma\Delta p$, where Φ is the incident flux density and σ is the cross-section. The energy transfer to the nucleus is

$$\Delta E = c(p - p') \simeq (1 - \cos\theta)\frac{p^2}{m} + \left[\cos\psi - \cos(\psi - \theta)\right]vp \ , \tag{5.5}$$

where $v = P/m$ is the velocity of the nucleus before collision. The order of magnitude of this energy transfer is given by $\Delta E = (p/m + v)p$.

The mean velocity of atoms in condensed matter at temperatures of the order of the room temperature $T = 300K$ is $v = \sqrt{T/m}$. For $m = 10^5 m_e$, where $m_e \simeq 10^{-27}g$ is the electron mass, we get $v \simeq 10^4 cm/s$ and the momentum $p = mv = 10^{-18} g \cdot cm/s$ ($1K = 1.38 \times 10^{-16} erg$). The corresponding energy of the incident particle is $E = cp = 3 \times 10^{-8} erg \simeq 20keV$ s. It follows that we can use $\Delta E = p^2/m$ for $E > 20keV$ and $E = vp$ for $E < 20keV$.

5.3 Classical scattering

Let us assume that the interaction arises independently from each atom (nucleus) placed at position $\boldsymbol{R}_i$ in the sample, such that we can write the interaction as

$$H_1 = \sum_i v(\boldsymbol{R} - \boldsymbol{R}_i) \ , \tag{5.6}$$

where $v(\boldsymbol{R} - \boldsymbol{R}_i)$ is the interaction at $\boldsymbol{R}$ generated by the i-th state (the same for all sites); the site label i goes from 1 to N, where N is the total number of sites in the sample. We call this scattering a classical scattering. We assume that the interaction $v(\boldsymbol{R} - \boldsymbol{R}_i)$ has an abrupt variation near the i-th site, like, *e.g.,* a contact interaction. The equation of motion for a particle interacting with a projectile is $\frac{d\boldsymbol{p}_{mn}}{dt} = -(gradv)_{mn}$, where $\boldsymbol{p}$ is the particle momentum and m, n denote the quantum states. Since $gradv$ is a highly-localized function on the site position, the matrix elements reduce to $(gradv)_{mn} \sim \varphi_m^*(\boldsymbol{R}_i)\varphi_n(\boldsymbol{R}_i)$. We can see that this potential gives rise to a quantum-mechanical motion associated to each particle in the target. The corresponding energy transfer is thermalized.

The rate of transition is given by the well-known Fermi's "golden rule"

$$dw = (2\pi/\hbar) \left| (H_1)_{fi} \right|^2 \delta(E_i - E_f) d\nu_f \ , \tag{5.7}$$

where $(H_1)_{fi}$ is the matrix element of the interaction between the final state (f) and the initial state (i), E_i is the initial energy, E_f is the final energy and $d\nu_f = V p_f^2 dp_f do/(2\pi\hbar)^3$ is the number of final states with momentum $\boldsymbol{p}_f$ in the range dp_f, in the volume V and the solid angle do ($\hbar = 10^{-27} erg \cdot s$ is Planck's constant).

The initial state of an incoming particle is $\frac{1}{\sqrt{V}}e^{i\boldsymbol{k}_i\boldsymbol{R}}$ and the final state of the scattered particle is $\frac{1}{\sqrt{V}}e^{i\boldsymbol{k}_f\boldsymbol{R}}$, where $\boldsymbol{k}_{i,f} = \boldsymbol{p}_{i,f}/\hbar$ are the corresponding wavevectors. The matrix element of the interaction becomes

$$(H_1)_{fi} = \frac{1}{V}\sum_i \int d\boldsymbol{R}v(\boldsymbol{R}-\boldsymbol{R}_i)e^{i(\boldsymbol{k}_i-\boldsymbol{k}_f)\boldsymbol{R}} \ . \tag{5.8}$$

We notice in this equation the lack of translational symmetry. The translational symmetry is recovered by using the center-of-mass coordinate $\boldsymbol{R}_c = \frac{1}{N}\sum_i \boldsymbol{R}_i$ and the internal coordinates $\boldsymbol{r}_i = \boldsymbol{R}_i - \boldsymbol{R}_c$, $\sum_i \boldsymbol{r}_i = 0$ (baricentric coordinates). By introducing $\boldsymbol{R} = \boldsymbol{r} + \boldsymbol{R}_c$, $\boldsymbol{R}_i = \boldsymbol{r}_i + \boldsymbol{R}_c$ in equation (5.8), we get

$$(H_1)_{fi} = \frac{1}{V}\sum_i \int d\boldsymbol{r}v(\boldsymbol{r}-\boldsymbol{r}_i)e^{i(\boldsymbol{k}_i-\boldsymbol{k}_f)\boldsymbol{r}}e^{i(\boldsymbol{k}_i-\boldsymbol{k}_f)\boldsymbol{R}_c} \ ; \tag{5.9}$$

it follows that R_c should be viewed as a dynamical variable, such that the matrix element is obtained by inserting the wavefunctions of the sample $\frac{1}{\sqrt{V}}e^{i\boldsymbol{K}_{i,f}\boldsymbol{R}_c}$, where $\boldsymbol{K}_{i,f}$ are the initial and final wavevectors of the sample (momenta $\boldsymbol{P}_{i,f} = \hbar\boldsymbol{K}_{i,f}$) and integrating over $\boldsymbol{R}_c$. The matrix element becomes

$$\begin{gathered}(H_1)_{fi} = \frac{1}{V^2}\sum_i \int d\boldsymbol{R}_c d\boldsymbol{r}v(\boldsymbol{r}-\boldsymbol{r}_i)\cdot \\ \cdot e^{i(\boldsymbol{k}_i-\boldsymbol{k}_f)\boldsymbol{r}}e^{i(\boldsymbol{k}_i-\boldsymbol{k}_f)\boldsymbol{R}_c}e^{i(\boldsymbol{K}_i-\boldsymbol{K}_f)\boldsymbol{R}_c} \ .\end{gathered} \tag{5.10}$$

Performing the integration over $\boldsymbol{R}_c$ we get the momentum conservation

$$(H_1)_{fi} = \frac{1}{V}\delta_{\boldsymbol{k}_i+\boldsymbol{K}_i,\boldsymbol{k}_f+\boldsymbol{K}_f}\sum_i \int d\boldsymbol{r}v(\boldsymbol{r}-\boldsymbol{r}_i)e^{i(\boldsymbol{k}_i-\boldsymbol{k}_f)\boldsymbol{r}} \ , \tag{5.11}$$

such that the translational symmetry is recovered. In the summation over i we should omit a site, due to the condition $\sum_i \boldsymbol{r}_i = 0$, which, for $N \gg 1$ is irrelevant.

According to equation (5.7), to the momentum conservation

$$\boldsymbol{k}_i + \boldsymbol{K}_i = \boldsymbol{k}_f + \boldsymbol{K}_f \tag{5.12}$$

we should add the energy conservation

$$\frac{\hbar^2 K_i^2}{2M} + c\hbar k_i = \frac{\hbar^2 K_f^2}{2M} + c\hbar k_f \ , \tag{5.13}$$

where M is the mass of the (macroscopic) sample. Since $\hbar k_{i,f}/Mc \ll 1$, these equations are satisfied for $k_i = k_f$ and $\boldsymbol{K}_f - \boldsymbol{K}_i = \boldsymbol{k}_i - \boldsymbol{k}_f$, *i.e.* the scattered particle changes the direction of its momentum, transfers the difference to the sample, while preserving its energy (elastic scattering). Henceforth, we tacitly understand the momentum and the energy conservation in the scattering formulae. Moreover, since $\delta(E_i - E_f)dp_f = dp_f/dE_f = 1/c$ (which amounts to an integration over the final energies), we get the rate of transitions per unit time in the solid angle

$$dw = \frac{2\pi E^2}{V\hbar c^3(2\pi\hbar)^3}\left|\sum_i \int d\boldsymbol{r} v(\boldsymbol{r} - \boldsymbol{r}_i)e^{i(\boldsymbol{k}_i - \boldsymbol{k}_f)\boldsymbol{r}}\right|^2 do \ , \tag{5.14}$$

where $E = E_f = cp_f = c\hbar k_f$ is the energy of the scattered particle (identical with the energy of the incident particle, $E_f = E_i = E$). Further on, we assume a contact interaction $v(\boldsymbol{r} - \boldsymbol{r}_i) = va^3\delta(\boldsymbol{r} - \boldsymbol{r}_i)$, where v is the strength of the interaction and a is the range of the spatial extension of the interaction (*e.g.*, the dimension of the atomic nucleus $a \simeq 10^{-13}cm$). The number of transitions per unit time becomes

$$dw = \frac{2\pi E^2 v^2 a^6}{V\hbar c^3(2\pi\hbar)^3}\left|\sum_i e^{i(\boldsymbol{k}_i - \boldsymbol{k}_f)\boldsymbol{r}_i}\right|^2 do \ ; \tag{5.15}$$

we note the occurrence of the form-factor of the sample

$$F(\boldsymbol{k}_i - \boldsymbol{k}_f) = \sum_i e^{i(\boldsymbol{k}_i - \boldsymbol{k}_f)\boldsymbol{r}_i} \ . \tag{5.16}$$

According to our normalization ($\frac{1}{\sqrt{V}}e^{i\boldsymbol{k}_i\boldsymbol{R}}$, a particle in the volume V), the flux density of the incoming particles is c/V (number of incident particles per unit time and unit area of cross-section). We get the cross-section

$$d\sigma = \frac{2\pi E^2 v^2 a^6}{\hbar c^4(2\pi\hbar)^3}\left|F(\boldsymbol{k}_i - \boldsymbol{k}_f)\right|^2 do = \sigma_0\left|\sum_i e^{i(\boldsymbol{k}_i - \boldsymbol{k}_f)\boldsymbol{r}_i}\right|^2 do \ , \tag{5.17}$$

where $\sigma_0 do$ is the cross-section of a site (*e.g.*, atomic nucleus), *i.e.* $\sigma_0 = 2\pi E^2 v^2 a^6/\hbar c^4 (2\pi\hbar)^3$.[4] For neutrinos the product va^3 corresponds to Fermi's constant G.

The number of transitions per unit time is

$$dw = \Phi d\sigma \ , \tag{5.18}$$

where Φ is the particle flux density (number of incoming particles per unit time and unit area). The momentum transfer to the sample per unit time (the force $\boldsymbol{df}$ acting upon the sample in the element of the solid angle) is

$$d\dot{\boldsymbol{P}} = d\boldsymbol{f} = \hbar d\dot{\boldsymbol{K}} == \hbar(\boldsymbol{k}_i - \boldsymbol{k}_f)dw = \Phi\hbar(\boldsymbol{k}_i - \boldsymbol{k}_f)d\sigma \ . \tag{5.19}$$

We note that the rate of change of the energy transfer in a given direction is $d\dot{E} = \boldsymbol{V} d\dot{\boldsymbol{P}} = \boldsymbol{V} d\boldsymbol{f}$, where $\boldsymbol{V}$ is the velocity acquired by the sample. For a macroscopic sample this is a very small velocity. We can use the parametrization $\boldsymbol{k}_i = k(0,0,1)$, $\boldsymbol{k}_f = k(\sin\theta\cos\varphi, \sin\theta\sin\varphi, \cos\theta)$, such that we get the total force (by integrating over the solid angle)

$$f = \Phi\hbar k \int (1 - \cos\theta) d\sigma \tag{5.20}$$

in the forward direction. The integral in equation (5.20) is the transport cross-section. For a uniform scattering the form-factor F does not depend on the momentum transfer, such that the transverse components of the force are zero and the forward force is

$$f = 4\pi\Phi\hbar k\sigma_0 |F|^2 = 4\pi n_i E\sigma_0 |F|^2 = n_i E\sigma \ , \tag{5.21}$$

where $n_i = \Phi/c$ is the density of the incoming particles and E is the energy of the incoming particle. We note that $n_i E = w_i$ is the energy density of the incoming particles (actually the density of $c\hbar k = cp$), such that $F = w_i\sigma$. This is precisely the result obtained in the X-Rays Diffraction chapter for the electromagnetic radiation.

It is worth noting that the above treatment of the scattering holds also for electromagnetic radiation, leading to the cross-section

$$d\sigma = \sigma_0 |F(\boldsymbol{k}_i - \boldsymbol{k}_f)|^2 do \ , \tag{5.22}$$

[4]We note that the momentum conservation is independent of the energy conservation, such that the formula is also valid for inelastic scattering.

where $\sigma_0 do$ is the cross-section of a single scatterer and F denotes the form-factor of the sample (equation (5.17)). Indeed, the interaction $(q^2/2mc^2)A^2$ of a pointlike charge q with the (reduced) mass m with the electromagnetic field with the vector potential $\boldsymbol{A}$ is a local interaction, associated to each site, such that the total interaction with the sample can be represented by equation (5.6). Similarly, the transition probability (equation (5.7)) can be replaced by its form in the second-order of the perturbation theory for the local interaction $-q\boldsymbol{Ap}/mc$, where $\boldsymbol{p}$ is the (relative) momentum of the charge; inelastic collisions can be treated in this case, with the conservation of the corresponding momentum and energy.[5] Moreover, equation (5.6) can be generalized to

$$H_1 = \int d\boldsymbol{R}' n(\boldsymbol{R}') v(\boldsymbol{R} - \boldsymbol{R}') \ , \tag{5.23}$$

where $n(\boldsymbol{R}')$ is the density of the particles in the sample (or the density of the physical quantities in the sample which generates the interaction). In this case we are able to compute the cross-section of the scattering by the inhomogeneities in the sample, or by its elementary excitations; of course, the form-factor changes in this case, including the momentum associated with these inhomogeneities, as well as the position dependence of the interaction.

5.4 Form-factor

The form-factor is calculated in the X-Rays Diffraction chapter. For randomly distributed nuclei the form-factor is

$$F(\boldsymbol{k}_i - \boldsymbol{k}_f) = N\delta_{\boldsymbol{k}_i, \boldsymbol{k}_f} \tag{5.24}$$

and the total cross-section is

$$\sigma = \sigma_0 N^2 \frac{(2\pi)^2}{k^2 V^{2/3}} = \sigma_0 N^{4/3} \frac{(2\pi)^2}{k^2 v^{2/3}} \ , \tag{5.25}$$

[5] For soft X-rays the atomic collisions are classical, but the cross-section preserves its form given by equation (5.22). This is so because the waves $e^{-i\omega t + i\boldsymbol{kr}}$ in the classical electromagnetic radiation can be viewed as wavefunctions of the photons. Even if the individual scattering may be treated classically, its nature is quantum-mechanical and, consequently, statistical.

where σ_0 is the cross-section of a single nucleus, N is the number of atoms (nuclei) in the sample with the macroscopic volume V, $k = k_i = k_f$ is the magnitude of the wavevector and $v = V/N$ is the volume associated to a nucleus. The cross-section exhibits one peak in the forward direction.

For a crystal $F(\boldsymbol{k}_i - \boldsymbol{k}_f) = N\delta_{\boldsymbol{k}_i+\boldsymbol{g},\boldsymbol{k}_f}$,where $\boldsymbol{g}$ is a reciprocal vector. The cross-section exhibits Bragg peaks. The cross-section of one peak is the same as in equation (5.25),

$$\sigma_{\boldsymbol{g}} = \sigma_0 N^{4/3} \frac{(2\pi)^2}{k^2 v^{2/3}} \ . \tag{5.26}$$

For a large number of peaks we may associate an area $(\Delta g)^2 = (2\pi)^2/v^{2/3}$ to each peak, such that the total number of peaks is $4\pi k^2/(4\pi^2/v^{2/3}) = k^2 v^{2/3}/\pi$. The total cross–section is

$$\sigma = 4\pi\sigma_0 N^{4/3} \ . \tag{5.27}$$

This is a coherent diffraction. The total number of peaks is of the order $(d/\lambda)^2$, where d is the mean separation distance between the nuclei (atoms) and λ is the wavelength of the particles. For an energy $1MeV$ the wavelength is $\lambda \simeq 10^{-10}cm$, so we can see that the number of peaks is large (of the order 10^5, the magnitude of d is a few Å's).

For $1mol$ the number of atoms (nuclei) is $N \simeq 6 \times 10^{23}$. Making use of $\sigma_0 = 10^{-44}cm^2$ (for neutrinos) we get a cross-section $\sigma \simeq 10^{-12}cm^2$. We can see that the condition $\sigma \ll A$, where A is the area of the sample, is satisfied (the sapphire crystal used in Weber's experiments has a volume $\simeq 6cm^3$ and mass $\simeq 24g$, density $\rho \simeq 4g/cm^2$ at room temperature).

According to equation (5.21) the force $f = \Phi E\sigma/c$ acting upon the crystal is $f \simeq 10^{-16}dyn$ for a flux density $\Phi = 10^{12}/cm^2 \cdot s$ and energy $E = 1MeV$. This is an extremely small, unmeasurable, force.

5.5 Temperature dependence

Let $U(\boldsymbol{r}_i - \boldsymbol{r}_j)$ be the interaction energy between two identical nearest-neighbours atoms (molecules, unit cells in crystals) placed at equilibrium positions $\boldsymbol{r}_{i,j}$. The total interaction energy is $\frac{1}{2}\sum_{ij} U(\boldsymbol{r}_i - \boldsymbol{r}_j)$, where the summation is performed over all the nearest-neighbours. Let $\boldsymbol{u}_{i,j}$ be small displacements of these atoms from their equilibrium positions. The hamiltonian of these atoms reads

$$H = \sum_i \frac{1}{2m} p_i^2 + \frac{1}{4} U \sum_{ij} (\boldsymbol{u}_i - \boldsymbol{u}_j)^2 \ , \tag{5.28}$$

where $\boldsymbol{p}_i$ is the momentum of the i-th atom with mass m and U is the second derivative of the interaction $U(\boldsymbol{r}_i - \boldsymbol{r}_j)$ at equilibrium positions. We limit ourselves to the second-order expansion in powers of the displacement difference (the first-order term is vanishing at equilibrium) and assume an isotropic interaction. We use the Fourier series expansion

$$\boldsymbol{u}_i = \frac{1}{\sqrt{N}} \sum_{\boldsymbol{k}} \boldsymbol{u}_{\boldsymbol{k}} e^{i\boldsymbol{k}\boldsymbol{r}_i} \ , \ \boldsymbol{u}_{\boldsymbol{k}} = \frac{1}{\sqrt{N}} \sum_i \boldsymbol{u}_i e^{-i\boldsymbol{k}\boldsymbol{r}_i} \tag{5.29}$$

where N is the number of atoms in the sample and the wavevectors $\boldsymbol{k}$ are restricted to the first Brillouin zone for a crystal. The displacements are real, such that $u^*_{-\boldsymbol{k}} = u_{\boldsymbol{k}}$. The hamiltonian becomes

$$H = \sum_{\boldsymbol{k}} \left[\frac{1}{2m} \boldsymbol{p}_{\boldsymbol{k}} \boldsymbol{p}_{-\boldsymbol{k}} + \frac{1}{2} m\omega^2(\boldsymbol{k}) \boldsymbol{u}_{\boldsymbol{k}} \boldsymbol{u}_{-\boldsymbol{k}} \right] \ , \tag{5.30}$$

where

$$\omega^2(\boldsymbol{k}) = \frac{U}{m} \sum_j (1 - \cos \boldsymbol{k}\boldsymbol{a}_j) \ , \tag{5.31}$$

$\boldsymbol{a}_j$ being the nearest-neighbours vectors. Since $\boldsymbol{k}\boldsymbol{a}_j < 1$, we may use $\omega^2(\boldsymbol{k}) \simeq \frac{1}{2}(U/m) \sum_j (\boldsymbol{k}\boldsymbol{a}_j)^2$.

The hamiltonian given by equation (5.30) is the hamiltonian of a sum of harmonic oscillators with frequency $\omega(\boldsymbol{k})$ and mass m. For each $\boldsymbol{k}$-mode there exist three polarizations, labelled by α. For each polarization we quantize this motion by

$$\begin{aligned} u_{\boldsymbol{k}\alpha} &= \sqrt{\frac{\hbar}{2m\omega(\boldsymbol{k})}} (a^*_{\boldsymbol{k}\alpha} + a_{-\boldsymbol{k}\alpha}) \ , \\ p_{\boldsymbol{k}\alpha} &= i\sqrt{\frac{\hbar m\omega(\boldsymbol{k})}{2}} (a^*_{\boldsymbol{k}\alpha} - a_{-\boldsymbol{k}\alpha}) \end{aligned} \tag{5.32}$$

(where $\hbar$ is Planck's constant); the hamiltonian becomes

$$H = \sum_{\boldsymbol{k}\alpha} \frac{1}{2}\hbar\omega(\boldsymbol{k}) \left(a^*_{\boldsymbol{k}\alpha} a_{\boldsymbol{k}\alpha} + a_{\boldsymbol{k}\alpha} a^*_{-\boldsymbol{k}\alpha} \right) \ . \tag{5.33}$$

These quanta of energy correspond to three branches of acoustic phonons (in the limit $\boldsymbol{k}\boldsymbol{a}_j \rightarrow 0$ the frequency is proportional to the magnitude of the wavevector, as for sound). We may include atoms in the unit cell, with different masses, and get optical phonons with non-zero frequencies in the long-wavelength limit. Also, $\hbar\boldsymbol{k}$ may be viewed as the phonon momentum.

The phonon occupation number at finite temperatures $T = 1/\beta$ is

$$n_{\boldsymbol{k}\alpha} = \frac{1}{e^{\beta\hbar\omega(\boldsymbol{k})} - 1} \ . \tag{5.34}$$

The number of phonon states is defined by the number of degrees of freedom,

$$3N = 3\sum_{\boldsymbol{k}} 1 = 3\frac{4\pi V}{3(2\pi)^3} k_D^3 \ , \tag{5.35}$$

where V is the volume of the sample and the wavevector cutoff k_D is the Debye wavevector. The inclusion of the optical branches of phonons (number of atoms in the unit cell) does not change this relationship. It is convenient to use an average over the vectors $\boldsymbol{a}_j$ and write the frequency of the acoustic phonons as $\omega(k) \simeq \frac{1}{2}(U/m)d^2k^2$, where d is approximately the distance between the nearest-neighbours. Then, equation (5.35) may be written as

$$N = \frac{4\pi V}{3(2\pi)^3 c_s^3} \omega_D^3 \ , \tag{5.36}$$

where $c_s = \sqrt{U/2md}$ is the sound velocity and ω_D is the Debye frequency. Noteworthy, this equation does not depend on the optical branches. We note that k_D is of the order $1/d$ and $\omega_D \simeq c_s/d$ ($V \simeq Nd^3$). The total energy of the phonons is given by

$$\begin{aligned} E &= 3\tfrac{4\pi V}{(2\pi)^3 c_s^3} \int_0^{\omega_D} d\omega \cdot \omega^2 \tfrac{\hbar\omega}{e^{\beta\hbar\omega} - 1} = \\ &= \tfrac{9NT^4}{\hbar^3\omega_D^3} \int_0^{\hbar\omega_D/T} dx \tfrac{x^3}{e^x - 1} \ . \end{aligned} \tag{5.37}$$

This equation has a certain dependence on the optical branches, which may be included in a variation of the parameter ω_D. Typical Debye temperatures are of the order $10-100K$, such that, at room temperature we may use the approximation x^2 for the integrand (for sapphire $\omega_D \simeq 10^3K$, $1eV = 1.1 \times 10^4K$, $1K = 1.38 \times 10^{-16}erg$).

The form-factor F (equation (5.16)) depends on the atomic positions $\boldsymbol{r}_i$ (this is why we call this scattering a classical scattering). These positions are affected by the thermal motion, so we need to average the squared form-factor over the thermal motion. The thermal average ($<...>$) of the squared form-factor is given by

$$
\begin{aligned}
<\mid F(\boldsymbol{q}) \mid^2>= \sum_{ij} e^{i\boldsymbol{q}(\boldsymbol{r}_i-\boldsymbol{r}_j)} < e^{i\boldsymbol{q}(\boldsymbol{u}_i-\boldsymbol{u}_j)} >= \\
= \sum_{ij} e^{i\boldsymbol{q}(\boldsymbol{r}_i-\boldsymbol{r}_j)} e^{-\frac{1}{2}<[\boldsymbol{q}(\boldsymbol{u}_i-\boldsymbol{u}_j)]^2>} \quad ,
\end{aligned}
\tag{5.38}
$$

where

$$
\begin{aligned}
< [\boldsymbol{q}(\boldsymbol{u}_i - \boldsymbol{u}_j)]^2 >= q_\alpha q_\beta(< u_{i\alpha}u_{i\beta} > + \\
+ < u_{j\alpha}u_{j\beta} > + < u_{i\alpha}u_{j\beta} > + < u_{j\alpha}u_{i\beta} >) \; .
\end{aligned}
\tag{5.39}
$$

From equations (5.29) and (5.32) we get

$$
< u_{i\alpha}u_{j\beta} >= \delta_{\alpha\beta}\frac{\hbar}{mN}\sum_{\boldsymbol{k}}\frac{\coth[\beta\hbar\omega(k)/2]}{\omega(k)}e^{i\boldsymbol{k}(\boldsymbol{r}_i-\boldsymbol{r}_j)} \tag{5.40}
$$

and

$$
\begin{aligned}
< [\boldsymbol{q}(\boldsymbol{u}_i - \boldsymbol{u}_j)]^2 >= \\
= \frac{6\hbar q^2}{mN}\sum_{\boldsymbol{k}}\frac{\coth[\beta\hbar\omega(k)/2]}{\omega(k)}\left[1 - e^{i\boldsymbol{k}(\boldsymbol{r}_i-\boldsymbol{r}_j)}\right] \; .
\end{aligned}
\tag{5.41}
$$

We introduce the notation

$$
f(k) = \frac{3\hbar q^2}{m}\frac{\coth[\beta\hbar\omega(k)/2]}{\omega(k)} \tag{5.42}
$$

and write

$$
< [\boldsymbol{q}(\boldsymbol{u}_i - \boldsymbol{u}_j)]^2 >= \frac{2}{N}\sum_{\boldsymbol{k}} f(k)\left[1 - e^{i\boldsymbol{k}(\boldsymbol{r}_i-\boldsymbol{r}_j)}\right] \; . \tag{5.43}
$$

and

$$e^{-\frac{1}{2}<[\boldsymbol{q}(\boldsymbol{u}_i-\boldsymbol{u}_j)]^2>\cdot}=e^{-\frac{1}{N}\sum_{\boldsymbol{k}}f(k)}e^{\frac{1}{N}\sum_{\boldsymbol{k}}f(k)e^{i\boldsymbol{k}(\boldsymbol{r}_i-\boldsymbol{r}_j)}}=$$
$$=e^{-\frac{1}{N}\sum_{\boldsymbol{k}}f(k)}\left[1+\frac{1}{N}\sum_{\boldsymbol{k}}f(k)e^{i\boldsymbol{k}(\boldsymbol{r}_i-\boldsymbol{r}_j)}+...\right]\ . \tag{5.44}$$

We use the notation

$$W=\frac{1}{N}\sum_{\boldsymbol{k}}f(k) \tag{5.45}$$

and write the squared form-factor as

$$<\mid F(\boldsymbol{q})\mid^2>=e^{-W}\cdot$$
$$\cdot\left[\sum_{ij}e^{\boldsymbol{q}(\boldsymbol{r}_i-\boldsymbol{r}_j)}+\frac{1}{N}\sum_{\boldsymbol{k}}f(k)\sum_{ij}e^{i(\boldsymbol{k}+\boldsymbol{q})(\boldsymbol{r}_i-\boldsymbol{r}_j)}+...\right]\ , \tag{5.46}$$

or

$$<\mid F(\boldsymbol{q})\mid^2>=N^2e^{-W}\sum_{\boldsymbol{g}}\left[\delta_{\boldsymbol{q},\boldsymbol{g}}+\frac{1}{N}f(\mid\boldsymbol{q}+\boldsymbol{g}\mid)+...\right] \tag{5.47}$$

for a crystal, where $\boldsymbol{q}+\boldsymbol{g}$ is restricted to the first Brillouin zone; for random atomic positions $\boldsymbol{g}=0$. e^{-W} is the Debye-Waller factor,[6] while the N-term is the diffuse scattering.[7] We can see that the Bragg peaks are preserved at finite temperatures, their intensity being diminished by the Debye-Waller factor.

For randomly distributed positions of the atoms $f(q)$ given by equation (5.42) varies between $f(q)\simeq T/mc_s^2$ for $q\rightarrow 0$ and $f(q)\simeq\hbar q/mc_s$ for large q (of the order of the wavevector of the incident particles). The energy mc_s^2 is of the order $10^{-10}erg$ ($c_s=10^6cm/s$), or 10^6K. Therefore, at room temperature and for $\hbar cq=1MeV$ the variation range of the function $f(q)$ is from 10^{-4} to 10^4.

For a crystal the summation over $\boldsymbol{g}$ in the diffuse-scattering term leads to a vector $\boldsymbol{q}+\boldsymbol{g}$ in the first Brillouin zone. We can estimate the contribution of the diffuse term by averaging the function f over the volume

[6]P. Debye, "Interferenz von Roentgenstrahlen und Waermebewegung", Ann. Phys. **43** 49 (1914); I. Waller, "Theoretische Studien zur Interferenz- und Dispersionstheorie der Roentgenstrahlen". Dissertation, Uppsala (1925).

[7]W. H. Zachariasen, "A theoretical study of the diffuse scattering of X-rays by crystals", Phys. Rev. **57** 597 (1940).

$\frac{4\pi}{3}\omega_D^3/c_s^3$, with the result $\overline{f} = q^2T/m\omega_D^2$. The diffuse-scattering term has a large interval of variation, according to the variation of the momentum transfer $\boldsymbol{q} = \boldsymbol{k}_i - \boldsymbol{k}_f$.

The exponent of the Debye-Waller factor from equation (5.45) is given by

$$W = \frac{3\hbar q^2}{m}\frac{4\pi V}{(2\pi)^3 N c_s^3}\int_0^{\omega_D} d\omega \cdot \omega \coth(\beta\hbar\omega/2) =$$
$$= \frac{3\hbar q^2}{m}\frac{4\pi V}{(2\pi)^3 N c_s^3}\left(\frac{2}{\beta\hbar}\right)^2 \int_0^{\beta\hbar\omega_D/2} dx \cdot x\coth(x) \ . \tag{5.48}$$

At room temperature $\beta\hbar\omega_D/2$ is less than unity, such that we may approximate the integrand in equation (5.48) by 1. We get

$$W \simeq \frac{(qd)^2 T}{mc_s^2} \ . \tag{5.49}$$

At room temperature $T/mc_s^2 \simeq 10^{-4}$, such that the maximum value of W is unity, attained for $q \simeq 10^{10}cm^{-1}$. The Debye-Waller factor affects little the intensity of the peaks.

5.6 Coherent *vs* incoherent scattering

The interaction $v(\boldsymbol{R} - \boldsymbol{R}_i)$ in equation (5.6) may include internal coordinates of the atoms (nuclei), like, for instance, the interaction $-q\boldsymbol{A}_i\boldsymbol{p}_i/mc$ of a pointlike charge q with mass m placed at the i-th site with the electromagnetic radiation with the vector potential $\boldsymbol{A}_i$; a factor $a^3\delta(\boldsymbol{R} - \boldsymbol{R}_i)$ can be included, where a is of the order of the atomic dimension. The momentum $\boldsymbol{p}_i$ of the charge depends on internal degrees of freedom. Consequently, the matrix elements of the interaction v should include the internal wavefunctions (even if the treatment may be classical, see the X-rays Diffraction chapter)). The wavefunctions are defined up to phase factors, such that each site contributes a phase factor $e^{i\chi_i}$. The form-factor (equation (5.16)) becomes

$$F(\boldsymbol{q}) = \sum_i e^{i\chi_i} e^{i\boldsymbol{q}\boldsymbol{r}_i} \tag{5.50}$$

and its square is

$$\left|F(\boldsymbol{q})\right|^2 = \sum_{ij} e^{i(\chi_i-\chi_j)} e^{i\boldsymbol{q}(\boldsymbol{r}_i-\boldsymbol{r}_j)} \ . \tag{5.51}$$

We should average over these phase factors.

The interaction of an incident beam of particles with a macroscopic sample may proceed with the wavefunction of the entire sample. We call this scattering a (macroscopic) quantum scattering. If so, the average over the phase factors implies $\chi_i = \chi_j$ and the squared form-factor is

$$\left|F(\boldsymbol{q})\right|^2 = \sum_{ij} e^{i\boldsymbol{q}(\boldsymbol{r}_i-\boldsymbol{r}_j)} = N^2 \sum_{\boldsymbol{g}} \delta_{\boldsymbol{q},\boldsymbol{g}} \ , \tag{5.52}$$

where N is the number of sites in the sample and $\boldsymbol{g}$'s are reciprocal vectors (for a crystal; for randomly distributed sites $\boldsymbol{g} = 0$). We can see that the squared form-factor is proportional to the square of the number of sites. This is a coherent scattering. The cross-section is obtained by multiplying by the solid-angle width of each scattering peak, which reduces the N^2-dependence to a $N^{4/3}$-dependence. This is a coherent diffraction. If, for some reasons, the sites are not quantum-mechanically (and statistically) correlated, *i.e.* they are independent (randomly distributed), then the average over the phase factors implies $\overline{e^{i(\chi_i-\chi_j)}} = \delta_{ij}$, and the squared form-factor

$$\left|F(\boldsymbol{q})\right|^2 = \sum_{ij} \overline{e^{i(\chi_i-\chi_j)}} e^{i\boldsymbol{q}(\boldsymbol{r}_i-\boldsymbol{r}_j)} = N \tag{5.53}$$

and the cross-section is proportional to the number of sites; this is an incoherent scattering. Coherence domains may appear, each characterized by its own phase factor, the distinct phase factors being uncorrelated.

5.7 Neutrino coherent scattering

According to equation (5.18), the number of scatterings per unit time, is $w = \Phi\sigma$. It follows the time interval between two collisions

$$\tau = \frac{1}{\Phi\sigma} = \frac{1}{4\pi\Phi\sigma_0 N^{4/3}} \tag{5.54}$$

(for a coherent diffraction); we call this time the "collision" time. According to the numerical estimations given above ($\Phi = 10^{12}/cm^2 \cdot s$, $\sigma = 10^{-12}cm^2$; $N = 6 \times 10^{23}$, $\sigma_0 = 10^{-44}cm^2$) we get

$$\tau \simeq 1s \ . \tag{5.55}$$

The time needed for a neutrino to pass through $1cm$ is $t_0 \simeq 3\times 10^{-11}s$. We can see that the neutrino collisions are extremely rare.

In an elementary act of collision a neutrino transfers to the atom (nucleus) a momentum of the order $\Delta p = \hbar k$. For $k \simeq 10^{10}cm^{-1}$ (energy $1MeV$) we get $\Delta p = 10^{-17}g \cdot cm/s$. The displacement of the atom is of the order $\simeq \hbar/\Delta p \simeq 10^{-10}cm$. At room temperature T an atom has a thermal velocity of the order $v = \sqrt{T/m}$, where m is the mass of the atom. This velocity is of the order $v = \left(10^{-14}/10^5 \cdot 10^{-27}\right)^{1/2} \simeq 10^4 cm/s$, and the momentum of the atom is $mv \simeq 10^{-18}g \cdot cm/s$. We take the mass of the atom (nucleus) equal to $10^5 m_e$, where $m_e \simeq 10^{-27}g$ is the electron mass. We can see that $\hbar k$ is greater than the momentum of the atom. It follows that the perturbation energy of an atom is of the order $\Delta E = p^2/m \simeq 10^{-12}erg$. The time needed for that atom to recover its equilibrium is $\Delta t_{eq} \simeq \hbar/\Delta E \simeq 10^{-15}s$.[8] Therefore, the atoms (nuclei) recover quickly their equilibrium after each collision with the neutrinos. It follows that the incident beam of neutrinos sees the atoms (nuclei) in their equilibrium positions, thermal motion included (the thermal time is $\hbar/T \simeq 10^{-14}s$ at room temperature). The scattering should include the wavefunction of the entire sample. We expect a coherent neutrino scattering.

We note that the momentum transfer in a elementary act of collision may be smaller than $\hbar k$ (*e.g.*, for a forward scattering), which would lead to a longer recovery time, but this is only a fraction of all the collisions. Also, a slightly longer recovery time is obtained at lower temperatures.

Let us assume that the scattering is coherent. We may use a cross-section $\sigma = 4\pi\sigma_0 N^2$ which is of the order $\sigma \simeq 10^4 cm^3$ for $1mol$;

[8] At quantum level a wavefunction $e^{-\frac{i}{\hbar}Et}$ is associated to the particle with energy E. This wavefunction varies between 1 and -1, passing through 0. It follows that it may lose (or gain) the energy E in a time of the order $\Delta t_{eq} = \hbar/E$.

it would lead to a "collision" time $\tau \simeq 10^{-16}s$, which is smaller than $\Delta t_{eq} \simeq 10^{-15}s$, all the other numerical data being the same. Actually, the figure $\sigma \simeq 10^4 cm^3$ is unphysical, such that we may still expect a coherent scattering. In general, for a coherent scattering we should have $\tau = 1/\Phi\sigma > \Delta t_{eq} = \hbar/\Delta E$, where ΔE is the mean energy transfer. For $\Delta E = 10^{-12}erg$, we get $\sigma < \Delta E/\hbar\Phi$, which leads to $\sigma < 10^3 cm^2$ for our numerical data. We may conclude that the coherent scattering for neutrinos may be self-consistent.

Similar considerations apply to the scattering of the electromagnetic radiation. The electrons in atoms have a high velocity, which, together with their low mass, make them to recover very quickly their equilibrium in atoms, or in their states in condensed matter. The momentum transfer from the radiation is taken by the entire atom. For lower energies ($E < 20keV$) the energy transfer is of the order $\Delta E = \sqrt{T/mp}$.

However, the one-particle cross-section of the scattering of the electromagnetic radiation is of the order $\sigma_0 \simeq 10^{-25}cm^2$. The cross-section in these cases increases, approximately, by a factor $10^{44} \cdot 10^{-25} = 10^{19}$, which invalidates the condition $\sigma < \Delta E/\hbar\Phi$ for coherent scattering. In this case the "collision" time is of the order $10^{-20}s$ (for $\sigma = 4\pi\sigma_0 N^{4/3}$), which is much shorter than the equilibrium time $\Delta t_{eq} \simeq 10^{-15}s$, such that we may expect an incoherent scattering, or a coherent diffraction. On the other hand, such a cross-section is larger than the area of the sample, which is unphysical. An incoherent scattering gives a cross-section $\sigma = 4\pi\sigma_0 N \simeq 1cm^2$ for $1mol$, which increases the "collision" time to $\tau \simeq 10^{-12}s$ and invalidates the condition for an incoherent scattering. We conclude that in these cases there exists a mixture of incoherent and coherent scattering, arising from coherence domains. We can see that the ultimate reason for a coherent scattering valid for neutrinos in the whole crystal is their extremely small cross-section.

5.8 Coherent-incoherent interplay

Let us assume that a beam of incident particles with wavevector $\boldsymbol{k}$ is scattered by a macroscopic sample to a wavevector $\boldsymbol{k}'$, such that the energy conservation leads to $k = k'$. The general form of the cross-section is

$$d\sigma = \sigma_0 \left|F(\boldsymbol{q})\right|^2 do \ , \tag{5.56}$$

where σ_0 is the cross-section of the scattering by a single particle in the sample, do is the solid angle in the direction of $\boldsymbol{k}'$ and $F(\boldsymbol{q})$, $\boldsymbol{q} = \boldsymbol{k}' - \boldsymbol{k}$, is the form-factor (equation (5.17)). The cross-section σ_0 may depend on $\cos\theta$, where θ is the angle between $\boldsymbol{k}$ and $\boldsymbol{k}'$; we write $\sigma_0(\cos\theta)$. The form-factor is given by

$$F(\boldsymbol{q}) = \sum_i e^{i\boldsymbol{q}\boldsymbol{r}_i} \ , \tag{5.57}$$

where $\boldsymbol{r}_i$ is the position of the i-th particle in the sample. Let us assume that the scattering is coherent within domains with volumes v_n and incoherent for distinct domains, such that the form-factor is written as

$$F(\boldsymbol{q}) = \sum_n e^{i\chi_n} \sum_{i\in v_n} e^{i\boldsymbol{q}\boldsymbol{r}_i} \ , \tag{5.58}$$

where $e^{i\chi_n}$ are incoherent phase factors. We assume that all v_n have the same value $v = l^3$, where l is the dimension of the domains.

For a crystal the summation over $\boldsymbol{r}_i$ in equation (6.58) is determined up to $\boldsymbol{q} = \boldsymbol{g}$, where $\boldsymbol{g}$ is any reciprocal vector; for randomly distributed atoms $\boldsymbol{g} = 0$. For $\boldsymbol{q} = \boldsymbol{g}$ in equation (6.58) we get $\sum_{i\in v_n} e^{i\boldsymbol{q}\boldsymbol{r}_i} = l^3/d^3$, where d is the mean separation distance between the atoms. For small $\boldsymbol{q}$'s in the vicinity of a reciprocal vector we may pass from summation to an integral in equation (6.58), with the result

$$\sum_{i\in v} e^{i\boldsymbol{q}\boldsymbol{r}_i} = \frac{1}{d^3}\int_v d\boldsymbol{r} e^{i\boldsymbol{q}\boldsymbol{r}} = \frac{4\pi}{q^3 d^3}(\sin ql - ql\cos ql) \tag{5.59}$$

(actually the domain radius is $(3/4\pi)^{1/3}l$). This oscillating function decreases with increasing q and has a peak $4\pi l^3/3d^3$ for $q = 0$ with a width $\Delta q \simeq 2\pi/l$. The extrema of the function $\sin ql - ql\cos ql$ are placed at $ql = \pi n$, where n is any integer. At these points the function given by equation (5.59) is $4\pi(-1)^{n+1}l/q^2d^3$. Since $ql \gg 1$, the maximum from $q = 0$ dominates. Therefore, we may use the approximation

$$\sum_{i \in v} e^{i\boldsymbol{q}\boldsymbol{r}_i} \simeq \frac{4\pi l^3}{3d^3} \; , \; -\pi/l < \Delta q < \pi/l \; . \tag{5.60}$$

We note that for a single domain this approximation corresponds to a coherent scattering. For multiple domains this coherent scattering is considerably reduced. On the other hand, the envelope of the square of this function goes approximately as $\sim l^2/q^4d^6$, which is considerably enhanced in comparison with a coherent scattering (for a δ-function this contribution is zero). Therefore, we conclude that the coherent domains reduce considerably the intensity of the Bragg peaks and enhance the diffuse scattering. This is in agreement with the experimental observation that the Bragg peaks of X-rays scattering are broader than expected.

The average over the phase factors in $\sum_{n,m} e^{i(\chi_n - \chi_m)}$ gives V/l^3, where V is the volume of the sample. Consequently, the squared form-factor is

$$|F(\boldsymbol{q})|^2 = \frac{V}{l^3}\frac{(4\pi l^3)^2}{(3d^3)^2} \tag{5.61}$$

within the solid-angle range given by equation (5.60) about each reciprocal vector; this solid angle is $\Delta o = \pi(\pi/l)^2/k^2$. It follows that the cross-section for one peak is

$$\sigma_{\boldsymbol{g}} = \sigma_0(\cos\theta_{\boldsymbol{g}})\frac{V}{l^3}\frac{(4\pi l^3)^2}{(3d^3)^2}\frac{\pi^3}{l^2k^2} \; . \tag{5.62}$$

The vectors $\boldsymbol{k}'$ are on the Ewald sphere with radius k (for elastic scattering). The Bragg condition gives $g = 2k\sin\theta/2$. This condition defines the angle $\theta_{\boldsymbol{g}}$ in equation (5.62). The total cross-section is obtained by summing $\sigma_{\boldsymbol{g}}$ over all the reciprocal vectors on the Ewald sphere. We associate the area $(2\pi/d)^2$ to each reciprocal vector, such that the number of reciprocal vectors on the Ewald sphere is $4\pi k^2/(2\pi/d)^2 = k^2d^2/\pi$. This number is approximately $(d/\lambda)^2$, where λ is the wavelength of the incident particles. For a large energy of the incident particles the number of peaks is large, such that the solid angle $4\pi^2/k^2d^2$ for each $\boldsymbol{g}$ is small. We may pass to integration over the solid angles of the reciprocal vectors, and get the total cross-section as

$$\sigma = \frac{V}{l^3}\frac{(4\pi l^3)^2}{(3d^3)^2}\frac{\pi^3}{l^2k^2}\cdot\frac{k^2d^2}{4\pi^2}2\pi\int d\theta\sin\theta\sigma_0(\cos\theta) = \\ = \frac{8\pi^4}{9}\frac{Vl}{d^4}\int d\theta\sin\theta\sigma_0(\cos\theta)\ . \tag{5.63}$$

We can see that the cross-section of coherent diffraction $\sim V^{4/3}$ is reduced by the multiple coherent domains by a factor $R = l/V^{1/3}$, where l is the dimension of the domains. In the limit $l = d$ corresponding to a monoatomic "domain" we recover the incoherent cross-section $\sim V$. (The numerical factor $8\pi^4/9$ should be replaced by 2π for a corrected domain radius.) The domain dimension l remains a fitting parameter. This reduction is also valid for inelastic collisions.

The domains may be viewed formally as "defects". If f is the fraction of defects the dimension of a domain is $l = d/f^{1/3}$ and the reduction factor is $R = d/f^{1/3}V^{1/3} = 1/(fN)^{1/3}$, where N is the number of atoms in the sample.

There exists a limitation of the coherence domains. We assume that the phase factor is the same within each domain. This amounts to a continuous (uniform) approximation. However, each atom, with volume $v_0 = a_0^3$, where a_0 is the atomic dimension, has v/v_0 possibilities to induce a discontinuity, in all directions, where $v = a^3$ and a is the mean separation distance between the atoms. It follows that the maximal radius of non-uniformities is $(v/v_0)a$; the sphere with this radius include $N_d = (v/v_0)^3$ atoms.[9] This is the minimal range over which we may view a uniform, continuous domain. For $a_0 = 1$Å and $a = 4$Å we get a radius $\simeq 256$Å and a number of atoms in the domain of the order $N_d \simeq 10^6$. On the other hand, domains of this order of magnitude may also be viewed as being non-uniform. The only element of continuity remains their surface, which defines the domain. It follows that such domains are coherent over a number $N_d^{2/3}$ of atoms only. This leads to a further reduction in the coefficient R. It is worth emphasizing that domains with a number of atoms smaller than $\simeq 10^6$ are completely incoherent, and N_d should be replaced by 1 ($f = 1$).

[9] M. Apostol, *Statistical Physics*, Cambridge Scholars Publishing, Newcastle upon Tyne (2021).

5.9 Quantum scattering

5.9.1 Introduction

A sufficient condition of coherent scattering is a transfer wavelength much longer than the dimension of the target-particle assembly; for large targets this implies an extremely small momentum and energy transfer.[10] Also, classical scattering, like Thomson scattering or elastic X-ray diffraction, is an instance of coherent scattering. For neutrino scattering by macroscopic samples these conditions are not met. For a finite momentum transfer the form-factor may reduce appreciably the scattering rate, and an incoherent scattering may appear. This is the counter-argument[11] to Weber's suggestion of a coherent scattering of neutrinos by a large, stiff, perfect crystal.[12] However,

[10] L. Landau and E. M. Lifshitz, *Course of Theoretical Physics*, vol. 4 (V. B. Berestetskii, E. M. Lifshitz and L. P. Pitaevskii, *Quantum Electrodynamics*), Butterworth-Heinemann, Oxford (1982); Daniel Z. Freedman, "Coherent effects of a weak neutral current", Phys. Rev. **D9** 1389 (1974); D. Akimov et al, COHERENT Collaboration, "Observation of coherent elastic neutrino-nucleus scattering", Science **357** 1123 (2017); E. Akhmedov, G. Arcadi, M. Lindner and S. Vogl, "Coherent scattering and macroscopic coherence: implications for neutrino, dark matter and axion detection", JHEP (2018), doi: 10.10007/JHEP10 (2018)045.

[11] G. F. Bertsch and Sam M. Austin, "Comment on "Method for observation of neutrinos and antineutrinos"", Phys. Rev. **C34** 361 (1986); R. Casella, "On the theory of coherent neutrino scattering from crystals", Nuovo Cimento **94** 42 (1986); T. H. Ho, "Comments on the "Method for observation of neutrinos and antineutrinos", Phys, Lett. **B166** 295 (1986); M. N. Butler, "Comment on "Method for observation of neutrinos and antineutrinos"", Phys. Rev. **C35** 1164 (1987); T. Y. Aharanov, F. T. Avignon III, A. Casher and S. Nussinov, "Constraints on anomalous scattering of neutrinos from crystals", Phys. Rev. Lett. **58** 1173 (1987); H. J. Lipkin, "N dependence of coherent radiation from crystals", Phys. Rev. Lett. **58** 1176 (1987).

[12] J. Weber, "Method for observation of neutrinos and antineutrinos", Phys. Rev. **C31** 1468 (1985); "Apparent observation of abnormally large coherent scattering cross sections using keV and MeV range antineutrinos, and solar neutrinos", Phys. Rev. **D38** 32 (1988). See also J. D. Fronson and B. C. Jacobs, "Null result for enhanced neutrino scattering in crystals", Phys. Rev. **A46** 2235 (1992); Y. Su, B. R. Heckel, E. G. Adelberger, J. H. Gundlach, M. Harris, G. L. Smith and H. E. Swanson, "New tests of the universality of free fall", Phys. Rev. **D50** 3614 (1994); J. Luo, X. Chen, J. G. Li, S. H. Fan and F. G. Li, "Test of abnormaly large coherent scattering cross section using solar-neutrino", Chin. Phys. Lett. **12** 136 (1995); G. Q. Feng, S. Q. Yang, L.

a perfect crystal may undergo quantum transitions between states of definite crystal quasi-momentum, where the crystal is translated as a whole by a lattice vector.[13] We show below that such transitions lead to a quantum coherent scattering. The interplay between coherent and incoherent scattering is discussed.

In the current treatments of various scattering processes (like X-ray, neutron, Raman scattering, the Moessbauer effect) the wavefunction of the particle assembly is considered irrelevant. As shown above in the Neutrino coherent scattering section, in the particular case of the neutrino scattering, at least, the multi-particle wavefunction of the target is relevant, and leads to quantum coherent scattering.

5.9.2 Center of mass

Let us consider a set of N particles, labelled by $i, j, ... = 1, 2, ...N$, with positions $\boldsymbol{r}_i$ and mass m_i. By using the identity

$$\boldsymbol{r}_i = \boldsymbol{R} + \sum_j \frac{m_j}{M}(\boldsymbol{r}_i - \boldsymbol{r}_j) \ , \tag{5.64}$$

where $M = \sum_i m_i$ is the total mass and $\boldsymbol{R} = \sum_i \frac{m_i}{M}\boldsymbol{r}_i$ is the position of the center of mass, we may pass from the variables $\boldsymbol{r}_i$ to the variables $\boldsymbol{R}$ and $\boldsymbol{r}_i - \boldsymbol{r}_j$; the number of the latter is C_N^2 (since $\boldsymbol{r}_i - \boldsymbol{r}_j = -(\boldsymbol{r}_j - \boldsymbol{r}_i)$). This change of variables may account for a two-particle interaction (which depends on $\boldsymbol{r}_i - \boldsymbol{r}_j$), but for interacting particles the pairs ij (and the variables $\boldsymbol{r}_i - \boldsymbol{r}_j$) are not independent.[14]

C. Tu and J. Luo, "Improvement of test of solar neutrino coherent scattering with torsion pendullum", Chin. Phys. Lett. **23** 2052 (2006); I. Cruceru, G. Nicolescu and O. G. Duliu, "A dedicated torsion balance to detect neutrinos by coherent scattering on high Debye temperature monocrystals", Int. J. Mod. Phys. **A23** 1719 (2008); M. Cruceru, G. Nicolescu, O. G. Duliu and I. Cruceru, "Detection of solar neutrinos by coherent scattering in high-Debye temperature monocrystals", Int. J. Mod. Phys. **A26** 2773 (2011); G. Nicolescu, "A heuristic approach to the detection of solar neutrinos", J. Phys. **G40** 055201 (2013).

[13] R. E. Peierls, "Zur kinetischen Theorie der Waermeleitung in Kristallen", Ann. Phys. **3** 1055 (1929).

[14] Joseph B. Aviles, Jr., "Separation of the center-of-mass in independent particle systems", Ann. Phys. **50** 393 (1968) and References therein; Michal

For free, independent particles we can use the change of variables from $\boldsymbol{r}_i$ to $\boldsymbol{R}$ and $\boldsymbol{\xi}_i = \sum_j \frac{m_j}{M}(\boldsymbol{r}_i - \boldsymbol{r}_j) = \boldsymbol{r}_i - \boldsymbol{R}$ (baricentric coordinates). The number of variables $\boldsymbol{\xi}_i$ is $N-1$, since $\sum_i m_i \boldsymbol{\xi}_i = 0$. The new variables are independent. The wavefunction of this assembly can be written as $e^{i\boldsymbol{K}_0\boldsymbol{R}} F(\boldsymbol{\xi}_1, \boldsymbol{\xi}_2, ...)$, where $\hbar\boldsymbol{K}_0$ is the total momentum and $F(\boldsymbol{\xi}_1, \boldsymbol{\xi}_2, ...)$ is the internal wavefunction of the assembly. Indeed, a change $\boldsymbol{r}_i \longrightarrow \boldsymbol{r}_i + \boldsymbol{a}$, with an arbitrary translation $\boldsymbol{a}$, leaves the internal coordinates $\boldsymbol{\xi}_i$ unchanged and introduces a phase factor $e^{i\boldsymbol{K}_0\boldsymbol{a}}$, which is the definition of the translational symmetry for a homogeneous space and the corresponding (total) momentum. By such a translation the assembly acquires a total momentum $\hbar\boldsymbol{K}_0$. Additional internal degrees of freedom, with corresponding variables, may be included, besides the coordinates $\boldsymbol{\xi}_i$. Since $\boldsymbol{\xi}_i = \boldsymbol{r}_i - \boldsymbol{R}$, we may also write the wavefunction as $e^{i\boldsymbol{K}_0\boldsymbol{R}} F(\boldsymbol{r}_1, \boldsymbol{r}_2, ...)$, with the condition $\sum_i m_i \boldsymbol{r}_i = M\boldsymbol{R}$. This condition indicates that one internal degree of freedom of the assembly is superfluous (it is accounted for by the center of mass), such that, for a large number of particles, this condition is irrelevant.

It is convenient to use a particle field, instead of a multi-particle wavefunction $F(\boldsymbol{r}_1, \boldsymbol{r}_2, ...)$. For a field, the variables $\boldsymbol{r}_i$ are viewed as labelling states, and to each particle a wavefunction $\varphi(\boldsymbol{r} - \boldsymbol{r}_i)$ is associated. In general, the one-particle wavefunctions φ may be distinct, and additional one-particle degrees of freedom may be included. For identical particles the wavefunctions φ are the same for all particles. Besides the total-momentum factor, the field is written as a superposition of one-particle wavefunctions

$$\psi(\boldsymbol{r}) = e^{i\boldsymbol{K}_0\boldsymbol{R}} \sum_i c_i \varphi(\boldsymbol{r} - \boldsymbol{r}_i) \ , \tag{5.65}$$

where the coefficients c_i are such that

$$\begin{aligned} \int d\boldsymbol{r} \left|\psi(\boldsymbol{r})\right|^2 = \sum_{ij} c_i^* c_j \int d\boldsymbol{r} \varphi^*(\boldsymbol{r} - \boldsymbol{r}_i)\varphi(\boldsymbol{r} - \boldsymbol{r}_j) = \\ = \sum_i c_i^* c_i = N \ ; \end{aligned} \tag{5.66}$$

Svrcek, "Centre-of-mass separation in quantum mechanics: Implications for the many-body treatment in quantum chemistry and solid state physics", arXiv: 1008.3968v1 (physics.chem-ph) (2010); doi: 10.1007/978-94-007-2076-3.28.

the integral may include a general scalar product and the wavefunctions $\varphi(\boldsymbol{r}-\boldsymbol{r}_i)$ are orthonormal. A convenient representation is to view these wavefunctions as localized on the positions $\boldsymbol{r}_i$, over a distance of the order of the dimension of the particle. From equation (5.66) we can see that $|\psi(\boldsymbol{r})|^2$ is the density of particles. Also, $\psi(\boldsymbol{r})/\sqrt{N}$ may be viewed as a one-particle wavefunction associated to the particle assembly. The set of the wavefunctions $\varphi(\boldsymbol{r}-\boldsymbol{r}_i)$ is a complete set $(\frac{1}{N}\sum_i \varphi^*(\boldsymbol{r}-\boldsymbol{r}_i)\varphi(\boldsymbol{r}'-\boldsymbol{r}_i)=\delta(\boldsymbol{r}-\boldsymbol{r}'))$. The total momentum $\hbar\boldsymbol{K}_0$ can be added as another label, besides the labels $\boldsymbol{r}_i$, the orthogonality and the completeness are preserved. The superposition involved in the definition of the field and the superposition coefficients c_i may account for correlations. Since the wavefunctions $\varphi(\boldsymbol{r}-\boldsymbol{r}_i)$ are given, such a description of particle assemblies by means of fields is applicable to the internal motion inside the particles and to states associated to the positions $\boldsymbol{r}_i$.

Since the functions $\varphi(\boldsymbol{r}-\boldsymbol{r}_i)$ are localized, the coefficients c_i in equation (5.65) should be phase factors. Indeed,

$$\begin{aligned} |\psi(\boldsymbol{r})|^2 &= \sum_{ij} c_i^* c_j \varphi^*(\boldsymbol{r}-\boldsymbol{r}_i)\varphi(\boldsymbol{r}-\boldsymbol{r}_j) = \\ &= \sum_i |c_i|^2 |\varphi(\boldsymbol{r}-\boldsymbol{r}_i)|^2 = \\ &= |c_j|^2 |\varphi(\boldsymbol{r}-\boldsymbol{r}_j)|^2 = |\varphi(\boldsymbol{r}-\boldsymbol{r}_j)|^2 \ , \end{aligned} \tag{5.67}$$

for $\boldsymbol{r}$ in the immediate vicinity of any $\boldsymbol{r}_j$; equation (5.67) shows that c_i must be phase factors. In the limitig case where the correlations are absent, these phase factors, denoted by $c_i = e^{i\chi_i}$, are randomly distributed. They do not depend on the wavefunctions $\varphi(\boldsymbol{r}-\boldsymbol{r}_i)$. This situation is applicable to gases, liquids or amorphous solids, as long as the quantum-mechanical description is appropriate, as, for instance, electrons in amorphous solids.[15] Also, the coefficients c_i may be viewed as particle destruction operators (by the definition of the density given by equation (5.66)); their commutation/anticommutation relations imply quantum-mechanical correlations.

[15] N. F. Mott, "Electrons in disordered structures", Adv. Phys. **16** 49 (1967); "Conduction in non-crystalline systems", Phil. Mag. **22** 7 (1970); M. H. Cohen, "Electronic structure and transport in covalent amorphous semiconducting alloys", J. Non-Crystalline Solids **2** 432 (1970).

For arbitrary equilibrium positions $\boldsymbol{r}_i^0$ the field is written as

$$\psi(\boldsymbol{r}) = e^{i\boldsymbol{K}_0\boldsymbol{R}}\sum_i e^{i\chi_i}\varphi(\boldsymbol{r}-\boldsymbol{r}_i^0-\boldsymbol{u}_i) \ , \tag{5.68}$$

where the phase factors $e^{i\chi_i}$ are randomly distributed and $\boldsymbol{u}_i$ denote the small displacements of the particles from equilibrium positions. These displacements are functions of $\boldsymbol{r}_i^0$. They consist of a uniform displacement $\boldsymbol{u}_0$ and lattice vibrations $\boldsymbol{v}_i$, such that $\boldsymbol{u}_i = \boldsymbol{u}_0 + \boldsymbol{v}_i$, subject to the condition $\sum_i \boldsymbol{v}_i = 0$. The degrees of freedom corresponding to $\boldsymbol{u}_0$ and $\boldsymbol{v}_i$ are distinct; $\boldsymbol{u}_0$ corresponds to a center-of-mass uniform displacement, while $\boldsymbol{v}_i$ are vibrations. Therefore, $\boldsymbol{u}_0$ may be included in the phase factor $e^{i\boldsymbol{K}_0\boldsymbol{R}}$, which becomes $e^{i\boldsymbol{K}_0(\boldsymbol{R}_0+\boldsymbol{u}_0)}$, where $\boldsymbol{R}_0 = \frac{1}{N}\sum_i \boldsymbol{r}_i^0$. We may omit $\boldsymbol{u}_0$ and write down the field as

$$\psi(\boldsymbol{r}) = e^{i\boldsymbol{K}_0\boldsymbol{R}_0}\sum_i e^{i\chi_i}\varphi(\boldsymbol{r}-\boldsymbol{r}_i^0-\boldsymbol{v}_i) \ . \tag{5.69}$$

We call this field an incoherent field, due to the presence of the random phase factors $e^{i\chi_i}$. We recall that $\varphi(\boldsymbol{r}-\boldsymbol{r}_i^0-\boldsymbol{v}_i)$ are functions localized on $\boldsymbol{r}_i^0+\boldsymbol{v}_i$. We note that we have N degrees of freedom, due to the condition $\sum_i \boldsymbol{v}_i = 0$.

If the fixed, equilibrium positions $\boldsymbol{r}_i^0$ of N identical particles are arranged in a crystal, then the set of wavefunctions $\varphi(\boldsymbol{r}-\boldsymbol{r}_i^0)$ is invariant under a lattice displacement, in contrast to the set of wavefunctions $\varphi(\boldsymbol{r}-\boldsymbol{r}_i^0-\boldsymbol{v}_i)$ (where not all $\boldsymbol{v}_i$ are vanishing). It follows that we need to separate these two sets of wavefunctions and, in addition, the superposition

$$e^{i\boldsymbol{K}_0\boldsymbol{R}_0}\sum_i c_i\varphi(\boldsymbol{r}-\boldsymbol{r}_i^0) \tag{5.70}$$

should acquire a phase factor under the operation of lattice translation. This is possible only if $c_i = e^{i\boldsymbol{K}\boldsymbol{r}_i^0}$, where $\hbar\boldsymbol{K}$ is the crystal quasi-momentum. The superposition corresponding to $\boldsymbol{v}_i$ preserves its incoherent character. If we limit ourselves to internal motion, we may omit the phase factor $e^{i\boldsymbol{K}_0\boldsymbol{R}_0}$ of the total momentum, which amounts to fixing the position of the center of mass at zero ($\boldsymbol{R}_0 = 0$). Consequently, we define the coherent field

$$\Phi(\boldsymbol{r}) = \sum_i e^{i\boldsymbol{K}\boldsymbol{r}_i^0}\varphi(\boldsymbol{r}-\boldsymbol{r}_i^0) \ . \tag{5.71}$$

We note that $\Phi(\boldsymbol{r})$ describes the translation of the whole crystal by a lattice vector; this is why we call it a coherent field. Also,

$$\int d\boldsymbol{r}\,|\Phi(\boldsymbol{r})|^2 = N \tag{5.72}$$

and

$$\int d\boldsymbol{r}\Phi^*(\boldsymbol{r})\sum_i e^{i\chi_i}\varphi(\boldsymbol{r}-\boldsymbol{r}_i^0-\boldsymbol{v}_i) = 0 \tag{5.73}$$

(for extended lattice vibrations). Moreover,

$$\sum_i e^{i\boldsymbol{K}\boldsymbol{r}_i^0} = N\delta_{\boldsymbol{K},0} \ , \tag{5.74}$$

while $\sum_i e^{i\chi_i} = 0$. This difference induces the main distinction between a coherent field and an incoherent one. If Φ is labelled by $\boldsymbol{K}$, the coherent fields are orthogonal to each other for distinct wavevectors $\boldsymbol{K}$. Also, we note that $\sum_{\boldsymbol{K}} e^{i\boldsymbol{K}\boldsymbol{r}_i^0} = N\delta_{\boldsymbol{r}_i^0,0}$. If the crystal moves rigidly by a lattice vector, it acquires a quasi-momentum $\hbar\boldsymbol{K}$ (the condition $\sum_i \boldsymbol{v}_i = 0$ ensures the total number $N-1$ of internal degrees of freedom). The quasi-momentum $\hbar\boldsymbol{K}$ is defined up to a reciprocal vector of the lattice (which defines umklapp processes in crystals[16]).

5.9.3 Crystal field

The crystal field is a superposition of the coherent field given by equation (5.71) and the incoherent field given by equation (5.69). We write this superposition as

$$\Psi(\boldsymbol{r}) = \sqrt{1-g^2}\Phi(\boldsymbol{r}) + g\psi(\boldsymbol{r}) \ , \tag{5.75}$$

where the weight factor g is proportional to the square root of the relative number of vibration states of the lattice, properly normalized. This relative number of phonon states is proportional to

$$I = \frac{1}{\omega_D^3}\int_0^{\omega_D} d\omega \frac{\omega^2}{e^{\hbar\omega/T}-1} \ , \tag{5.76}$$

[16] R. E. Peierls, "Zur kinetischen Theorie der Waermeleitung in Kristallen", Ann. Phys. **3** 1055 (1929).

where ω_D is the Debye frequency, ω is the phonon frequency and T denotes the temperature; the Debye temperature is $\Theta = \hbar\omega_D$. For $T/\Theta \ll 1$ this integral is $I \simeq 2.4(T/\Theta)^3$, while for $T/\Theta \gg 1$ we get $I \simeq \frac{1}{2}T/\Theta$. A normalized expression for the relative number of states is $4.8(T/\Theta)^2$ for $T/\Theta \ll 1$ and 1 for $T/\Theta \gg 1$. An interpolation formula is $I \simeq 4.8(T/\omega_D)^2/\left[1+4.8(T/\omega_D)^2\right]$, such that

$$g \simeq \frac{2.2(T/\omega_D)}{\left[1+4.8(T/\omega_D)^2\right]^{1/2}} \; . \tag{5.77}$$

We can see that for high Debye temperatures the main contribution to the crystal field comes from the coherent field ($g \ll 1$), while for low Debye temperatures the main contribution to the crystal field comes from the incoherent field ($g \to 1$). The coefficient g is the square root of a diagonal density matrix. Equation (5.75) is valid at thermal equilibrium.

5.9.4 Photon emission (absorption)

Let us consider a set of N identical particles, like atoms, molecules, atomic nuclei, arranged in a solid, capable of emitting or absorbing electromagnetic radiation. The emission (absorbtion) of a photon is made by each particle by transition from a state with the wavefunction $\varphi^{a(b)}$ to a state with (orthogonal) wavefunction $\varphi^{b(a)}$. The incoherent fields $\psi^{a(b)}(\boldsymbol{r})$ are given by equation (5.69), the coherent fields $\Phi^{a(b)}_{\boldsymbol{K}}(\boldsymbol{r})$ are given by equation (5.71) and the crystal fields $\Psi^{a(b)}_{\boldsymbol{K}}$ are given by equation (5.72).

In order to estimate the photon emission (absorption) rates we need the matrix elements of the interaction between states described by $\psi^a(\boldsymbol{r})$ and $\psi^b(\boldsymbol{r})$ for incoherent emission (absorption) and between states described by $\Phi^a_{\boldsymbol{K}}(\boldsymbol{r})$ and $\Phi^b_{\boldsymbol{K}'}(\boldsymbol{r})$ for coherent emission (absorption). The interaction hamiltonian is

$$H = -\frac{1}{c}\sum_i \boldsymbol{j}(\boldsymbol{r}-\boldsymbol{r}_i)\boldsymbol{A}(\boldsymbol{r}) \; , \tag{5.78}$$

where $\boldsymbol{j}(\boldsymbol{r}-\boldsymbol{r}_i)$ is the electromagnetic current inside the i-th particle and $\boldsymbol{A}$ is the (transverse) vector potential of the electromagnetic field

($div\boldsymbol{A} = 0$, c denotes the speed of light in vacuum). The longitudinal components of the electromagnetic field do not participate in interaction. The vector potential is represented by

$$\boldsymbol{A} = \sum_{\boldsymbol{k}} c\alpha_k \left(\boldsymbol{e_k} a_{\boldsymbol{k}} e^{-i\omega_k t + i\boldsymbol{kr}} + \boldsymbol{e_k} a^*_{\boldsymbol{k}} e^{i\omega_k t - i\boldsymbol{kr}} \right) , \tag{5.79}$$

where $\alpha_k = \sqrt{2\pi\hbar/\omega_k V}$, $a_{\boldsymbol{k}}$, $a^*_{\boldsymbol{k}}$ are the annihilation/creation operators of photons with frequency $\omega_k = ck$ and polarization $\boldsymbol{e_k}$ ($\boldsymbol{e_k k} = 0$), V denotes the volume and $\hbar$ is Planck's constant ($\boldsymbol{k} \neq 0$).

Let us compute first the matrix elements H^{ba} by using the incoherent fields given by equation (5.69). For the emission of a $\boldsymbol{k}$-photon (in vacuum) we get

$$H^{ba} = -\sum_i \sqrt{\frac{2\pi\hbar}{\omega_k V}} \left(\boldsymbol{j}^{ba} \boldsymbol{e_k} \right) e^{-i\boldsymbol{k}(\boldsymbol{r}_i^0 + \boldsymbol{v}_i)} e^{i(\omega_{ba} + \omega_k)t} , \tag{5.80}$$

where $\boldsymbol{j}^{ba}$ is the matrix elements of the current, the frequency $\omega_{ba} = (E_b - E_a)/\hbar$ corresponds to the difference between the two energies $E_{a,b}$ of the two states a, b and we assume that the electromagnetic potential is practically uniform over the particle volume (dipolar approximation); the time factors are included in equation (5.80). We note that the current $\boldsymbol{j}(\boldsymbol{r} - \boldsymbol{r}_i)$ is localized inside the i-th particle. The displacements $\boldsymbol{v}_i$ are very small in comparison with $\boldsymbol{r}_i^0$, such that we may use the expansion

$$e^{-i\boldsymbol{k}(\boldsymbol{r}_i^0 + \boldsymbol{v}_i)} = e^{-i\boldsymbol{k}\boldsymbol{r}_i^0}(1 - i\boldsymbol{k}\boldsymbol{v}_i) ; \tag{5.81}$$

it is easy to see that the summation over i leads to

$$H^{ba} = i\sum_i \sqrt{\frac{2\pi\hbar}{\omega_k V}} \left(\boldsymbol{j}^{ba} \boldsymbol{e_k} \right) e^{-i\boldsymbol{k}\boldsymbol{r}_i^0} (\boldsymbol{k}\boldsymbol{v}_i) e^{i(\omega_{ba} + \omega_k)t} . \tag{5.82}$$

Since the displacement $\boldsymbol{v}_i$ can be written as

$$\boldsymbol{v}_i = \frac{1}{\sqrt{N}} \sum_{\boldsymbol{q}} \boldsymbol{v_q} e^{i\boldsymbol{q}\boldsymbol{r}_i^0} , \tag{5.83}$$

we get a factor $\delta_{\boldsymbol{kq}}$ in equation (5.82). Therefore, we can see that the photon momentum $\hbar\boldsymbol{k}$ is compensated by the absorption of a quantum of lattice vibration with momentum $\hbar\boldsymbol{q} = \hbar\boldsymbol{k}$ (or emission of a vibration quantum with momentum $-\hbar\boldsymbol{k}$). The matrix element given above becomes

$$H^{ba} = i\sqrt{\frac{2\pi\hbar N}{\omega_k V}}\left(\boldsymbol{j}^{ba}\boldsymbol{e_k}\right)(\boldsymbol{k}\boldsymbol{v_k})e^{i(\omega_{ba}+\omega_k\pm\Omega_k)t} , \tag{5.84}$$

where Ω_k is the frequency of the quantum of lattice vibrations. The number of transitions per unit time (emission rate) is

$$w^{ba}_{incoh} = \frac{4\pi^2 N}{\omega_k V}\left|\boldsymbol{j}^{ba}\boldsymbol{e_k}\right|^2 |\boldsymbol{k}\boldsymbol{v_k}|^2 \delta(E_b - E_a + \hbar\omega_k \pm \hbar\Omega_k) . \tag{5.85}$$

For the total emission rate the above expressions are added, to account for the two vibration processes ($\pm\hbar\Omega_k$), then a factor $Vk^2dkdo/(2\pi)^3$ is introduced, in order to get the emission rate in the energy interval $d(\hbar\omega_k) = c\hbar dk$ and the solid angle do; an average over polarization can also be used. Also, the emission rate implies a thermal average $\overline{|\boldsymbol{k}\boldsymbol{v_k}|^2}$ over the vibration displacements. If the vibration energy $\hbar\Omega_k$ is much lower than the (room) temperature T (*e.g.*, the Debye temperature smaller than the room temperature), we may use a classical harmonic oscillator with mass m for each particle, which leads to $\overline{|\boldsymbol{k}\boldsymbol{v_k}|^2} = k^2T/m\Omega_k^2$. This factor is reminiscent of the Debye-Waller factor. We can see from equation (5.85) that the emission rate of N particles is N times the emission rate of one particle, as expected for an incoherent emission. We note that the incoherent emission is associated with phonon excitations (momentum transfer $\hbar\boldsymbol{k} \neq 0$).

For a perfect crystal we need to compute the matrix elements of the interaction between states $\Psi^a_{\boldsymbol{K}}$ and $\Psi^b_{\boldsymbol{K}'}$ given by equation (5.75). The matrix element of the interaction between the coherent state $\Phi^a_{\boldsymbol{K}}(\boldsymbol{r})$ and the incoherent state ψ^b is vanishing, as a consequence of distinct localization, such that we are left with a matrix element H^{ba} given by equation (5.84) and the coherent matrix element

$$H^{ba}_{\boldsymbol{K}'\boldsymbol{K}} = -\sum_i \sqrt{\frac{2\pi\hbar}{\omega_k V}} e^{i(\boldsymbol{K}-\boldsymbol{K}'-\boldsymbol{k})\boldsymbol{r}_i^0}\left(\boldsymbol{j}^{ba}\boldsymbol{e_k}\right)e^{i(\omega_{ba}+\omega_k)t} \tag{5.86}$$

for the emission of a photon. We can see that the photon momentum $\hbar\boldsymbol{k}$ is emitted by the whole crystal on account of the change of the crystal quasi-momentum $\hbar\boldsymbol{K} \longrightarrow \hbar\boldsymbol{K}'$. The crystal moves as a whole by a lattice vector. The emission rate consists of two distinct parts: a coherent emission given by

$$w_{coh}^{ba} = \frac{4\pi^2 N^2}{\omega_k V} \left| \boldsymbol{j}^{ba} \boldsymbol{e_k} \right|^2 \delta(E_b - E_a + \hbar\omega_k) \tag{5.87}$$

and a much smaller incoherent part given by equation (5.85). Each contribution should be multiplied by its weight factor, such that we get the total rate

$$w^{ba} = (1 - g^2) w_{coh}^{ba} + g^2 w_{incoh}^{ba} \ . \tag{5.88}$$

The coherent part of the emission rate is proportional to N^2 and implies a change of the quasi-momentum of the crystal, while the incoherent part is proportional to N and implies the excitation of a lattice vibration. The change in the crystal quasi-momentum was recognized in the Moessbauer effect.[17]

5.9.5 Coherent scattering

The scattering rate is given by

$$w_{fi} = \frac{2\pi}{\hbar} \left| \sum_n \frac{H_{fn} H_{ni}}{E_n - E_i - i0^+} \right|^2 \delta\left(E_f - E_i\right) \ , \tag{5.89}$$

where f is the final state, i is the initial state and n denotes the intermediate state.[18] Let us assume that a $\boldsymbol{k}$-photon is absorbed and a $\boldsymbol{k}'$-photon is emitted (in vacuum), where the initial/final state is labelled by a/b. For a crystal the field in the inital state is

$$\begin{aligned} \Psi_{\boldsymbol{K}}(\boldsymbol{r}) = (1 - g^2)^{1/2} \textstyle\sum_i e^{i\boldsymbol{K}\boldsymbol{r}_i^0} \varphi^a(\boldsymbol{r} - \boldsymbol{r}_i^0) + \\ + g \textstyle\sum_i e^{i\chi_i} \varphi^a(\boldsymbol{r} - \boldsymbol{r}_i^0 - \boldsymbol{v}_i) \ , \end{aligned} \tag{5.90}$$

[17] P. Guetlich, E. Bill and A. X. Trautwein, "Moessbauer Spectroscopy and Transition Metal Chemistry", Springer, Heidelberg (2011).

[18] For electromagnetic interaction the term $\frac{q^2}{2mc^2} A^2$ should be added, for a point-like charge q with mass m, which gives a contribution in the first-order of the perturbation theory.

where the first contribution on the right in equation (5.90) corresponds to the coherent field $\Phi^a_{\boldsymbol{K}}$, while the second contribution corresponds to the incoherent field ψ^a. Similar expressions hold for the final state with $\boldsymbol{K} \longrightarrow \boldsymbol{K}'$ and $a \longrightarrow b$. Due to the localization of the wavefunctions $\varphi^{a,b}(\boldsymbol{r}-\boldsymbol{r}_i)$, the intermediate states are $\varphi^c(\boldsymbol{r}-\boldsymbol{r}_i^0)$, or $\varphi^c(\boldsymbol{r}-\boldsymbol{r}_i^0-\boldsymbol{v}_i)$. The matrix element of the interaction between a coherent state and an incoherent one is vanishing, for the same reason. For a coherent scattering the scalar product gives a factor

$$\sum_i e^{i(-\boldsymbol{K}'-\boldsymbol{k}'+\boldsymbol{K}+\boldsymbol{k})\boldsymbol{r}_i^0} = N\delta_{\boldsymbol{K}'+\boldsymbol{k}',\boldsymbol{K}+\boldsymbol{k}} \tag{5.91}$$

(in the dipolar approximation). The variation of the photon momentum $\hbar(\boldsymbol{k}'-\boldsymbol{k})$ is absorbed by the crystal quasi-momentum $\hbar(\boldsymbol{K}-\boldsymbol{K}')$. Therefore, the coherent scattering rate is

$$w_{coh} = N^2 w \ , \tag{5.92}$$

where w is the scattering rate for a particle. The radiation is scattered in all directions of the wavevector $\boldsymbol{k}'$. The average change in the quasi-momentum $\Delta(\hbar\boldsymbol{K}) = \hbar\boldsymbol{k} - \hbar\overline{\boldsymbol{k}'} = \hbar\boldsymbol{k}$ is the inital momentum $\hbar\boldsymbol{k}$ of the incident radiation. The corresponding change in the energy is taken by the center of mass; since the mass of the crystal is very large, this energy is extremely small.

For an incoherent scattering the scalar product gives a factor

$$\begin{aligned}\sum\nolimits_i e^{i(-\chi_i+\chi_i)+i(-\boldsymbol{k}'+\boldsymbol{k})(\boldsymbol{r}_i^0+\boldsymbol{v}_i)} = \\ = \sum\nolimits_i e^{i(-\boldsymbol{k}'+\boldsymbol{k})\boldsymbol{r}_i^0}[1+i(\boldsymbol{k}-\boldsymbol{k}')\boldsymbol{v}_i+..] \ .\end{aligned} \tag{5.93}$$

The first term in this equation should be omitted, because it corresponds to a coherent displacement $\boldsymbol{v}_i = 0$; the remaining term leads to

$$\frac{1}{\sqrt{N}} i(\boldsymbol{k}-\boldsymbol{k}')\boldsymbol{v}_{\boldsymbol{q}} \sum_i e^{i(-\boldsymbol{k}'+\boldsymbol{k}+\boldsymbol{q})\boldsymbol{r}_i^0} = \sqrt{N} i(\boldsymbol{k}-\boldsymbol{k}')\boldsymbol{v}_{\boldsymbol{q}}\delta_{\boldsymbol{k}'-\boldsymbol{k},\boldsymbol{q}} \ ; \tag{5.94}$$

the change in the photon momentum is compensated by the momentum of a lattice vibration. The scattering rate acquires a factor N with respect to the scattering of a particle and, in addition, the energy conservation implies the energy of the vibration quantum $(\delta(E_b + \hbar\omega_{k'} - E_a - \hbar\omega_k \pm \hbar\Omega_q))$.[19]

For the electromagnetic interaction given by equation (5.78) the coherent scattering rate is

$$w_{fi}^{coh} = \frac{(2\pi)^3 \hbar N^2}{\omega_k \omega_{k'} V^2} \left| \sum_c \frac{(\boldsymbol{j}^{bc}\boldsymbol{e}_{\boldsymbol{k}'})(\boldsymbol{j}^{ca}\boldsymbol{e}_{\boldsymbol{k}})}{E_c - E_a - \hbar\omega_k - i0^+} \right|^2 \cdot \\ \cdot\delta(E_b + \hbar\omega_{k'} - E_a - \hbar\omega_k) \ , \tag{5.95}$$

while the incoherent scattering rate is

$$w_{fi}^{incoh} = \frac{(2\pi)^3 \hbar N}{\omega_k \omega_{k'} V^2} \left| \sum_c \frac{(\boldsymbol{j}^{bc}\boldsymbol{e}_{\boldsymbol{k}'})(\boldsymbol{j}^{ca}\boldsymbol{e}_{\boldsymbol{k}})(\boldsymbol{q}\boldsymbol{v}_{\boldsymbol{q}})}{E_c - E_a - \hbar\omega_k - i0^+} \right|^2 \cdot \\ \cdot\delta(E_b + \hbar\omega_{k'} - E_a - \hbar\omega_k \pm \hbar\Omega_q) \ , \tag{5.96}$$

where $\boldsymbol{q} = \boldsymbol{k}' - \boldsymbol{k}$. The crystal scattering rate is

$$w^{fi} = (1 - g^2) w_{fi}^{coh} + g^2 w_{fi}^{incoh} \ . \tag{5.97}$$

The total rate is obtained by integrating over the final states. For $\omega_k \neq \omega_{k'}$ ($\boldsymbol{k} \neq \boldsymbol{k}'$) we have a Raman scattering. In the above matrix elements we assume that the photon is absorbed in the initial state and emitted in the final state (Stokes line, excitation from the ground state). For $E_a = E_b$, $\omega_k = \omega_{k'}$ we have an elastic scattering (Rayleigh, Thomson). It may happen that $E_c = E_a + \hbar\omega_k$. This is a resonant scattering (resonance fluorescence). Then, the infinitesimal term $i0^+$ is replaced by the finite width of the intermediate energy level (which brings the main contribution).

5.9.6 Cross-section

The scattering cross-section is obtained by integrating the scattering rate over the number of final states $Vk'^2 dk' do'/(2\pi)^3$ and dividing by the incoming particle flux density Φ. We get the total cross-section for the coherent scattering (equation (5.95))

$$\sigma_{coh} = N^2 \sigma_0 \ , \tag{5.98}$$

[19] I. E. Tamm, "Ueber die Quantentheorie der molekularen Lichtzerstreuung in festen Koerpern", Z. Phys. **60** 345 (1930).

where

$$\sigma_0 = \frac{4\pi}{c^3 SV} \overline{\left| \sum_c \frac{(\boldsymbol{j}^{bc}\boldsymbol{e}_{\boldsymbol{k}'})(\boldsymbol{j}^{ca}\boldsymbol{e}_{\boldsymbol{k}})(\boldsymbol{q}\boldsymbol{v}_{\boldsymbol{q}})}{E_c - E_a - \hbar\omega_k - i0^+} \right|^2} \frac{\omega_k - \omega_{ba}}{\omega_k} \tag{5.99}$$

is the cross-section for one particle ($\omega_k > \omega_{ba}$); the average is performed over all the orientations of the wavevectors $\boldsymbol{j}$ and $\boldsymbol{k}'$ (unpolarized target and photons).

For the incoherent scattering (equation (5.96)) we note that the phonon frequency Ω_q in the function δ may be neglected, since the phonon velocity is much smaller than the photon velocity ($\boldsymbol{q} = \boldsymbol{k}' - \boldsymbol{k}$); for elastic scattering ($\omega_{ba} = 0$, $\omega_k = \omega_{k'}$, $k = k'$) the main contribution comes from the forward direction, and $q \to 0$ ($q = 2k \sin\frac{\theta}{2}$, where θ is the angle between $\boldsymbol{k}$ and $\boldsymbol{k}'$), which makes the cross-section very small, in view of the factor $(\boldsymbol{q}\boldsymbol{v}_q)^2$. The factor $(\boldsymbol{q}\boldsymbol{v}_q)^2$ should be thermally averaged, which leads to

$$\overline{(\boldsymbol{q}\boldsymbol{v}_q)^2} = \frac{\hbar q^2}{2m\Omega_q} \coth \frac{\hbar\Omega_q}{2T} , \tag{5.100}$$

where m is the particle mass and T denotes the temperature. The highest value of this factor is $q^2 T/m\Omega_q^2 = T/mu^2$, where u is a mean phonon velocity. This value is obtained for $\hbar\Omega_q \ll 2T$, or $\theta \ll T$, *i.e.* for usual, "soft" solids. In the opposite limit the incoherent cross-section is very low. Therefore, the incoherent scattering contributes little to the scattering in stiff, perfect crystals. Under these conditions we get the total incoherent cross-section

$$\sigma_{incoh} = N(T/mu^2)\sigma_0 . \tag{5.101}$$

We note that the Bragg peaks (the diffraction) disappear in the coherent scattering, since a form-factor is absent. The coherent scattering described herein is a purely quantum-mechanical effect, where the quasi-momentum conservation precludes the occurrence of the reciprocal lattice vectors.

5.10 Neutrino scattering

The calculations presented above are performed for photons, by using the interaction hamiltonian given by equation (5.78). Actually, a coherent scattering in the whole crystal does not occur for photons. Indeed, for soft X-rays the equilibrium time is long; it may be longer than the "collision" time. For energetic photons, like gamma rays with energy $E = 1MeV$, the perturbation energy of an atom is $\Delta E = p^2/m$, where p is the photon momentum and m is the mass of the atom. According to the above discussion, the momentum transfer is of the order $\Delta p = p = 10^{-17}g \cdot cm/s$ for an energy $1MeV$. The energy transfer is $\Delta E = 10^{-12}erg$ and the equilibrium time is of the order $\Delta t_{eq} = \hbar/\Delta E = 10^{-15}s$. On the other hand, the "collision" time is $\tau = 1/\Phi\sigma$ (equation (5.54)), where Φ is the flux density and σ is the cross-section. For a coherent scattering $\sigma = \sigma_0 N^2$, where σ_0 is of the order $10^{-25}cm^2$. For $1mol$ we get $\sigma \simeq 10^{22}cm^2$, and for a flux density $\Phi = 10^{12}/cm^2 \cdot s$ the "collision" time is of the order $\tau = 10^{-34}s$. This time is much shorter than the equilibrium time, so we expect the photons to see independent atomic targets, and the scattering to be incoherent. For an incoherent scattering $\sigma = \sigma_0 N \simeq 10^{-1}cm^2$ and the "collision" time is $\tau = 10^{-11}s$, which is longer than the equilibrium time. For a coherent diffraction the cross-section is $\sigma = \sigma_0 N^{4/3} \simeq 10^7 cm^2$ and $\tau \simeq 10^{-19}s < \Delta t_{ex}$. We can see that either the cross-sections are larger than the area of the crystal, or the incoherent regime is inconsistent. These estimations are made for the crystal as a whole. Very likely, we have an admixture of coherent and incoherent scattering, but definitely not a quantum coherent scattering in the whole crystal.

The condition for a coherent scattering is $1/\Phi\sigma_0 N^2 > \Delta t_{eq}$, *i.e.* $N < 1/(\Phi\sigma_0\Delta t_{eq})^{1/2}$. On the other hand, the condition $N < (A/\sigma_0)^{1/2}$ should also be fulfilled, where A is the area of the sample. For $\sigma_0 = 10^{-25}cm^2$ we get $N < 10^{14}$ ($\Phi = 10^{12}/cm^2 \cdot s$, $\Delta t_{eq} = 10^{-15}s$) and $N < 3\times 10^{13}A^{1/2}$. We can see that these conditions cannot be fulfilled for a macroscopic sample. On the other hand, for $\sigma_0 = 10^{-44}cm^2$ (neutrinos) these conditions read $N < 3 \times 10^{23}$ and $N < 10^{22}A^{1/2}$. Such conditions can be fulfilled for neutrinos. The force acting upon the crystal is $f = \frac{\Phi}{c}E\sigma$ and, since $\sigma < A < 1/\Phi\Delta t_{eq}$, the force satisfies the condition $f < \frac{\Phi}{c} EA \simeq 5 \times 10^{-5} A\, dyn$. This limiting force is measurable for a crystal of the order $1mol$. It is worth noting that a larger crystal does not improve the condition of coherent scattering. For an incoherent scattering the cross-section is $\sigma = \sigma_0 N$, and the condition of incoherent scattering gives $N > 1/\Phi\sigma_0\Delta t_{eq}$, which is not satisfied for $\sigma_0 = 10^{-25} cm^2$ ($\Phi = 10^{12}/cm^2 \cdot s$, $\Delta t_{eq} = 10^{-15} s$). For a coherent diffraction $\sigma = \sigma_0 N^{4/3}$, and the conditions read $N > 1/(\Phi\sigma_0\Delta t_{eq})^{3/4}$ and $N < (A/\sigma_0)^{3/4}$. With our numerical data ($\sigma_0 = 10^{-25} cm^2$, $\Phi = 10^{12}/cm^2 \cdot s$, $\Delta t_{eq} = 10^{-15} s$) these conditions lead to $N > 10^{21}$ and $N < 10^{18} A^{3/4}$. The second condition is not satisfied. We expect that the cross-section is reduced in this case by multiple coherence domains, such that an admixture of scatterings is possible.

Let us assume an incident neutrino wavefunction $\frac{1}{\sqrt{V}} e^{i\boldsymbol{kr}}$ and an emergent (scattered) neutrino wavefunction $\frac{1}{\sqrt{V}} e^{i\boldsymbol{k}'\boldsymbol{r}}$, where V is the volume of the sample and $\boldsymbol{k}$, $\boldsymbol{k}'$ are the incident/emergent wavevectors. Further on, we assume an interaction hamiltonian

$$H = a^3 h \sum_i \delta(\boldsymbol{r} - \boldsymbol{r}_i) \ , \tag{5.102}$$

where h depends on the internal coordinates of the target particle with dimension a placed at position $\boldsymbol{r}_i$ (contact interaction). The wavefunction of the whole target is

$$\begin{aligned} \Psi_{\boldsymbol{K}}(\boldsymbol{r}) = (1-g^2)^{1/2} \textstyle\sum_i e^{i\boldsymbol{K}\boldsymbol{r}_i^0} \varphi(\boldsymbol{r} - \boldsymbol{r}_i^0) + \\ + g \textstyle\sum_i e^{i\chi_i} \varphi(\boldsymbol{r} - \boldsymbol{r}_i^0 - \boldsymbol{v}_i) \ , \end{aligned} \tag{5.103}$$

where g is the weight factor given by equation (5.77), $\boldsymbol{K}$ is the quasi-momentum of the crystal, $\boldsymbol{r}_i^0$ denote the equilibrium positions of the atoms in the crystal, $\boldsymbol{v}_i$ are the displacements of the atoms from these positions, $\varphi(\boldsymbol{r} - \boldsymbol{r}_i)$ are single-particle wavefunctions localized on the positions $\boldsymbol{r}_i$ ($\boldsymbol{r}_i = \boldsymbol{r}_i^0$, or $\boldsymbol{r}_i = \boldsymbol{r}_i^0 + \boldsymbol{v}_i$) and $e^{i\chi_i}$ are random phase factors. The number of transitions per unit time is given by

$$w = \frac{2\pi}{\hbar} |H_{fi}|^2 \delta(E_f - E_i) \ , \tag{5.104}$$

where $E_{f,i}$ are the energies of the final/initial states and H_{fi} are the matrix elements of the interaction between these states. We assume that the initial state and the final state are internal states of the particles. We compute first the matrix elements of the interaction H between the incoherent states, *i.e.* the states which have the phase factors $e^{i\chi_i}$ in equation (5.103). These are the states denoted by $\psi(\boldsymbol{r})$ in equation (5.69) (with $\boldsymbol{R}_0 = 0$). For these states the positions $\boldsymbol{r}_i$ in equation (5.102) are $\boldsymbol{r}_i = \boldsymbol{r}_i^0 + \boldsymbol{v}_i$. The normalization to unity of the initial wavefunction $\frac{1}{\sqrt{V}} e^{i\boldsymbol{kr}}\psi(\boldsymbol{r})$ requires a factor $\sqrt{a^3}$ for the functions $\varphi(\boldsymbol{r} - \boldsymbol{r}_i)$, which amounts to $|\varphi(\boldsymbol{r}_j - \boldsymbol{r}_i)|^2 = \delta_{ij}$. We get immediately

$$H_{fi} = \frac{a^3 h_{fi}}{V} \sum_i e^{i(\boldsymbol{k}-\boldsymbol{k}')\boldsymbol{r}_i^0} \left(1 + i(\boldsymbol{k} - \boldsymbol{k}')\boldsymbol{v}_i + ...\right) \ , \tag{5.105}$$

where

$$\boldsymbol{v}_i = \frac{1}{\sqrt{N}} \sum_{\boldsymbol{q}} e^{i\boldsymbol{q}\boldsymbol{r}_i^0} \boldsymbol{v}_{\boldsymbol{q}} \tag{5.106}$$

is the phonon field. The first term in equation (5.105) corresponds to a coherent scattering, so it should be left aside. We get the matrix element

$$H_{fi} = -\frac{a^3 h_{fi}}{V} \sqrt{N} \left(i\boldsymbol{q}\boldsymbol{v}_{\boldsymbol{q}}\right) \delta_{\boldsymbol{k}',\boldsymbol{k}+\boldsymbol{q}} \ . \tag{5.107}$$

We can see that the scattering excites phonons, which ensure the conservation of the energy and the momentum, *i.e.* $E_f = \varepsilon_f + \hbar\omega_{k'} \pm \hbar\Omega_{\boldsymbol{q}}$, $E_i = \varepsilon_i + \hbar\omega_k$, $\boldsymbol{k}' = \boldsymbol{k} + \boldsymbol{q}$, where $\varepsilon_{i,f}$ are the internal energies of a particle in the target, $\omega_k = ck$, $\omega_{k'} = ck'$ are the frequencies of the incoming and emergent neutrino and Ω_b is the frequency of the phonon. A phonon wavevector $\boldsymbol{q}$ may arise either from the creation of a phonon with wavevector $\boldsymbol{q}$ or from the destruction of a phonon with wavevector $-\boldsymbol{q}$, whence the two phonon frequencies $\pm\Omega_{\boldsymbol{q}}$. The transition rate becomes

$$w = \frac{2\pi}{\hbar} \frac{\left|a^3 h_{fi}\right|^2}{V^2} N \left|(\boldsymbol{q}\boldsymbol{v}_{\boldsymbol{q}})\right|^2 \delta(\varepsilon_f + \hbar\omega_{k'} \pm \hbar\Omega_{\boldsymbol{q}} - \varepsilon_i - \hbar\omega_k) \ , \tag{5.108}$$

where N is the number of atoms in the sample and $\boldsymbol{q}$ is determined by $\boldsymbol{q} = \boldsymbol{k}' - \boldsymbol{k}$. Since $\omega_k \gg \Omega_{\boldsymbol{q}}$ we may omit the phonon frequencies in the energy conservation in equation (5.108). Also, we may limit ourselves to elastic scattering, $\varepsilon_i = \varepsilon_f$. The factor $\left|(\boldsymbol{q}\boldsymbol{v}_{\boldsymbol{q}})\right|^2$ should be averaged over thermal phonon states. For $T \ll \Theta_D$, where T is the temperature and Θ_D is the Debye temperature, the average is very small. For $T \gg \Theta_D$ the average is of the order $q^2T/m\Omega_q^2 = T/mu^2$, where u is the mean phonon velocity. Also, we need to multiply by the number of final states $Vk^2dkdo/(2\pi)^3$ in the solid angle do, and integrate over these states. The cross-section is obtained by dividing to the flux density c/V. We get the incoherent cross-section

$$\sigma_{incoh} = \left|a^3h_{fi}\right|^2 N\overline{|(\boldsymbol{q}\boldsymbol{v}_{\boldsymbol{q}})|^2}\frac{E^2}{\pi\hbar^4c^4} \; ; \tag{5.109}$$

it can be written as

$$\sigma_{incoh} = N\sigma_{0incoh} \; , \tag{5.110}$$

where σ_{0incoh} is the single-particle cross-section. This incoherent cross-section differs from the single-particle cross-section σ_0 by the thermal factor $\overline{|(\boldsymbol{q}\boldsymbol{v}_{\boldsymbol{q}})|^2}$.

The matrix elements of the interaction between coherent and incoherent states in equation (5.103) are zero. We are left with the matrix elements between coherent states, *i.e.* the states with the phase factors $e^{i\boldsymbol{K}\boldsymbol{r}_i^0}$ in equation (5.103). The matrix elements of the interaction are

$$\begin{aligned} H_{fi} &= \tfrac{a^3h_{fi}}{V}\textstyle\sum_i e^{i(\boldsymbol{K}-\boldsymbol{K}')\boldsymbol{r}_i^0}e^{i(\boldsymbol{k}-\boldsymbol{k}')\boldsymbol{r}_i^0} = \\ &= \tfrac{a^3h_{fi}}{V}N\delta_{\boldsymbol{K}'+\boldsymbol{k}',\boldsymbol{K}+\boldsymbol{k}} \; ; \end{aligned} \tag{5.111}$$

we can see that the difference in the neutrino momentum is compensated by the quasi-momentum of the crystal. The energies are $E_f = \hbar^2K'^2/2M + \hbar\omega_{k'} + \varepsilon_f$, $E_i = \hbar^2K^2/2M + \hbar\omega_k + \varepsilon_i$, where M is the mass of the crystal. Since this mass is very large, we may neglect the crystal energy in the energy conservation. Also, we limit ourselves to elastic scattering (although the coherence considerations are also valid for inelastic scattering). By integrating over the final states we get the total transition rate

$$w = \frac{\left|a^3h_{fi}\right|^2}{V}N^2\frac{k^2}{\pi\hbar^2c} \tag{5.112}$$

and the total coherent cross-section (by dividing to the flux density

$\Phi = c/V$)

$$\sigma_{coh} = \left|a^3 h_{fi}\right|^2 N^2 \frac{E^2}{\pi\hbar^4 c^4} \ ; \tag{5.113}$$

it can be written as

$$\sigma_{coh} = N^2 \sigma_0 \ , \tag{5.114}$$

where σ_0 is the (coherent) cross-section of a single particle.

The total cross-section is

$$\sigma = (1 - g^2)\sigma_{coh} + g^2 \sigma_{incoh} \ . \tag{5.115}$$

The force acting upon the crystal (in the forward direction) is

$$f = \frac{\Phi}{c} \sigma E \tag{5.116}$$

(equation (5.21)).

We note that the coherent cross-section given by equation (5.114) does not exhibit Bragg peaks, in contrast to the coherent diffraction. This is due to the use of the quantum-mechanical wavefunction (field) of the target.[20] Also, except for the weight factor $1 - g^2$, it does not depend on the temperature.

The numerical estimations are done for a sapphire crystal of high purity with a volume $\simeq 6cm^3$ (mass $\simeq 24g$). The volume of the unit cell is $\simeq 7 \times 10^2 \text{Å}^3$, such that the number of unit cells is $N \simeq 10^{22}$ (less than $0.1mol$). The Debye temperature is $\Theta_D \simeq 10^3 K$, such that, at room temperature, $1 - g^2 \simeq 0.67$ (equation (5.77)). Making use of $\sigma_0 = 10^{-44} cm^2$, we get a coherent cross-section $\sigma = (1 - g^2)\sigma_{coh} \simeq 0.67cm^2$. We note that this cross-section is smaller than the area of the sample. Also, the collision time $\tau = 1/\Phi\sigma$ ($\simeq 10^{-12}s$, $\Phi = 10^{12}/cm^2 \cdot s$) is longer than the equilibrium time Δt_{eq} ($\simeq 10^{-15}s$), such that the condition of coherent scattering is preserved. For $\Phi = 10^{12}/cm^2 \cdot s$ and $E = 1MeV$ the force acting upon the crystal is $f \simeq 3.4 \times 10^{-5} dyn$. This is a measurable force. The contribution of the incoherent scattering is much smaller. The cross-section (and the force) can be larger, due to the number of Al atoms in the unit cell

[20] M. Apostol, "Weber's coherent scattering and neutrino detection", Roum. Reps. Phys. **60** 315 (2008).

of the sapphire crystal. However, it cannot be larger than the area of the sample. Small deviations in the above input data may lead to unmeasurable forces. For instance, if $T/\Theta_D = 2$ the cross-section becomes $\sigma \simeq 0.03cm^2$ and the force $f \simeq 10^{-6}dyn$, which cannot be easily measured. Also, we note that a large unit cell may increase the cross-section σ_0. The input data used here are close to those employed in Weber's experiments and the experiments which confirmed Weber's results.[21]

5.11 Coherence domains

A larger cross-section σ than the area A of the target is an unphysical situation. The target is out of equilibrium. The equilibrium is preserved by developing scattering domains, characterized by random phase factors. This way, the entropy increases, under given conditions. The increase should be minimal, such that the new cross-section $\overline{\sigma}$ is as large as possible. We call these domains coherence domains.[22] Therefore, the condition

$$\overline{\sigma} < A \tag{5.117}$$

should be satisfied, with a cross-section as large as possible ($\overline{\sigma} = A$).

We may have two kinds of scattering inside each domain. For one kind the target particles may behave independently; for a crystal they give a coherent diffraction. If the atoms are randomly distributed the diffraction reduces to a forward peak. The time between two successive collisions is shorter than the equilibrium time of the target

[21] J. Weber, "Method for observation of neutrinos and antineutrinos", Phys. Rev. **C31** 1468 (1985); "Apparent observation of abnormally large coherent scattering cross sections using keV and MeV range antineutrinos, and solar neutrinos", Phys. Rev. **D38** 32 (1988); I. Cruceru, G. Nicolescu and O. G. Duliu, "A dedicated torsion balance to detect neutrinos by coherent scattering on high Debye temperature monocrystals", Int. J. Mod. Phys. **A23** 1719 (2008); M. Cruceru, G. Nicolescu, O. G. Duliu and I. Cruceru, "Detection of solar neutrinos by coherent scattering in high-Debye temperature monocrystals", Int. J. Mod. Phys. **A26** 2773 (2011); G. Nicolescu, "A heuristic approach to the detection of solar neutrinos", J. Phys. **G40** 055201 (2013).

[22] M. Apostol, "Coherence domains in matter interacting with radiation", Phys. Lett. A**373** 379 (2009); M. Apostol and M. Ganciu, "Coherent polarization driven by external electromagnetic fields", Phys. Lett. **A374** 4848 (2010).

particles. The former time is given by $\overline{\tau} = 1/\Phi\overline{\sigma}$, where Φ is the incident flux density; the latter is of the order $\Delta t_{eq} = \hbar/\Delta E$, where $\Delta E = (p/m + v)p$, v being the thermal velocity of the particles, m is the mass of the atom and p is the momentum of the incident particle. The condition for a coherent diffraction is

$$\overline{\tau} < \Delta t_{eq} \ . \tag{5.118}$$

The second type of scattering is the coherent scattering. In this case the opposite condition

$$\overline{\tau} > \Delta t_{eq} \tag{5.119}$$

is satisfied.

Let us assume that the number of domains is n_d, each domain including a number N_d of target particles; we assume identical domains, as a statistical mean. We have $n_d = N/N_d$, where N is the total number of particles in the sample. It is convenient to introduce the fraction $f = 1/N_d$, by analogy with the purity fraction of the monocrystals. This fraction varies between $f = 1/N$, when we have only one domain, *i.e.* the whole target, and $f = 1$, when the whole target is "fractured" in atomic domains.

The above conditions depend on the numerical data. We adopt a typical situation where $\Phi = 10^{12}/cm^2 \cdot s$, an incident energy $E = 1MeV$, an incident momentum $p = E/c = 5 \times 10^{-17} g \cdot cm/s$, a particle mass $m = 10^5 \cdot 10^{-27} g$ and the room temperature $T = 300K$. We get $v \simeq 10^4 cm/s$, $\Delta E = 2.5 \times 10^{-11} erg$ and $\Delta t_{eq} \simeq 4 \times 10^{-17} s$.

In the Coherent-incoherent interplay section it is shown that the coherent diffraction cross-section, written as $\sigma = \sigma_0 N^{4/3}$, is reduced by the coherence domains to

$$\overline{\sigma} = \frac{1}{(fN)^{1/3}} \sigma \ , \ \sigma = \sigma_0 N^{4/3} \ , \tag{5.120}$$

where σ_0 is the single-particle cross-section. The same formula is valid for the forward scattering peak in a sample with random positions of the particles, by using the corresponding σ, which is reduced by the number of peaks. We can see that for $f = 1$ this formula recovers the cross-section of a total incoherent scattering ($\overline{\sigma} \sim N$). The coherent cross-section $\sigma = \sigma_0 N^2$ is reduced by the coherence domains to

$$\overline{\sigma} = \sigma_0 N N_d = \frac{1}{fN}\sigma \ , \ \sigma = \sigma_0 N^2 \ ; \tag{5.121}$$

we can see again that for $f = 1$ this formula gives the total incoherent scattering. Therefore, the total incoherence is already contained in these cross-section formulae.

In view of the maximal cross-section condition (under given constraints), we start the examination of these formulae with the coherent scattering (for the same σ_0). We use, as a typical example, $N = 10^{22}$. Let us take first the neutrinos, with $\sigma_0 = 10^{-44}cm^2$ and $\overline{\sigma} = \sigma/fN$, $\sigma = 1cm^2$. The condition given by equation (5.117) leads to $f > 10^{-22}/A$, while the condition given by equation (5.119) gives $f > 4 \times 10^{-27}$. Since $\overline{\sigma}$ should be as large as possible, with $1/N < f < 1$ satisfying these conditions, we get $f = 1/N$ for a macroscopic target with $A > 1cm^2$; this means that we have a total coherent scattering in the whole target. It is worth noting that f increases, and the cross-section decreases, for a smaller area of the target. For $f = 1/N$ the cross-section is $\overline{\sigma} = \sigma = \sigma_0 N^2 = 1cm^2$ and the force acting upon the crystal is $\frac{\Phi}{c}\overline{\sigma}E = 5 \times 10^{-5}\, dyn$. These quantities are reduced by the weight factor $1 - g^2$, as shown above.

Let us consider now the coherent scattering for $\sigma_0 = 10^{-25}cm^2$ (X-, gamma rays). The two conditions given by equations (5.117) and (5.119) amount to $f > 10^{-3}/A$ and $f > 4 \times 10^{-8}$. We choose $f = 10^{-3}/A$, which maximize the cross-section (and is in the range $1/N$ to 1). Therefore, we have $n_d = fN = 10^{19}/A$ coherence domains, which reduce appreciably the cross-section to $\overline{\sigma} = A$. Such a cross-section may lead to a measurable force ($\simeq 3 \times 10^{-5} A\, dyn$), though it is difficult to achieve experimentally the conditions of $1MeV$ gamma rays with a flux density $10^{12}/cm^2 \cdot s$. We note that $\overline{\sigma} = \sigma_0 N/f$ with $f = 10^{-3}/A$ is close to an incoherent scattering ($\overline{\sigma} = 10^3 A\sigma_0 N \sim N$). However, it is worth noting that the number of atoms $N_d = 10^3 A$ in each coherence domain is much smaller than the critical number of atoms which defines a minimal domain. Therefore, the number of atoms in each domain is in fact $N_d = 1$ ($f = 1$), which makes the cross-section $\overline{\sigma} = \sigma_0 N = 10^{-3}cm^2$ ($< A$) and an unmeasurable force $\simeq 5 \times 10^{-8}\, dyn$. The scattering is incoherent in the whole target. A slight increase may arise from larger domains, where N_d is replaced by $N_d^{3/2}$ (finite-size (surface) effects).

Let us examine the formation of domains for the coherent diffraction of electromagnetic radiation. The cross-section is $\sigma = \sigma_0 N^{4/3}$ and $\overline{\sigma} = \sigma/(fN)^{1/3} = \sigma_0 N/f^{1/3}$. The first inequality is $\overline{\sigma} < A$, *i.e.* $f^{1/3} > \sigma_0 N/A$. The "collision" time is $\overline{\tau} = 1/\Phi\overline{\sigma} = f^{1/3}/\Phi\sigma_0 N$; the second inequality is $\overline{\tau} < \Delta t_{eq}$, *i.e.* $f^{1/3} < \Phi\sigma_0 N\Delta t_{eq}$. Therefore, it is necessary to have $\sigma_0 N/A < \Phi\sigma_0 N\Delta t_{eq}$, *i.e.* $\Delta t_{eq} > 1/\Phi A$. We get $E < \sqrt{\Phi A\hbar mc^2}$ or $E < \Phi A\hbar c/v$, since $\Delta t_{eq} = \hbar m/p^2$, $\hbar c/vE(E = cp)$. Since $\Phi A\hbar c/v < \sqrt{\Phi A\hbar mc^2}$, we choose $E < \Phi A\hbar c/v$. For $\Phi = 10^{12}/cm^2 \cdot s$ and $v = 10^4 cm/s$ we get $E < 2A\,keV$. For such energies the scattering of the electromagnetic radiation is practically classical. Since the cross-section is maximal we choose $f^{1/3} = \sigma_0 N/A$ (which is in the range $1/N^{1/3}$ to 1) and $\overline{\sigma} = A$. The force exerted upon the target is $\frac{\Phi}{c}AE$. For our numerical data it is smaller than $\simeq 10^{-7}A^2\,dyn$. In usual experimental conditions, for any reasonably large area A and flux density Φ, it is difficult to achieve these conditions ($E < \Phi A\hbar c/v$) and to measure this force. It is worth noting that the number of atoms in each domain $N_d = A^3/(\sigma_0 N)^3$ is $10^9 A^3$, which is larger than the critical value for minimal domains.

Let us summarize the above discussion. The cross-section should satisfy the condition

$$\overline{\sigma} < A \; . \tag{5.122}$$

On the other hand we have an equilibrium time $\Delta t_{eq} = \hbar/\Delta E$ and a "collision" time $\overline{\tau} = 1/\Phi\overline{\sigma}$. For a coherent scattering $\overline{\sigma} = \sigma_0 N/f$ and we should have the inequality $\overline{\tau} > \Delta t_{eq}$. These two conditions lead to

$$f > \frac{\sigma_0 N}{A} \; , \; f > \frac{\Phi\hbar}{\Delta E}\sigma_0 N \; , \tag{5.123}$$

where $\Delta E = (p/m + v)p$. Let us assume $\sigma_0 N/A > \frac{\Phi\hbar}{\Delta E}\sigma_0 N$ (which means $\Delta E > 10^{-3}AeV$). Then, $f > \sigma_0 N/A$. If, on the contrary, $\sigma_0 N/A < \frac{\Phi\hbar}{\Delta E}\sigma_0 N$ (very low energies), then $\Delta E = vp$ and $f > \frac{\Phi\hbar c}{vE}\sigma_0 N > \sigma_0 N/A$. In both cases $f > \sigma_0 N/A$. For neutrinos, this inequality means $f > 10^{-22}/A$, which is much smaller than $1/N$. Therefore, $f = 1/N$ (since the cross-section should be maximal within given constraints) and $\overline{\sigma} = \sigma_0 N^2$. We have a coherent scattering in the whole crystal, and a measurable force acting upon the crystal. For electromagnetic radiation the inequality $f > \sigma_0 N/A$ means $f > 10^{-3}/A$, *i.e.* $f = 10^{-3}/A$ (which is in the range $1/N$ to 1) and domains with a number of atoms $N_d = 1/f < 10^3 A$. These domains are inconsistent; they reduce to $N_d = 1$ and $f = 1$, which means an incoherent scattering with a cross-section $\overline{\sigma} = \sigma_0 N$ and an unmeasurable force acting upon the crystal.

For a coherent diffraction $\overline{\sigma} = \sigma_0 N/f^{1/3}$ and we should have the inequality $\overline{\tau} < \Delta t_{eq}$; together with equation (5.122) it leads to

$$f^{1/3} > \frac{\sigma_0 N}{A} \ , \ f^{1/3} < \frac{\Phi\hbar}{\Delta E}\sigma_0 N \ . \qquad (5.124)$$

Therefore, $\sigma_0 N/A < \frac{\Phi\hbar}{\Delta E}\sigma_0 N$, *i.e.* the energies should be very small. Therefore, $\Delta E = Ev/c$ and $E < \frac{\Phi\hbar c}{v} A = 2A\,keV$. For neutrinos this inequality does not apply. It follows $f^{1/3} = \sigma_0 N/A$ for electromagnetic radiation, which is in the range $1/N < f < 1$. The cross-section is $\overline{\sigma} = A$ (which is higher than the incoherent cross-section $\sigma_0 N$) and the force acting upon the crystal is very small (unmeasurable), as a consequence of the small energy. The coherent diffraction is favoured in comparison to incoherent scattering, since it has a higher cross-section.

We conclude that neutrinos may suffer a coherent scattering in the whole sample, and produce a measurable force on the target. Energetic electromagnetic radiation may suffer a completely incoherent scattering, producing a force which is unmeasurable. Soft X-rays suffer a coherent diffraction within incoherent domains, with an unmeasurable force.

6 Neutron Scattering

6.1 Single-particle scattering

Neutrons are produced by radioactive sources and nuclear reactors with a wide range of energies E, from fractions of eV's (thermal neutrons) to MeV's, and a typical flux density $\Phi = 10^{12}/cm^2 \cdot s$. They are massive (mass m of the order $10^3 m_e$, electron mass $m_e = 10^{-27}g$) and non-relativistic, in contrast to the electromagnetic radiation, or neutrinos. The neutrons interact, by a contact interaction, with atomic nuclei and by their magnetic moment with atomic electrons (their interaction depends on spin). Consequently, they generate a quantum-mechanical motion of the atomic targets, as expected. They are able to probe not only the positions of the atomic nuclei (and atoms), but also elementary excitations like phonons, magnons, etc, by inelastic collisions.

The neutron interaction with the nuclei is strong, so the Born approximation or a first-order perturbation scattering is not appropriate. However, for contact interaction (a δ-potential) we have an exact scattering formula, given in the Scattering, Collisions chapter. Specifically, the cross-section is given by

$$\sigma_0 = 4\pi \left| \frac{1}{-\frac{1+u}{u}a^{-1} - ik} \right|^2 , \tag{6.1}$$

where $u = ma^2 U_0 / 2\pi\hbar^2$, U_0 is the strength of the interaction, a is the range of the potential and k is the wavevector ($\hbar = 10^{-27} erg \cdot s$ is Planck's constant). This formula is valid for $ak \lesssim 1$ (resonant scattering); otherwise, k is missing in equation (6.1). $au/(1+u)$ is the scattering length (Wigner's formula[1]). The cross-section is uniform.

[1]See M. Apostol, *Theory of Quanta*, Nova, NY (2019) and the References therein.

The interaction extends over a range of the order of the nucleus dimension, *i.e.* a is a few fermi ($1fm = 10^{-13}cm$), the localization energy is $\hbar^2/ma^2 \simeq 10^{-5}eV$ and $U_0 \gg \hbar^2/ma^2$. The reduced interaction strength is much larger than unity, such that $\sigma_0 \simeq 10^{-24}$. For small u the length a is larger, and the cross-section has the same order of magnitude. $U_0 a^3 \delta(\boldsymbol{r})$ is a pseudopotential, originating at $\boldsymbol{r} = 0$.[2]

6.2 Scattering parameters

Let us consider the scattering of a neutron with momentum $\boldsymbol{p}$ with a nucleus with mass M and momentum $\boldsymbol{P}$. After collision the neutron has a momentum $\boldsymbol{p}'$ and the atom has a momentum $\boldsymbol{P}'$. The momentum and energy conservation reads

$$\begin{gathered} \boldsymbol{P}' + \boldsymbol{p}' = \boldsymbol{P} + \boldsymbol{p} \ , \\ \frac{P'^2}{2M} + \frac{p'^2}{2m} = \frac{P^2}{2M} + \frac{p^2}{2m} \ . \end{gathered} \tag{6.2}$$

These are four equations with six unknowns ($\boldsymbol{P}'$ and $\boldsymbol{p}'$). Besides the azimuthal angles we introduce two free parameters, namely the angle θ between $\boldsymbol{p}$ and $\boldsymbol{p}'$ and the angle ψ between $\boldsymbol{p}$ and $\boldsymbol{P}$; the angle between $\boldsymbol{p}'$ and $\boldsymbol{P}$ is $\psi - \theta$. From the first equation (6.2) we get

$$P'^2 = P^2 + p^2 + p'^2 + 2Pp\cos\psi - 2pp'\cos\theta - 2Pp'\cos(\psi - \theta) \ , \tag{6.3}$$

which, introduced in the second equation (6.2), leads to

$$\begin{gathered} \left(1 + \frac{m}{M}\right) p'^2 - 2\frac{m}{M}\left[p\cos\theta + P\cos(\psi - \theta)\right] p' + \\ + 2\frac{m}{M} Pp\cos\psi - \left(1 - \frac{m}{M}\right) p^2 = 0 \ . \end{gathered} \tag{6.4}$$

Since $m/M \ll 1$, we expand the solution in powers of m/M, such that

$$p' = p - \frac{m}{M}(1 - \cos\theta)p + \frac{m}{M}\left[\cos(\psi - \theta) - \cos\psi\right] P + \dots \ . \tag{6.5}$$

[2] E. Fermi, "Motion of neutrons in hydrogeneous substances", Ricerca Scientifica **7** 13 (1936). Higher-order theory of perturbation generates pseudopotentials which depend on the incident energy.

We can see that $p' \simeq p$ and, from equation (6.4), $P' \simeq 2p\sin\frac{\theta}{2}$ for $p \gg P$ and $P' \simeq P$ for $p \ll P$. The momentum transfer to the atom along the direction of the projectile is $\Delta p = (1-\cos\theta)p$, such that the force is $F = \Phi\sigma\Delta p$, where Φ is the incident flux density and σ is the cross-section. Since Δp is of the order p, the force is $F = \Phi\sigma\sqrt{2mE}$, where $E = p^2/2m$ is the energy of the neutron.

The energy transfer to the nucleus is

$$\Delta E = \frac{p^2}{2m} - \frac{p'^2}{2m} \simeq (1-\cos\theta)\frac{p^2}{M} + [\cos\psi - \cos(\psi-\theta)]\, vp \ , \quad (6.6)$$

where $v = P/M$ is the velocity of the atom before collision. The order of magnitude of this energy transfer is given by $\Delta E = (p/M + v)p$.

The mean velocity of atoms in condensed matter at temperatures of the order of the room temperature $T = 300K$ is $v = \sqrt{T/M}$. For $M = 10^5 m_e$, where $m_e \simeq 10^{-27}g$ is the electron mass, we get $v \simeq 10^4 cm/s$ and the momentum $P = Mv = 10^{-18} g \cdot cm/s$ ($1K = 1.38 \times 10^{-16} erg$). A comparison between p/M and v leads to an energy $E = p^2/m = P^2/m$ of the incident particle, *i.e.* $E = 10^{-12} erg \simeq 1eV$. It follows that we can use $\Delta E = p^2/M$ for $E > 1eV$ and $\Delta E = vp$ for $E < 1eV$.

6.3 Macroscopic target. Quantum scattering

By a macroscopic target we understand a sample of condensed matter with $N \gg 1$ "atoms" (atomic nuclei, molecules, unit cells). The atoms have a thermal velocity $v \simeq \sqrt{T/M}$, where T is the temperature and M is the mass of the atom. In an elementary act of collision the atom receives a momentum of the order of the momentum $p = \sqrt{mE}$ of the incident neutron, where m is the neutron mass and E is the neutron energy. The atom is perturbed by an energy $\Delta E \simeq p^2/M + vp$, such that it recovers its thermal equilibrium in time $\Delta t_{eq} \simeq \hbar/\Delta E$. For instance, at room temperature $= 300K$ ($1K = 1.38 \times 10^{-16} erg$, $1eV = 1.16 \times 10^4 K$) the velocity of the atom is $v \simeq 10^4 cm/s$ ($M \simeq 10^5 \times 10^{-27} g$). For $E = 1MeV$ the neutron momentum is $p \simeq 10^{-15} g \cdot cm/s$, and its wavelength is $\lambda \simeq 10^{-12} cm$ ($1eV = 1.6 \times 10^{-12} erg$). The equilibrium time is $\Delta t_{eq} \simeq 10^{-19} s$ ($\Delta E \simeq 10^{-8} erg$).

On the other hand, there exist $\Phi\sigma$ collisions per unit time, where Φ is the flux density of incident particles and σ is the cross-section of the target. The time between two succesive collisions is $\tau = 1/\Phi\sigma$ (collision time). If $\tau > \Delta t_{eq}$, the atoms recover their equilibrium before the next collisions, such that the incident particles see the target as a whole. In that case we need to use the wavefunction of the entire sample. We call it a (macroscopic) quantum scattering. On the contrary, if $\tau < \Delta t_{eq}$, the incident particles see each atom individually, as independent targets, each with its own wavefunction. We call that a classical scattering.

Let us assume first the case of a quantum scattering $\tau > \Delta t_{eq}$, *i.e.*

$$\frac{1}{\Phi\sigma} > \frac{\hbar}{\Delta E} \ . \tag{6.7}$$

We need to compute the cross-section σ. Of course,

$$\sigma < A \ , \tag{6.8}$$

where A is the area of the target. Indeed, the number of incident particles per unit time is $\simeq \Phi A$, while the number of scattered particles per unit time is $\Phi\sigma$. The conservation of the number of particles requires $\Phi\sigma < \Phi A$; the difference (if any) is the number of transmitted particles.

Let us consider first a crystal, with atomic positions $\boldsymbol{r}_i^0, = 1, 2, ...N$, and displacements from these positions $\boldsymbol{v}_i$. The wavefunction of a crystal is

$$\Psi_{\boldsymbol{K}}(\boldsymbol{r}) = \sqrt{1-g^2}\Phi_{\boldsymbol{K}}(\boldsymbol{r}) + g\psi(\boldsymbol{r}) \ , \tag{6.9}$$

where

$$\Phi_{\boldsymbol{K}}(\boldsymbol{r}) = \textstyle\sum_i e^{i\boldsymbol{K}\boldsymbol{r}_i^0}\varphi(\boldsymbol{r}-\boldsymbol{r}_i^0) \tag{6.10}$$

and

$$\psi(\boldsymbol{r}) = \sum_i e^{i\chi_i}\varphi(\boldsymbol{r}-\boldsymbol{r}_i^0-\boldsymbol{v}_i) \ ; \tag{6.11}$$

$\boldsymbol{K}$ is the quasi-momentum of the crystal, $e^{i\chi_i}$ are random phase factors and $\varphi(\boldsymbol{r}-\boldsymbol{r}_i)$, $\boldsymbol{r}_i = \boldsymbol{r}_i^0$ or $\boldsymbol{r}_i = \boldsymbol{r}_i^0+\boldsymbol{v}_i$, are (orthonormal) wavefunctions localized on the positions $\boldsymbol{r}_i$ (see the Neutrino Scattering chapter);

these wavefunctions may include internal degrees of freedom. The weight factor g is given by

$$g \simeq \frac{2.2(T/\omega_D)}{\left[1+4.8(T/\omega_D)^2\right]^{1/2}} \ , \tag{6.12}$$

where ω_D is the Debye temperature. The wavefunction $\Phi_{\boldsymbol{K}}(\boldsymbol{r})$ corresponds to a motion of the crystal as a whole, with atoms in fixed, regular, relative positions; we call it the coherent wavefunction. The wavefunction $\psi(\boldsymbol{r})$, which includes the random phase factors $e^{i\chi_i}$, is an incoherent wavefunction. These two wavefunctions are orthogonal to each other ($\boldsymbol{v_i} \neq 0$).

The interaction hamiltonian is

$$H = a^3 h \sum_i \delta(\boldsymbol{r}-\boldsymbol{r}_i) \ , \tag{6.13}$$

where h depends on the internal coordinates of the target particle with dimension a placed at position $\boldsymbol{r}_i$ (contact interaction). The number of transitions per unit time is given by

$$w = \frac{2\pi}{\hbar} \left|H_{fi}\right|^2 \delta(E_f - E_i) \ , \tag{6.14}$$

where $E_{f,i}$ are the energies of the final/initial states and H_{fi} are the matrix elements of the interaction between these states. We assume an incident neutron wavefunction $\frac{1}{\sqrt{V}}e^{i\boldsymbol{kr}}$ and an emergent (scattered) neutron wavefunction $\frac{1}{\sqrt{V}}e^{i\boldsymbol{k}'\boldsymbol{r}}$, where V is the volume of the crystal and $\boldsymbol{k}$, $\boldsymbol{k}'$ are the incident/emergent wavevectors. The normalization to unity of the initial wavefunction $\frac{1}{\sqrt{V}}e^{i\boldsymbol{kr}}\Phi_{\boldsymbol{K}}(\boldsymbol{r})$, or $\frac{1}{\sqrt{V}}e^{i\boldsymbol{kr}}\psi$, requires a factor $\sqrt{a^3}$ for the functions $\varphi(\boldsymbol{r}-\boldsymbol{r}_i)$, which amounts to $\left|\varphi(\boldsymbol{r}_j-\boldsymbol{r}_i)\right|^2 = \delta_{ij}$.

The matrix element of the interaction between the wavefunctions $\frac{1}{\sqrt{V}}e^{i\boldsymbol{kr}}\Phi_{\boldsymbol{K}}(\boldsymbol{r})$ and $\frac{1}{\sqrt{V}}e^{i\boldsymbol{k}'\boldsymbol{r}}\Phi_{\boldsymbol{K}'}(\boldsymbol{r})$ (coherent scattering) is

$$\begin{aligned} H_{fi} &= \tfrac{a^3 h_{fi}}{V} \textstyle\sum_i e^{i(\boldsymbol{K}-\boldsymbol{K}')\boldsymbol{r}_i^0} e^{i(\boldsymbol{k}-\boldsymbol{k}')\boldsymbol{r}_i^0} = \\ &= \tfrac{a^3 h_{fi}}{V} N \delta_{\boldsymbol{K}'+\boldsymbol{k}',\boldsymbol{K}+\boldsymbol{k}} \ . \end{aligned} \tag{6.15}$$

We can see that the difference in the neutron momenta is compensated by the quasi-momentum of the crystal. The energies are $E_f = \hbar^2 K'^2/2M_0 + \hbar^2 k'^2/2m$, $E_i = \hbar^2 K^2/2M_0 + \hbar^2 k^2/2m$, where M_0 is the mass of the crystal (neglecting inelastic processes). Since the mass M_0 is very large, we may neglect the crystal energy in the energy conservation. By integrating over the final states we get the total transition rate

$$w = \frac{\left|a^3 h_{fi}\right|^2}{V} N^2 \frac{2mk}{\pi\hbar^3} \tag{6.16}$$

and the total coherent cross-section (by dividing to the flux density $\Phi = \hbar k/mV$)

$$\sigma_{coh} = \left|a^3 h_{fi}\right|^2 N^2 \frac{2m^2}{\pi\hbar^4} \ ; \tag{6.17}$$

it can be written as

$$\sigma_{coh} = N^2 \sigma_0 \ , \tag{6.18}$$

where σ_0 is the cross-section of a single particle. It corresponds to $u \ll 1$ in Fermi's equation (6.1), as expected (first order of the perturbation theory). For higher-orders of the perturbation theory the structure of the cross-section ($\sim N^2$) is preserved (and inelastic collisions may be included). Therefore, we may use equation (6.18) with $\sigma_0 = 10^{-24} cm^2$.

The matrix elements of the interaction between the wavefunctions $\frac{1}{\sqrt{V}} e^{i\boldsymbol{kr}} \psi(\boldsymbol{r})$ and $\frac{1}{\sqrt{V}} e^{i\boldsymbol{k}'\boldsymbol{r}} \psi(\boldsymbol{r})$ is

$$H_{fi} = \frac{a^3 h_{fi}}{V} \sum_i e^{i(\boldsymbol{k}-\boldsymbol{k}')\boldsymbol{r}_i^0} \left(1 + i(\boldsymbol{k}-\boldsymbol{k}')\boldsymbol{v}_i + ...\right) \ , \tag{6.19}$$

where

$$\boldsymbol{v}_i = \frac{1}{\sqrt{N}} \sum_{\boldsymbol{q}} e^{i\boldsymbol{q}\boldsymbol{r}_i^0} \boldsymbol{v}_{\boldsymbol{q}} \tag{6.20}$$

is the phonon field. The first term in equation (6.19) corresponds to a coherent scattering ($\boldsymbol{v}_i = 0$, already included in equation (6.15)), so it should be left aside. We get the matrix element

$$H_{fi} = -\frac{a^3 h_{fi}}{V} \sqrt{N} \left(i\boldsymbol{q}\boldsymbol{v}_{\boldsymbol{q}}\right) \delta_{\boldsymbol{k}',\boldsymbol{k}+\boldsymbol{q}} \ . \tag{6.21}$$

We can see that the incoherent scattering excites phonons, which ensure the conservation of the energy and the momentum, *i.e.* $E_f = \hbar^2 k'^2/2m \pm \hbar\Omega_{\boldsymbol{q}}$, $E_i = \hbar^2 k^2/2m$, $\boldsymbol{k}' = \boldsymbol{k} + \boldsymbol{q}$, where $\Omega_{\boldsymbol{q}}$ is the frequency of the phonon. A phonon wavevector $\boldsymbol{q}$ may arise either from the creation of a phonon with wavevector $\boldsymbol{q}$ or from the destruction of a phonon with wavevector $-\boldsymbol{q}$, whence the two phonon frequencies $\pm\Omega_{\boldsymbol{q}}$. The transition rate becomes

$$w = \frac{2\pi}{\hbar} \frac{\left|a^3 h_{fi}\right|^2}{V^2} N \left|(\boldsymbol{q}\boldsymbol{v}_{\boldsymbol{q}})\right|^2 \delta(\hbar^2 k'^2/2m \pm \hbar\Omega_{\boldsymbol{q}} - \hbar^2 k^2/2m) \ , \quad (6.22)$$

where $\boldsymbol{q}$ is determined by $\boldsymbol{q} = \boldsymbol{k}' - \boldsymbol{k}$. The factor $\left|(\boldsymbol{q}\boldsymbol{v}_{\boldsymbol{q}})\right|^2$ should be averaged over thermal phonon states. It is reminiscent of the Debye-Waller factor (see the Neutrino Scattering chapter). For $T \ll \Theta_D$, where T is the temperature and Θ_D is the Debye temperature, the average $\overline{\left|(\boldsymbol{q}\boldsymbol{v}_{\boldsymbol{q}})\right|^2}$ is very small. For $T \gg \Theta_D$ the average is of the order $q^2 T/M\Omega_q^2 = T/Mc_s^2$, where c_s is the mean phonon velocity. Also, we need to multiply by the number of final states $Vk'^2 dk' do/(2\pi)^3$ in the solid angle do and the wavevector range k' to $k' + dk'$, and integrate over these states. The cross-section is obtained by dividing by the flux density $\hbar k/mV$. We get the incoherent cross-section, which is proportional to N. It can be written as

$$\sigma_{incoh} = N\sigma_0^{ph} \ , \quad (6.23)$$

where σ_0^{ph} is the single-particle incoherent cross-section arising from phonons. It is reduced by the factor $\overline{\left|(\boldsymbol{q}\boldsymbol{v}_{\boldsymbol{q}})\right|^2}$ in comparison to σ_0. The total cross-section of the crystal is

$$\sigma = (1 - g^2)\sigma_{coh} + g^2\sigma_{incoh} \ , \quad (6.24)$$

where we may neglect the incoherent cross-section in comparison to the coherent cross-section. The total force acting upon the crystal (in the forward direction) is

$$F = \Phi\sigma_0 p = \Phi\sigma_0\sqrt{2mE} \ . \quad (6.25)$$

For a target with atomic positions placed randomly (*e.g.*, amorphous solids, liquids, gases) the coherent wavefunction $\Phi_{\boldsymbol{K}}(\boldsymbol{r})$ is absent in equation (6.9); we are left with the incoherent wavefunction $\psi(\boldsymbol{r})$ and incoherent scattering arising from phonons.

6.4 Coherence domains. The principle of maximal cross-section

The coherent quantum scattering cross-section $\sigma = N^2\sigma_0$ for $\sigma_0 = 10^{-24}cm^2$ (neutrons) is larger than any reasonable area A of macroscopic targets (inequality (6.8) is not satisfied). The scattering is a statistical process. The increase of the cross-section is a non-equilibrium process. The target responds to this increase by developing uncorrelated coherence domains, which increase the entropy. The increase of entropy should be minimal, under given conditions, so the cross-section should be maximal.

Let us assume that n_d domains occur in the target, each with a (mean) volume v and N_d atoms, such that $n_d = N/N_d$. By analogy with the high-purity crystals we use the fraction $f = 1/N_d$. It varies between $f = 1/N$, when we have only one domain in the whole target, and $f = 1$, when the target is fractured in N "atomic" domains.

A constant phase factor over each domain is equivalent with the assumption of continuous, uniform domains. This assumption is valid as long as the dimension of the domain is larger than $(v/v_0)d$, where $v = d^3$ is the volume associated to the mean inter-atomic distance d and $v_0 = a_0^3$ is the volume of the atom with dimension a_0. Therefore, the number N_d of atoms in each domain should be larger than $(v/v_0)^3$ $(N_d > (v/v_0)^3)$. For instance, for $d = 4\text{Å}$ and $a_0 = 1\text{Å}$, the inequality $N_d \gtrsim 10^6$ should be satisfied. If not, the domains reduce to atoms (see the Neutrino Scattering chapter).[3]

[3] The variation of the volume due to the atomic extent of the particles is

$$\frac{\delta V}{V} = \frac{a_0^3}{a^3} ,$$

where a_0 is the atomic dimension and a is the mean separation distance between the particles; it follows $\delta V = (a_0/a)^3 a^3 N$, where N is the number of particles. On the other hand, the volume variation of a domain is $\delta V = Sa = N^{2/3}a^3$, where S is the area of the surface which encloses the domain. These two relations lead to $N^{1/3} = (a/a_0)^3$, which indicates a minimal dimension $R = N^{1/3}a = (a/a_0)^3 a$ for a domain. Making use of typical values $a_0 = 1\text{Å}$ and $a = 4\text{Å}$, we get $R = 64a = 256\text{Å}$ and $N = (R/a)^3 \simeq 10^5$. This is an estimation of the size of the domains (M. Apostol, *Statistical Physics*, Cambridge Scholars Publishing, Newcastle upon Tyne (2021)).

The scattering amplitude is

$$S=\sum_{a=1}^{n_d} e^{i\chi_a}\left(H_{fi}\right)_a \quad , \tag{6.26}$$

where $e^{i\chi_a}$ are random phase factors. By averaging the squared scattering amplitude over the phase factors, we get the coherent cross-section

$$\overline{\sigma}_{coh}=n_d N_d^2\sigma_0=N_d N\sigma_0=\frac{N\sigma_0}{f} \ . \tag{6.27}$$

For $f=1/N$ we get the coherent cross-section $N^2\sigma_0$ of the whole crystal, for $f=1$ we get the incoherent cross-section $N\sigma_0$. In this latter case the cross-section reduces to the incoherent cross-section arising from phonons. A similar procedure leaves the incoherent cross-section arising from phonons unchanged ($\overline{\sigma}_{incoh}=n_d N_d\sigma_0^{ph}=N\sigma_0^{ph}=\sigma_{incoh}$).

Let us examine the conditions of validity of a coherent neutron scattering in a crystal (we leave aside the weight factor $\sqrt{1-g^2}$). The inequalities (6.7) and (6.8) should be satisfied with $\overline{\sigma}_{coh}$. This leads to

$$f>\frac{N\sigma_0}{A} \ , \ f>\frac{\hbar\Phi}{\Delta E}N\sigma_0 \ . \tag{6.28}$$

Let us first assume $N\sigma_0/A>\hbar\Phi N\sigma_0/\Delta E$, *i.e.* $\Delta E>10^{-15}Aerg$ ($\Phi=10^{12}/cm^2\cdot s$); then, according to the principle of the maximal cross-section, $f=N\sigma_0/A$ and $\overline{\sigma}_{coh}=A$. If, on the contrary, $N\sigma_0/A<\hbar\Phi N\sigma_0/\Delta E$, we may take $\Delta E=vp$; then $f=\hbar\Phi N\sigma_0/\sqrt{mET/M}>N\sigma_0/A$ (and $\overline{\sigma}_{coh}<A$). In both cases, $N_d=1/f\leq A/N\sigma_0$. For large N, this is a small quantity. For instance, $0.1mol$ with $N=10^{22}$ gives $N_d\leq 10^2A$. The domains reduce in this case to atoms, and $f=1$. We are left with an incoherent scattering arising from phonons ($\sigma_{incoh}=N\sigma_0^{ph}$). A coherent scattering of neutrons is not possible in a large crystal. A similar situation occurs in targets with atoms placed randomly. The phonon single-particle cross-section is $\sigma_0^{ph}=\overline{|(\boldsymbol{q}\boldsymbol{v}_{\boldsymbol{q}})|^2}\sigma_0$. The largest value of the thermal average $\overline{|(\boldsymbol{q}\boldsymbol{v}_{\boldsymbol{q}})|^2}$ is of the order T/Mc_s^2. For $T=300K$ and $c_s=10^6cm/s$ (mean sound velocity) we get $T/Mc_s^2\simeq 10^{-4}$ and $\sigma_0^{ph}\simeq 10^{-28}cm^2$ ($\sigma_0=10^{-24}cm^2$). For $N=10^{22}$ the incoherent cross-section is $\sigma_{incoh}=10^{-6}cm^2$ and the

force acting upon the crystal (equation ((6.4)) is $F = 10^{-9} dyn$ for $\Phi = 10^{12}/cm^2 \cdot s$ and $E = 1MeV$. This is an unmeasurable force (for lower energies the force is smaller). The numerical data used above correspond to Weber's experiments,[4] and other similar experiments,[5] with neutrino scattering by a sapphire crystal.

6.5 Classical scattering

If the collision time τ is shorter than the equilibrium time Δt_{eq},

$$\tau < \Delta t_{eq} \ , \ \frac{1}{\Phi\sigma} < \frac{\hbar}{\Delta E} \tag{6.29}$$

the incident particles (neutrons) see independent target particles (atomic nuclei, atoms, molecules, unit cells), each with its own wavefunction. This is a classical scattering. We note that it is valid for low energies ($\Delta E < \hbar\Phi\sigma$).

Besides internal degrees of freedom, the initial wavefunction is $\frac{1}{\sqrt{V}}e^{i\boldsymbol{kR}}\frac{1}{\sqrt{V}}e^{i\boldsymbol{KR}_c}$ and the final wavefunction is $\frac{1}{\sqrt{V}}e^{i\boldsymbol{k}'\boldsymbol{R}}\frac{1}{\sqrt{V}}e^{i\boldsymbol{K}'\boldsymbol{R}_c}$, where $\boldsymbol{k}$, $\boldsymbol{k}'$ are the initial/final neutron wavevector and $\boldsymbol{K}$, $\boldsymbol{K}'$ are the initial/final wavevector of the center of mass of the target; $\boldsymbol{R}_c$ denotes the coordinate of the center of mass. The matrix elements of the interaction

$$H = a^3 h \sum_i \delta(\boldsymbol{R} - \boldsymbol{R}_i) \ , \tag{6.30}$$

(equation (6.13)), where $\boldsymbol{R}_i$ are the coordinates of the atoms, are

$$H_{fi} = \frac{a^3 h_{fi}}{V} \sum_i e^{i(\boldsymbol{k}-\boldsymbol{k}')\boldsymbol{r}_i} \delta_{\boldsymbol{K}'+\boldsymbol{k}',\boldsymbol{K}+\boldsymbol{k}} \ , \tag{6.31}$$

[4] J. Weber, "Method for observation of neutrinos and antineutrinos", Phys. Rev. **C31** 1468 (1985); J. Weber, "Apparent observation of abnormally large coherent scattering cross sections using keV and MeV range antineutrinos, and solar neutrinos", Phys. Rev. **D38** 32 (1988).

[5] I. Cruceru, G. Nicolescu and O. G. Duliu, "A dedicated torsion balance to detect neutrinos by coherent scattering on high Debye temperature monocrystals", Int. J. Mod. Phys. **A23** 1719 (2008); M. Cruceru, G. Nicolescu, O. G. Duliu and I. Cruceru, "Detection of solar neutrinos by coherent scattering in high-Debye temperature monocrystals", Int. J. Mod. Phys. **A26** 2773 (2011); G. Nicolescu, "A heuristic approach to the detection of solar neutrinos", J. Phys. **G40** 055201 (2013).

where $\boldsymbol{r}_i = \boldsymbol{R}_i - \boldsymbol{R}_c$ are the baricentric coordinates. We can see that the momentum is conserved, including the momentum of the center of mass. Also, the form-factor

$$F = \sum_i e^{i(\boldsymbol{k}-\boldsymbol{k}')\boldsymbol{r}_i} \tag{6.32}$$

appears in this equation. For a crystal the form-factor is $F = N\sum_{\boldsymbol{g}} \delta_{\boldsymbol{k}',\boldsymbol{k}+\boldsymbol{g}}$, where $\boldsymbol{g}$ are the reciprocal vectors of the crystal. The scattering exhibits diffraction (Bragg) peaks (for elastic scattering; the coherence considerations are valid also for inelastic scattering). For atoms placed randomly $\boldsymbol{g} = 0$; we have only a central (forward) diffraction peak. The diffraction peaks are placed on the Ewald sphere, $g = 2k\sin\frac{\theta}{2}$, where θ is the scattering angle.

The number of transitions per unit time for a $\boldsymbol{g}$-peak, in the wavevector interval $k' + dk'$, k' and the solid angle do, is

$$\begin{gathered} dw_{\boldsymbol{g}} = \frac{2\pi}{\hbar}\left(\frac{a^3 h_{fi}}{V}\right)^2 N^2\delta(\hbar^2 K'^2/2M_0 + \hbar^2 k'^2/2m- \\ -\hbar^2 K^2/2M_0 - \hbar^2 k^2/2m)\frac{Vk'^2 dk' do}{(2\pi)^3} \quad , \end{gathered} \tag{6.33}$$

where M_0 is the mass of the target and $\boldsymbol{k}' = \boldsymbol{k}+\boldsymbol{g}$ (elastic scattering); we can see that it is proportional to N^2; it is a coherent diffraction. Since $M_0 \gg m$, we may neglect the target energy in equation (6.33). The integration over the solid angle extends to $\Delta o = (2\pi/L)^2/k^2$, where L is the dimension of the sample. Integrating over k' and dividing by the incident flux density $\hbar k/mV$, we get the cross-section of one peak

$$\sigma_{\boldsymbol{g}} = N^{4/3}(a^3 h_{fi})^2\frac{m^2}{2\pi^2\hbar^4}\cdot\frac{4\pi^2}{d^2k^2} \quad , \tag{6.34}$$

where d is the mean inter-atomic distance. The cross-section of the coherent diffraction goes in fact as $N^{4/3}$. The number of peaks is $4\pi k^2/(2\pi/d)^2 = d^2k^2/\pi \gg 1$, such that the total diffraction cross-section is

$$\sigma_d = N^{4/3}(a^3 h_{fi})^2\frac{2m^2}{\pi\hbar^4} \tag{6.35}$$

(for $d^2k^2/\pi \gg 1$); it can be written as

$$\sigma_d = N^{4/3}\sigma_0 \quad , \tag{6.36}$$

where σ_0 is the single-particle cross-section (equation (6.17)). For the central peak σ_0 is multiplied by the factor $\pi/d^2k^2 \simeq (\lambda/d)^2$, where d is the mean inter-atomic distance and λ is the wavelength of the neutron.

This coherent-diffraction cross-section is affected by the Debye-Waller factor and the diffuse scattering (temperature effects, see the Neutrino Scattering chapter), which diminish the cross-section.

For uncorrelated domains of coherent diffraction the cross-section is

$$\overline{\sigma}_d = n_d N_d^{4/3} \sigma_0 = \frac{N\sigma_0}{f^{1/3}} ; \tag{6.37}$$

it goes from $\overline{\sigma}_d = N^{4/3}\sigma_0$ $(f = 1/N)$, for a one-domain target, to the incoherent cross-section $\overline{\sigma}_d = N\sigma_0$ $(f = 1)$, for atomic domains.

Let us examine the validity conditions for a coherent diffraction. From inequality (6.29) with $\overline{\sigma}_d$ and $\overline{\sigma}_d < A$ we get

$$\frac{N\sigma_0}{A} < f^{1/3} < \frac{\hbar\Phi}{\Delta E} N\sigma_0 , \tag{6.38}$$

which requires $\Delta E < \hbar\Phi A$ (low energies). Consequently, $f^{1/3} = N\sigma_0/A$ (in the validity range $1/N$ to 1) and $\overline{\sigma}_d = A$. The number of atoms in each domain $N_d = 1/f = 10^6 A^3$ (for $N = 10^{22}$ and $\sigma_0 = 10^{-24} cm^2$), which requires $A \gg 1$, for the domains to be consistent. Under such conditions, we conclude that coherent diffraction may be valid for low-energy neutrons in large crystals. The same is valid for the single peak in targets with randomly placed atoms, where the cross-section is smaller, so these conditions are also fulfilled. We note that the coherent diffraction is favoured in comparison to the incoherent scattering (at low energies), since the cross-section $\overline{\sigma}_d = A$ is larger than $\sigma_{incoh} = N\sigma_0^{ph}$. The force acting upon the crystal is $F = \Phi A p$ (equation (6.4)). Since $\Delta E < \hbar\Phi A$, we may take $\Delta E = vp$, which implies $F < \hbar\Phi^2 A^2/v$ $(E < (\Phi\hbar A)^2 M/2mT)$. For $\Phi = 10^{12}/cm^2 \cdot s$ and $T = 300K$, this inequality is $F < 10^{-7} A^2\, dyn$. For any reasonably large area A and flux density Φ it is difficult to achieve the condition $E < (\Phi\hbar A)^2 M/2mT$ and to measure this force.

6.6 Macroscopic quantum-mechanical scattering

6.6.1 Introduction

In two papers published in 1985 and 1988 Weber claimed that neutrinos (antineutrinos), from various sources like tritium, nuclear reactors and Sun, could be detected by their coherent scattering by a perfect, stiff, sapphire monocrystal with a high Debye temperature (mounted on a torsion balance and equilibrated by a lead dummy).[6] The coherent cross-section would be $\sigma = N^2\sigma_0$, where N is the number of unit cells in the target and σ_0 is the cross-section of a single unit cell (particle, *e.g.*, atomic nucleus). Such a highly enhanced cross-section $\sim N^2$ would give rise to a measurable force upon a torsion balance. Weber's claims have been criticized both on theoretical and experimental grounds, the main objection being that the form-factor would reduce appreciably the cross-section, and, on the other hand, such a coherence effect is not observed in $X-$, gamma rays or neutron scattering.[7] A discussion of the theoretical objections and negative

---

[6] J. Weber, "Method for observation of neutrinos and antineutrinos", Phys. Rev. **C31** 1468 (1985); "Apparent observation of abnormally large coherent scattering cross sections using keV and MeV range antineutrinos, and solar neutrinos", Phys. Rev. **D38** 32 (1988).

[7] G. F. Bertsch and S. M. Austin, "Comment on "Method for observation of neutrinos and antineutrinos"", Phys. Rev. **C34** 361 (1986); R. Casella, "On the theory of coherent neutrino scattering from crystals", Nuovo Cimento **94** 42 (1986); T. H. Ho, "Comments on the "Method for observation of neutrinos and antineutrinos""Phys. Lett. **B166** 295 (1986); M. N. Butler, "Comment on "Method for observation of neutrinos and antineutrinos"", Phys. Rev. **C35** 1164 (1987); T. Y. Aharanov, F. T. Avignon III, A. Casher and S. Nussinov, "Constraints on anomalous scattering of neutrinos from crystals", Phys. Rev. Lett. **58** 1173 (1987); H. J. Lipkin, "N dependence of coherent radiation from crystals", Phys. Rev. Lett. **58** 1176 (1987); J. D. Fronson and B. C. Jacobs, "Null result for enhanced neutrino scattering in crystals", Phys. Rev. **A46** 2235-2239 (1992); Y. Su, B. R. Heckel, E. G. Adelberger, J. H. Gundlach, M. Harris, G. L. Smith and H. E. Swanson, "New tests of the universality of free fall", Phys. Rev. **D50** 3614 (1994); J. Luo, X. Chen, J. G. Li, S. H. Fan and F. G. Li, "Test of abnormaly large coherent scattering cross section using solar-neutrino", Chin. Phys. Lett. **12** 136 (1995); G. Q. Feng, S. Q. Yang, L. C. Tu and J. Luo, "Improvement of test of solar neutrino coherent scattering with torsion pendullum", Chin. Phys. Lett. **23** 2052 (2006). See

experiments was given by Nicolescu, who presented a positive experiment;[8] indeed, an experiment by Cruceru et al exists, which confirms Weber's prediction for solar neutrinos.[9] The problem is still controversial. We show herein that a coherent scattering of neutrinos may appear in the conditions formulated by Weber and company, as a consequence of a quantum-mechanical treatment of the crystal as a whole (a macroscopic quantum-mechanical scattering).[10] This is a distinctive condition of the neutrino scattering, which is not fulfilled by other projectiles, with a higher cross-section. The main reason for such a behaviour is precisely the extremely small cross-section σ_0 ($10^{-44}cm^2$) of the neutrinos.

For $\sigma_0 = 10^{-44}cm^2$ and $N = 10^{22}$ (less than $0.1\,mol$) the coherent cross-section is $\sigma = 1cm^2$. For a neutrino flux density $\Phi = 10^{12}/cm^2 \cdot s$ the time between two collisions is $\tau = 1/\Phi\sigma = 10^{-12}s$. On the other hand, an atom in thermal equilibrium at room temperature has a velocity of the order $10^4cm/s$. In an elementary act of collision the atom is perturbed from its equilibrium state and receives a momentum of the order $p = E/c$, on average, where E is the energy of the neutrino projectile (and $c = 3\times10^{10}cm/s$ is the speed of light). For $E = 1MeV$ the momentum transfer is of the order $p = 5 \times 10^{-17}g \cdot cm/s$. Consequently, the energy perturbation of the atom with mass M is of the order $\Delta E = p^2/M = 2.5 \times 10^{-11}erg$ for $M = 10^5m_e$, where $m_e \simeq 10^{-27}g$ is the electron mass ($vp = 5 \times 10^{-13}erg \ll p^2/2m$). The time needed for this atom to recover its equilibrium is of the order $\Delta t_{eq} = \hbar/\Delta E \simeq 4 \times 10^{-17}s$ (where $\hbar \simeq 10^{-27}erg \cdot s$ is Planck's constant). We can see that $\tau \gg \Delta t_{eq}$. We can say that the atoms re-

also D. Akimov et al, COHERENT Collaboration, "Observation of coherent elastic neutrino-nucleus scattering", Science **357** 1123 (2017); E. Akhmedov, G. Arcadi, M. Lindner and S. Vogl, "Coherent scattering and macroscopic coherence: implications for neutrino, dark matter and axion detection", JHEP (2018), doi: 10.10007/JHEP10 (2018)045.

[8] G. Nicolescu, "A heuristic approach to the detection of solar neutrinos", J. Phys. G: Nucl. Part. Phys. **40** 055201 (2013).

[9] I. Cruceru, G. Nicolescu and O. G. Duliu, "A dedicated torsion balance to detect neutrinos by coherent scattering on high Debye temperature monocrystals", Int. J. Mod. Phys. **A23** 1719 (2008); M. Cruceru, G. Nicolescu, O. G. Duliu and I. Cruceru, "Detection of solar neutrinos by coherent scattering in high-Debye temperature monocrystals", Int. J. Mod. Phys. **A26** 2773 (2011).

[10] M. Apostol, "Macroscopic quantum-mechanical scattering. Coherent scattering of neutrinos", J. Theor. Phys. **334** (2021).

cover quickly their equilibrium state between two sucessive collisions, and the incident neutrino beam sees the crystal as a whole. Therefore, we need to adopt a quantum-mechanical treatment for the entire crystal. It is worth noting that if the cross-section increases to, say, $\sigma_0 = 10^{-24}cm^2$, as for X-, gamma rays or neutrons, the "collision" time decreases to $\tau = 10^{-32}s$, which is much shorter than the equilibrium time Δt_{eq}, all the other conditions remaining the same. In that case the incident projectile beam sees the crystal as consisting of distinct, independent atoms, such that a coherent scattering ($\sigma = N^2\sigma_0$) for the entire crystal is not possible. The particularity of a coherent scattering suffered by neutrinos in the whole crystal resides precisely in their extremely small cross-section σ_0. On the other hand, a single-particle cross-section $\sigma_0 = 10^{-24}cm^2$ increases considerably the total cross-section, such that we need to reconsider the scattering in this case.

6.6.2 Macroscopic quantum-mechanical scattering

Let us assume a macroscopic target consisting of $N \gg 1$ identical "atoms" (atomic nuclei, molecules, unit cells in a crystal). The interaction with an incident beam of particles can be written as

$$H = a^3 h(\xi) \sum_{i=1}^{N} \delta(\boldsymbol{r} - \boldsymbol{r}_i) \ , \tag{6.39}$$

where a is the range of the single-particle interaction $h(\xi)$, ξ denotes the internal coordinates of the atoms and $\boldsymbol{r}_i$ are the atomic positions. The time between two successive collisions is $\tau = 1/\Phi\sigma$, where Φ is the incident flux density and σ is the total cross-section. In an elementary act of collision an atom receives a momentum of the order of the momentum p of the incident particle. The atom has a thermal velocity $v \simeq \sqrt{T/M}$, where T is the temperature and M is the mass of the atom. The atomic motion is perturbed by an energy of the order $\Delta E = p^2/M + vp$, so it needs a time $\Delta t_{eq} \simeq \hbar/\Delta E$ to recover its equilibrium. Let us assume

$$\tau > \Delta t_{eq} \ ; \tag{6.40}$$

then, the incident particles see the macroscopic target as a whole, and we need to work with the wavefunction of the whole, macroscopic target.

A perfect monocrystal suffers two kinds of motion. One kind is the motion of the crystal as a whole, where all the laticial positions of the atoms $\boldsymbol{r}_i^0$ move by the same distance. The wavefunction corresponding to this motion is

$$\Phi_{\boldsymbol{K}}(\xi;\boldsymbol{r}) = \sum_{i=1}^{N} e^{i\boldsymbol{K}\boldsymbol{r}_i^0}\varphi(\xi;\boldsymbol{r}-\boldsymbol{r}_i^0) \ , \tag{6.41}$$

where $\boldsymbol{K}$ is the quasi-wavevector of the crystal (quasi-momentum $\hbar\boldsymbol{K}$) and $\varphi(\xi;\boldsymbol{r}-\boldsymbol{r}_i^0)$ are (orthonormal) wavefunctions localized on the positions $\boldsymbol{r}_i^0$. We can see that the wavefunction $\Phi_{\boldsymbol{K}}$ has the translational symmetry of the crystal. We call $\Phi_{\boldsymbol{K}}$ a coherent wavefunction. The other type of motion of the crystal corresponds to the thermal motion with atomic displacements $\boldsymbol{v}_i$ ($\boldsymbol{r}_i = \boldsymbol{r}_i^0 + \boldsymbol{v}_i$); the corresponding wavefunction is

$$\psi(\xi;\boldsymbol{r}) = \sum_{i=1}^{N} e^{i\chi_i}\varphi(\xi;\boldsymbol{r}-\boldsymbol{r}_i^0-\boldsymbol{v}_i) \ , \tag{6.42}$$

where $e^{i\chi_i}$ are random phase factors; we call ψ an incoherent wavefunction. Both these wavefunctions are normalized to N; they are orthogonal to each other. The wavefunction of the crystal is

$$\Psi_{\boldsymbol{K}} = \sqrt{1-g^2}\Phi_{\boldsymbol{K}} + g\psi \ , \tag{6.43}$$

where g is a weight coefficient. This coefficient is proportional to the square root of the relative number of vibration states of the crystal, properly normalized. This relative number of phonon states is proportional to

$$I = \frac{1}{\omega_D^3}\int_0^{\omega_D} d\omega \frac{\omega^2}{e^{\hbar\omega/T}-1} \ , \tag{6.44}$$

where ω_D is the Debye frequency and ω is the phonon frequency; the Debye temperature is $\Theta = \hbar\omega_D$. For $T/\Theta \ll 1$ the integral is $I \simeq 2.4(T/\Theta)^3$, while for $T/\Theta \gg 1$ the integral is $I \simeq \frac{1}{2}T/\Theta$. A normalized expression for the relative number of states is $4.8(T/\Theta)^2$

for $T/\Theta \ll 1$ and 1 for $T/\Theta \gg 1$. An interpolation formula is $I \simeq 4.8(T/\omega_D)^2/\left[1+4.8(T/\omega_D)^2\right]$, such that we can take for the weight coefficient

$$g \simeq \frac{2.2(T/\Theta)}{\left[1+4.8(T/\Theta)^2\right]^{1/2}} \ . \tag{6.45}$$

We can see that for high Debye temperatures the main contribution to $\Psi_{\boldsymbol{K}}$ comes from the coherent wavefunction ($g \ll 1$), while for low Debye temperatures the main contribution comes from the incoherent field ($g \to 1$).

Let us assume a wavefunction $\frac{1}{\sqrt{V}}e^{i\boldsymbol{kr}}$ for an incident particle, and an initial (i) wavefunction $\frac{1}{\sqrt{V}}e^{i\boldsymbol{kr}}\Psi_{\boldsymbol{K}}$, where V denotes the volume. The normalization to unity of this wavefunction requires a factor $\sqrt{a^3}$ in the wavefunctions φ, such that the scalar product is $\langle\varphi(\xi;\boldsymbol{r}_j-\boldsymbol{r}_i),\varphi(\xi;\boldsymbol{r}_j-\boldsymbol{r}_i)\rangle^2 = \delta_{ij}$. The matrix elements of the interaction between the wavefunctions $\Phi_{\boldsymbol{K}}$ and ψ are zero. The matrix elements of the interaction between two wavefunctions $\Phi_{\boldsymbol{K}}$ and $\Phi_{\boldsymbol{K}'}$ (coherent scattering) lead to the momentum conservation

$$H_{fi} \sim \sum_{i=1}^{N} e^{i(\boldsymbol{K}-\boldsymbol{K}')\boldsymbol{r}_i^0} e^{i(\boldsymbol{k}-\boldsymbol{k}')\boldsymbol{r}_i^0} = N\delta_{\boldsymbol{K}'+\boldsymbol{k}',\boldsymbol{K}+\boldsymbol{k}} \ , \tag{6.46}$$

where $\boldsymbol{k}'$, $\boldsymbol{K}'$ are the wavevectors of the final state (f). We can see that the difference in momentum of the incident particle is taken by the crystal, which moves as a whole. According to equation (6.46) the coherent cross-section is

$$\sigma_{coh} = N^2\sigma_0 \ , \tag{6.47}$$

where σ_0 is the single-particle cross-section. A similar calculation for the incoherent matrix elements (wavefunctions ψ) leads to

$$H_{fi} \sim \sum_{i=1}^{N} e^{i(\boldsymbol{k}-\boldsymbol{k}')\boldsymbol{r}_i^0}\left(1+i(\boldsymbol{k}-\boldsymbol{k}')\boldsymbol{v}_i+...\right) \ , \tag{6.48}$$

where

$$\boldsymbol{v}_i = \frac{1}{\sqrt{N}}\sum_{\boldsymbol{q}} e^{i\boldsymbol{q}\boldsymbol{r}_i^0}\boldsymbol{v}_{\boldsymbol{q}} \tag{6.49}$$

is the phonon field. The first term in equation (6.48) corresponds to a displacement of the crystal as a whole, so it is already included in the coherent scattering. We are left with

$$H_{fi} \sim \sqrt{N}\,(\boldsymbol{q}\boldsymbol{v}_{\boldsymbol{q}})\,\delta_{\boldsymbol{k}',\boldsymbol{k}+\boldsymbol{q}}\ . \tag{6.50}$$

We can see that the difference in momentum of the incident particle is taken by phonons; the incoherent scattering excites phonons. In addition, the incoherent cross-section is proportional to N. For the cross-section we need to average $(\boldsymbol{q}\boldsymbol{v}_{\boldsymbol{q}})^2$ over the thermal states. The maximum value of this average is of the order T/Mc_s^2 (at room temperature), where c_s is the mean phonon velocity (of the order $10^6 cm/s$). This thermal factor is reminiscent of the Debye-Waller factor (and the diffuse scattering). The total incoherent scattering can be written as

$$\sigma_{incoh} = N\sigma_0^{ph}\ , \tag{6.51}$$

where the single-particle cross-section σ_0^{ph}, arising from phonons, is smaller than σ_0 by the thermal factor.

The total cross-section of the crystal is

$$\sigma = \sqrt{1-g^2}\sigma_{coh} + g^2\sigma_{incoh}\ . \tag{6.52}$$

The incoherent cross-section is extremely small in comparison with the coherent cross-section. For atoms placed randomly (like in amorphous solids, liquids, etc) the coherent wavefunction $\Phi_{\boldsymbol{K}}$ is missing and the weight coefficient is $g = 1$.

6.6.3 Neutrino scattering

We adopt $N \simeq 10^{22}$ for the number of unit cells in the sapphire monocrystal ($\simeq 24g$, density $4g/cm^3$) used in Weber's experiments,[11]

[11] J. Weber, "Method for observation of neutrinos and antineutrinos", Phys. Rev. **C31** 1468 (1985); "Apparent observation of abnormally large coherent scattering cross sections using keV and MeV range antineutrinos, and solar neutrinos", Phys. Rev. **D38** 32 (1988).

and other similar experiments[12] (the volume of the unit cell of sapphire is large, $\simeq 10^3 \text{Å}^3$). Making use of $\sigma_0 = 10^{-44} cm^2$ we get a coherent cross-section $\sigma_{coh} \simeq 1cm^2$. For a Debye temperature $\Theta = 10^3 K$ the weight factor at room temperature is $\sqrt{1-g^2} \simeq 0.7$. The total coherent cross-section is $\sigma \simeq 0.7cm^2$. We note that this cross-section is smaller than the area of the crystal. For a flux density $\Phi = 10^{12}/cm^2 \cdot s$ the collision time is $\tau = 1/\Phi\sigma \simeq 10^{-12}s$. At room temperature the thermal velocity of an atom is $v \simeq 10^4 cm/s$. For a neutrino energy $E = 1MeV$ the momentum is $p = E/c \simeq 5 \cdot 10^{-17} g \cdot cm/s$. The perturbation energy is given by $vp \simeq 5 \times 10^{-13} erg$ and $p^2/M \simeq 2.5 \times 10^{-11} erg$ (for an atomic mass $M = 10^5 m_e$, where m_e is the electron mass). We can see that $\Delta E = p^2/M$ is higher, such that the equilibrium time is of the order $\Delta t_{eq} = \hbar/\Delta E \simeq 4 \times 10^{-17} s$. Since $\tau \gg \Delta t_{eq}$ the macroscopic quantum-scattering described above applies.

The force acting upon the target in the forward direction is $F = \Phi\sigma p \simeq 3.5 \times 10^{-5}\, dyn$. This is a measurable force. We note that it is sensitive to the values of the input parameters. For instance, a Debye temperature $\Theta = 100K$ leads to a weight coefficient $\sqrt{1-g^2} \simeq 0.15$ and a weaker force by a factor $\simeq 5$. Also, for an amorphous solid, although the conditions of a quantum-mechanical scattering may be fulfilled (the incoherent part), the force is extremely weak, as a consequence of the very small incoherent cross-section.

For solar neutrinos the single-particle cross-section may have the same order of magnitude ($\sigma_0 = 10^{-44} cm^2$), but the flux density is lower ($E \simeq 300keV$, $\Phi \simeq 10^{11}/cm^2 \cdot s$). The conditions of a coherent scattering are preserved, but the force is diminished by a factor $1/10$ ($\simeq 10^{-6} dyn$). A similar decrease appears for tritium neutrinos, though a higher σ_0 or a slightly greater number of unit cells N may compensate the decrease (while preserving the conditions of coherent scattering). We conclude that Weber-type experiments could exhibit a measurable

[12] I. Cruceru, G. Nicolescu and O. G. Duliu, "A dedicated torsion balance to detect neutrinos by coherent scattering on high Debye temperature monocrystals", Int. J. Mod. Phys. **A23** 1719 (2008); M. Cruceru, G. Nicolescu, O. G. Duliu and I. Cruceru, "Detection of solar neutrinos by coherent scattering in high-Debye temperature monocrystals", Int. J. Mod. Phys. **A26** 2773 (2011); G. Nicolescu, "A heuristic approach to the detection of solar neutrinos", J. Phys. G: Nucl. Part. Phys. **40** 055201 (2013).

force acting upon a sapphire crystal.

6.6.4 Other projectiles. Coherence domains

We adopt the value $\sigma_0 = 10^{-24}cm^2$ for other types of projectiles (like X-, gamma rays or neutrons). A coherent cross-section would be much larger than the area of the crystal. The crystal responds to this unphysical situation by developing coherence domains. Let us assume that n_d uncorrelated domains exist in the crystal, each with N_d unit cells (as a mean size), such that $n_d = N/N_d$. By a formal analogy with the high-purity crystals we use the fraction $f = 1/N_d$. This fraction varies between $f = 1/N$, when we have only one domain, *i.e.* the whole target, and $f = 1$, when the whole target is fragmented in "atomic" domains.

The scattering amplitude can be written as

$$S = \sum_{a=1}^{n_d} e^{i\chi_a} \left(H_{fi}\right)_a \ , \tag{6.53}$$

where $e^{i\chi_a}$ are random phase factors. By averaging the squared scattering amplitude over the phase factors, we get $\mid S \mid^2 = \sum_a \left|\left(H_{fi}\right)_a\right|^2$, such that the cross-section becomes

$$\overline{\sigma} = n_d \sigma_d \ , \tag{6.54}$$

where σ_d is the cross-section of a domain.

According to this equation, the coherent cross-section $\sigma_{coh} = \sigma_0 N^2$ is reduced by the coherence domains to

$$\overline{\sigma} = \sigma_0 n_d N_d^2 = \frac{\sigma_0 N}{f} \ ; \tag{6.55}$$

we can see that this formula gives the total coherent cross-section ($\sigma_0 N^2$) for $f = 1/N$ and the incoherent cross-section ($\sigma_0 N$) for $f = 1$. In this latter case σ_0 should be replaced by σ_0^{ph} (a similar procedure leaves the incoherent cross-section arising from phonons unchanged, $\overline{\sigma}_{incoh} = n_d N_d \sigma_0^{ph} = N\sigma_0^{ph} = \sigma_{incoh}$).

In order to have a quantum-mechanical scattering the conditions $\overline{\tau} > \Delta t_{eq}$ and $\overline{\sigma} < A$ should be satisfied, where A is the area of the target; they lead to

$$f > \frac{\hbar\Phi}{\Delta E}\sigma_0 N \ , \ f > \frac{\sigma_0 N}{A} \tag{6.56}$$

and a number of unit cells $N_d = 1/f < A/\sigma_0 N$ in each domain. For $\sigma_0 = 10^{-24}cm^2$ this number is too small for any macroscopic target ($N_d < 10^2 A$, $N = 10^{22}$); the domains are not well defined, such that f approaches unity and the scattering tends to an incoherent scattering. We conclude that the quantum-mechanical scattering cannot appear for large single-particle cross-sections, like $\sigma_0 = 10^{-24}cm^2$. We note that for neutrinos ($\sigma_0 = 10^{-44}cm^2$) $f > 10^{-22}/A$, $N_d < 10^{22}A$ and we may have one domain in the whole target. The coherent scattering occurs for neutrinos in a crystal precisely due to the small neutrino cross-section σ_0. The above considerations apply also to a polycrystalline target, where f is limited, in addition, by the size of the crystallites and, consequently, the cross-section is much diminished.

6.6.5 Classical scattering

If inequation (6.40) is not satisfied (*i.e.*, if $\tau < \Delta t_{eq}$), the incident particles see the target "atoms" (atomic nuclei, molecules, unit cells) as independent scatterers. We call this scattering a classical scattering. We can see that this condition implies low energies. The initial wavefunction is $\frac{1}{\sqrt{V}}e^{i\boldsymbol{kR}}\frac{1}{\sqrt{V}}e^{i\boldsymbol{KR}_c}$ (up to wavefunctions corresponding to the internal degrees of freedom), where $\boldsymbol{k}$ is the wavevector of the incident particle, $\boldsymbol{K}$ is the wavevector of the center of mass, $\boldsymbol{R}_c$ is the position of the center of mass and $\boldsymbol{R} = \boldsymbol{r} + \boldsymbol{R}_c$. The matrix elements of the interaction given by equation (6.39),

$$H_{fi} \sim \sum_{i=1}^{N} e^{i(\boldsymbol{k}-\boldsymbol{k}')\boldsymbol{r}_i}\delta_{\boldsymbol{K}'+\boldsymbol{k}',\boldsymbol{K}+\boldsymbol{k}} \ , \tag{6.57}$$

includes the form-factor

$$F(\boldsymbol{k}-\boldsymbol{k}') = \sum_{i=1}^{N} e^{i(\boldsymbol{k}-\boldsymbol{k}')\boldsymbol{r}_i} \ , \tag{6.58}$$

where $\boldsymbol{k}'$ is the wavevector of the scattered particle and $\boldsymbol{K}'$ is the final wavevector of the center of mass. We can see that the total momentum, including the momentum of the center of mass, is conserved. For a crystal $F(\boldsymbol{k}-\boldsymbol{k}') = N\delta_{\boldsymbol{k}',\boldsymbol{k}+\boldsymbol{g}}$, where $\boldsymbol{g}$ is a reciprocal vector of the lattice. For an amorphous target $\boldsymbol{g}=0$. It follows that we have diffraction peaks. For the cross-section of a peak we get $d\sigma_{\boldsymbol{g}} \sim N^2 do_{\boldsymbol{g}}$, where the solid angle $o_{\boldsymbol{g}}$ extends to the range $\Delta o_{\boldsymbol{g}} \simeq N^{-2/3}(2\pi/dk')^2$, where d is the mean distance between unit cells (scatterers). It follows $\sigma_{\boldsymbol{g}} \sim N^{4/3}/(dk')^2$. On the other hand the number of peaks is $\simeq (dk')^2 \gg 1$, such that the total cross-section is

$$\sigma = N^{4/3}\sigma_0 \ , \tag{6.59}$$

where σ_0 is the single-particle cross-section. For one peak σ should be divided by the number of peaks. As it is well known, this cross-section is affected by the Debye-Waller factor and diffuse scattering. According to equation (6.53) for $n_d = N/N_d = fN$ domains the cross-section is

$$\overline{\sigma} = n_d N_d^{4/3}\sigma_0 = \frac{\sigma_0 N}{f^{1/3}} \ . \tag{6.60}$$

For $f=1/N$ we recover the total cross-section $N^{4/3}\sigma_0$ of one domain, while for $f=1$ the cross-section reduces to $N\sigma_0$ of an incoherent scattering.

The conditions $\overline{\tau} = 1/\Phi\overline{\sigma} < \Delta t_{eq}$ and $\overline{\sigma} < A$ lead to

$$\frac{\sigma_0 N}{A} < f^{1/3} < \frac{\hbar\Phi}{\Delta E}\sigma_0 N \ , \tag{6.61}$$

which implies $N_d = 1/f < (A/\sigma_0 N)^3$ ($\overline{\sigma} < A$). For $\sigma = 10^{-24} cm^2$ and $N = 10^{22}$ the number of unit cells $N_d < 10^6 A^3$ in a domain may indicate well-defined domains for macroscopic targets. The force is bounded from above according to the inequality $F < \Phi A p$. The inequalities (6.61) imply $\Delta E < \hbar\Phi A$. This is a small energy transfer ($\Delta E < 10^{-15} A\, erg$), such that we may use $\Delta E = vp$ and $p < \frac{\hbar\Phi}{v}A$. The upper bound is given by $F < \frac{\hbar}{v}(\Phi A)^2 \simeq 10^{-7} A^2 dyn$ ($\Phi = 10^{12}/cm^2\cdot s$, $v = 10^4 cm/s$). For any reasonably large area A and flux density Φ it is difficult to satisfy these conditions ($p < \hbar\Phi A/v$) and to measure such a force in current experimental situations.

In conclusion, a quantum-mechanical scattering is identified in certain conditions in macroscopic targets, which may lead to a coherent cross-section in high-purity, stiff monocrystals. This coherent scattering may explain the Weber-type experiments of neutrino detection by using sapphire monocrystals. The coherent-scattering conditions are not fulfilled by other types of projectiles, with a higher single-particle cross-sections (like X-, gamma or neutrons). In these cases a classical diffraction may occur in crystals, which generates a weak force, at the limit of detection.

7 Moessbauer Effect

7.1 Resonance fluorescence

Let us assume that an excited atom decays by emitting an electromagnetic radiation with energy E (a photon). This energy is the difference between the energies of an excited atomic level and the ground-state. Its order of magnitude is $1eV$, corresponding to a frequency $\omega = 10^{15}s^{-1}$ ($1eV = 1.6 \times 10^{-12}erg$, Planck's constant $\hbar \simeq 10^{-27}erg \cdot s$). The excited energy levels have an energy uncertainty γ. For atomic levels it is of the order $\gamma \simeq \hbar\omega^2 a_0/c$, where $a_0 = e^2/m_e c^2 = 2.8 \times 10^{-13}cm$ is the electromagnetic radius of the electron with charge $e = -4.8 \times 10^{-10}esu$ and mass $m_e \simeq 10^{-27}g$ and $c = 3 \times 10^{10}cm/s$ is the speed of light in vacuum (see the Electromagnetic Radiation chapter). We get $\gamma \simeq 10^{-20}erg \simeq 10^{-8}eV$ (frequency $10^7 s^{-1}$). This is the natural breath of the atomic spectral line. We can see that the atomic spectral lines are extremely narrow.

Let us assume that the atom has the momentum $\boldsymbol{P}$ before the emission and the energy $P^2/2m + \hbar\omega$, where m is the mass of the atom. After emission, the atom has the momentum $\boldsymbol{P}'$, the photon has a momentum $\boldsymbol{p}$ and the total energy is $P'^2/2m + cp$. From the momentum conservation

$$\boldsymbol{P}' + \boldsymbol{p} = \boldsymbol{P} \tag{7.1}$$

and the energy conservation

$$\frac{P'^2}{2m} + cp = \frac{P^2}{2m} + \hbar\omega \tag{7.2}$$

we get

$$P'^2 = P^2 - 2pP\cos\theta + p^2 \ , \tag{7.3}$$

where θ is the angle made by $\boldsymbol{P}$ and $\boldsymbol{p}$, and

$$p^2 + 2(mc - P\cos\theta)p - 2m\hbar\omega = 0 \ . \tag{7.4}$$

The solutions of these equations are

$$p = -mc + P\cos\theta \pm \sqrt{(mc - P\cos\theta)^2 + 2m\hbar\omega}. \qquad (7.5)$$

The initial momentum is $\boldsymbol{P} = m\boldsymbol{v}$, where $\boldsymbol{v}$ is the velocity of the atom before emission. Since $v \ll c$ and $\hbar\omega \ll mc^2$ (rest energy of the electron $m_e c^2 \simeq 0.5MeV$), we get

$$p = \frac{\hbar\omega}{c}(1 + \frac{v}{c}\cos\theta - \hbar\omega/2mc^2 + ...) \ . \qquad (7.6)$$

The term $\hbar\omega/2mc^2$ is the non-relativistic correction to the rest energy of the atom (see equation (7.2)); the term $\frac{v}{c}\cos\theta$ is the Doppler correction. We can see that the photon frequency is changed into

$$\omega' = \omega(1 + \frac{v}{c}\cos\theta - \hbar\omega/2mc^2 + ...) \qquad (7.7)$$

due to the recoil of the atom. The energy exchanged by the atom is

$$\begin{aligned} \Delta E = P'^2/2m - P^2/2m = \hbar\omega - cp = \\ = -E\tfrac{v}{c}\cos\theta + E^2/2mc^2 \ . \end{aligned} \qquad (7.8)$$

We note that $E\frac{v}{c}\cos\theta \simeq \boldsymbol{pv}$ and $E^2/2mc^2 \simeq p^2/2m$, *i.e.* both corrections occur from $\Delta E = (\boldsymbol{P} - \boldsymbol{p})^2/2m - P^2/2m$. A similar atom may absorb the emitted photon, providing it has the initial momentum $\boldsymbol{P}'$. This is the resonance fluorescence.

In gases or liquids at temperature T (room temperature) the velocity $\boldsymbol{v}$ is a statistical variable. Consequently, for a given direction of the emitted photon, its frequency is spread from zero to infinity, due to the Doppler term $E\frac{v}{c}\cos\theta$ in equation (7.8). This is the Doppler broadening of the spectral line. It is dominant for small energies of the photon. The mean value of the velocity v (thermal velocity) is of the order $\overline{v} = \sqrt{T/m}$, with a spread of the same order of magnitude (the Boltzmann distribution is $\sim v^2 e^{-mv^2/2T}$). For an atomic mass $m = 10^5 m_e$ and at room temperature ($300K$), we get a thermal velocity $\overline{v} = 10^4 cm/s$ for atoms ($1K = 1.38\times10^{-16}erg$). The spectral line is shifted by $\Delta\omega \simeq \frac{\omega}{c}\sqrt{T/m}$, with a width of the same order of magnitude. For our numerical data the energy shift of the spectral line is of the order $\Delta E = \overline{v}E/c \simeq 10^{-18}erg$, with a similar width. These quantities are much smaller than the thermal energy $T \simeq 10^{-14}erg$ (at room temperature), but they are much larger than the natural width of the spectral line $\gamma \simeq 10^{-20}erg \simeq 10^{-8}eV$. It is worth noting that the change in energy ΔE implies a mean velocity $u = \Delta E/m\overline{v} \simeq 1cm/s$.

In solids the atom interacts with its neighbours, and moves as phonons, vibrations of the solid. The energy of N atoms (molecules, unit cells) in solids is $E_a \sim N\int^{\omega_D} d\omega\omega^2 \cdot \omega \coth(\hbar\omega/2T)$, where ω_D is the Debye frequency (see the Neutrino Scattering chapter). The function $\coth(\hbar\omega/2T)$ arises from the occupation number $(e^{\hbar\omega/T}-1)^{-1}$ plus the zero-point contribution 1/2. For typical Debye frequencies ($\omega_D = 10K - 100K$) and at room temperature $\coth(\hbar\omega/2T)$ may be approximated by $2T/\hbar\omega$, such that the zero-point energy brings no contribution. This approximation is valid even for higher Debye frequencies ($\omega_D = 1000K$). The distribution function $\sim \omega^2(e^{\hbar\omega/T}-1)^{-1}$ is similar to the Boltzmann distribution (the frequencies are cut off at the Debye frequency). Therefore, the mean energy of the atoms in solids (at room temperature) is $E_a \simeq NT$ and the thermal energy of an atom is $\simeq T$.

In solids the position of the atoms and their velocity are not determined. Indeed, the displacement of the atom is a Fourier superposition with amplitudes $\sim \sqrt{\hbar/m\omega}(a^*_{\boldsymbol{k}} + a_{-\boldsymbol{k}})$, where $a_{\boldsymbol{k}}$ ($a^*_{\boldsymbol{k}}$) are the destruction (creation) operators of phonons with momentum $\hbar\boldsymbol{k}$ and frequency ω. However, we can see that the displacement and the velocity are well defined (they are classical variables) as long as $\sqrt{\hbar/m\omega} \ll a$, where a is the mean separation distance between the atoms. This condition provides a lower bound for frequencies $\omega \gg \hbar/ma^2$. For a equal to a few $\mathring{A}'s$ ($1\mathring{A} = 10^{-8}cm$) we get $\omega \gg 10^{10}s^{-1}$ (wavelengths $\simeq 10^{-4}cm$, sound velocity is of the order $10^6 cm/s$). We note that typical Debye frequencies are $\omega_D = 10^{12}s^{-1} - 10^{13}s^{-1}$ ($10K - 100K$) and the thermal mean frequency is $\overline{\omega} \simeq T/\hbar = 10^{13}s^{-1}$. Therefore, we may use the (quasi-) classical approximation for phonons in solids, except for very low frequencies, which, however, have a very low statistical weight. It follows that the thermal velocity of the atoms in solids is $\overline{v} = \sqrt{T/m}$. At very low temperature the low frequency phonons bring the main contribution and these estimations are different. The spread of this distribution is of the same order of magnitude as the mean energy T (mean velocity $\overline{v} = \sqrt{T/m}$). Consequently, the Doppler shift and the Doppler broadening are the same in gases, liquids and solids (at room temperature).

The pertubation time produced by the energy $\Delta E = \overline{v}E/c$, *i.e.* the time needed for the atom to recover its equilibrium, is $\Delta t_{eq} = \hbar/\Delta E = \hbar c/\overline{v}E \simeq 2 \times 10^{-9}s$ (for $E = 1eV$). For comparison, the emission (absorption) time of the photon $\omega^{-1} = 10^{-15}s$ is much shorter; practically, the emission (absorption) proceeds instantaneously. In gases and liquids there exist collisions between the atoms, as well as sound waves (long wavelength phonons). In solids there exist only phonons. In gases the mean separation distance between the atoms is of the order $a \simeq 10\text{Å}$ (in normal conditions: room temperature $300K$ and pressure $10^6 dyn/cm^2$). For a mean velocity $\overline{v} = 10^4 cm/s$ the collision time is $\tau_{coll} = 10^{-11}s$. In liquids the mean separation distance between the atoms is of the order of a few Å's; the collision time in liquids is $\tau_{coll} = 10^{-12}s$. The phonon periods τ are such that $\tau > 1/\omega_D \simeq 10^{-12}s^{-1} - 10^{-14}s$ ($\omega_D = 10K - 1000K$). The thermal time is $\tau_{th} = \hbar/T \simeq 2.5 \times 10^{-14}s$ (at room temperature). Incidentally, the mean thermal displacement of the atom is $d = \overline{v}\tau_{th} \simeq 0.025\text{Å}$ (thermal wavelength).[1] We can see that during the equilibrium time Δt_{eq} both collisions and phonon excitations (emissions and absorptions) take place, and these processes can be thermalized. Therefore, the recoil energy of the atoms, appearing during emission or absorption of an atomic photon, is thermalized. We note that this recoil energy is both received and given by these processes (it is exchanged). The atomic photon emission or absorption is accompanied by collisions and phonon excitations.

Indeed, the perturbation energy $\Delta E = \overline{v}E/c \simeq 5 \times 10^{-19} erg$ corresponds to a frequency $\Delta E/\hbar \simeq 5 \times 10^8 s^{-1}$; this is also the order of magnitude of the rate γ of introducing this perturbation. The transition amplitude from/to a state with frequency ω (collisions, phonons) is proportional to

$$\int_{-\infty}^{t} dt' e^{i(\omega-\gamma)t'+\gamma t'} = e^{i(\omega-\gamma)t} \frac{e^{\gamma t}}{i(\omega-\gamma)+\gamma} ; \tag{7.9}$$

[1]See M. Apostol, *Physical Kinetics*, Cambridge Scholars Publishing, Cambridge (2020)).

the rate of transitions is proportional to

$$\frac{\gamma}{(\omega-\gamma)^2+\gamma^2}e^{2\gamma t} \quad , \tag{7.10}$$

whose main contribution is of the order $\simeq 1/\gamma$ in time γ^{-1}; for small γ the transitions may take place, in a long time, without disturbing the thermal equilibrium.

The situation is different for the nuclear gamma emission. A typical nuclear gamma energy is $E = 100keV$. Its natural breath Γ may be of the same order of magnitude as the atomic breath γ ($10^{-8}eV$). The Doppler term in equation (7.8) is $\Delta E = \overline{v}E/c \simeq 5\times 10^{-14}erg$ and the non-relativistic correction is $\Delta E = E^2/2mc^2 \simeq 1.6\times 10^{-13}erg$; these contributions are comparable. We take $\Delta E = 10^{-13}erg$. The shift and the Doppler spread in the gamma photon are much larger than for an atomic photon. They are of the same order of magnitude as the thermal energy $T \simeq 4\times 10^{-14}erg$ (at room temperature). We note that the non-relativistic correction gives only a shift in the spectral line, not a broadening. The additional mean velocity gained by the atom is of the order $u \simeq \sqrt{2\Delta E/m} \simeq 10^4 cm/s$ (as the thermal velocity $\overline{v}$). The equilibrium time of a recoiling nucleus is $\Delta t_{eq} \simeq 10^{-14}s$. We can see that this equilibrium time is too short to allow for collisions in gases or liquids, or to excite phonons. It is of the same order of magnitude as the thermal time τ_{th} at room temperature. The rate of transitions which exchange such a high energy is proportional to $1/\gamma \simeq 10^{-14}s$, which is very low. The frequency $\gamma = 10^{14}s^{-1}$ is higher than typical Debye frequencies $10^{12}s^{-1} - 10^{13}s^{-1}$ ($\omega_D = 10K - 100K$). After a sufficiently long time the recoil energy arising from a gamma photon is thermalized in gases or liquids, but in solids it is taken over by the entire solid. A similar situation is encountered with the capture of slow neutrons by atoms in a crystal,[2] and was suggested for collisions in a gas.[3] The resonance line is shifted, and the Doppler broadening is reduced.

The velocity V gained by the solid in the direction opposite to the emission direction is given by $Nmu = N_0 mV$, where N is the number

[2] W. E. Lamb, Jr., "Capture of slow neutrons by atoms in a crystal", Phys. Rev. **55** 190 (1939).

[3] R. H. Dicke, "The effect of collisions upon the Doppler width of spectral lines", Phys. Rev. **89** 472 (1953).

of emitting (absorbing) atoms and N_0 is the total number of atoms. Usually, N is smaller than N_0 by a few orders of magnitude, such that V is smaller than u ($u \simeq 10^4 cm/s$) by a few orders of magnitude. We can move the emitting source with a velocity of the order of the velocity V, such as to annihilate to a large extent the shift and the broadening of the spectral line and to get the resonance fluorescence. A similar situation is valid for the detector (receiver) of the gamma photon. By resonance fluorescence we can probe spectroscopically the local environment of the receiver. In amorphous solids there are localized high frequencies of vibrations, with short periods, so these vibrations may be excited, and the effect is poor. Among the suitable gamma active nuclei are Ir^{191}, Fe^{57}, etc. This is the Moessbauer effect.[4]

7.2 Photon emission

We assume an atom, or an atomic nucleus, with internal charges which may emit a photon. We call it a particle. Initially the particle has an internal wavefunction $\varphi_i(\xi)$, where ξ denotes the internal coordinates, with an energy $\hbar\omega_0$, of an excited state; finally, the particle decays to the groundstate, with a wavefunction $\varphi_f(\xi)$ and zero energy. The interaction hamiltonian of a charge q and mass m with the electromagnetic radiation (to the first non-relativistic approximation) is

$$H = -\frac{q}{mc}\boldsymbol{A}\boldsymbol{p}(\xi) \ , \tag{7.11}$$

where $\boldsymbol{A}$ is the vector potential and $\boldsymbol{p}(\xi)$ is the momentum of the charge (and $c = 3 \times 10^{10} cm/s$ is the speed of light in vacuum); in this approximation the momentum is $m\boldsymbol{v}$, where $\boldsymbol{v}$ is the velocity of the charge. The vector potential must be taken at the position of the charge. We consider in equation (7.11) the charge, the mass and the momentum as corresponding to an ensemble of internal charges. The

[4] R. L. Moessbauer, "Kernresonanzfluoreszenz von Gammastrahlung in Ir^{191}", Z. Phys. **151** 124 (1958); "Kernresonanzabsorption von γ-Strahlung in Ir^{191}", Naturwissenschaften **45** 538 (1958); "Kernresonanzabsorption von γ-Strahlung in Ir^{191}", Z. Naturforsch. **14a** 211 (1950); H. Frauenfelder, *The Moessbauer Effect*, Benjamin, NY (1963).

vector potential is

$$\boldsymbol{A}(\boldsymbol{R})=\frac{1}{\sqrt{V}}\sum_{\boldsymbol{k}}\alpha_k\boldsymbol{e}_{\boldsymbol{k}}\left(a_{\boldsymbol{k}}e^{i\boldsymbol{k}\boldsymbol{R}}+a^*_{\boldsymbol{k}}e^{-i\boldsymbol{k}\boldsymbol{R}}\right) \ , \tag{7.12}$$

where V is the volume, $\alpha_k = c\sqrt{2\pi\hbar/\omega_k}$, $\omega_k = ck$ is the photon frequency, $\boldsymbol{k}$ is the photon wavevector, $\boldsymbol{e}_{\boldsymbol{k}}$ is the polarization vector ($e_{\boldsymbol{k}} = 1$, $\boldsymbol{e}_{\boldsymbol{k}}\boldsymbol{k} = 0$) and $a_{\boldsymbol{k}}$, $a^*_{\boldsymbol{k}}$ are the photon destruction and creation operators (this is the occupation number, second quantization representation); a summation over polarization is tacitly understood. It is convenient to write the interaction as

$$H=-\frac{q\alpha_k}{mc}\boldsymbol{e}_{\boldsymbol{k}}\boldsymbol{p}(\xi) \tag{7.13}$$

for a $\boldsymbol{k}$-photon and use the photon wavefunction $\frac{1}{\sqrt{V}}e^{i\boldsymbol{k}\boldsymbol{R}}$. Since the photon wavelength is much longer than the dimension of the particle we adopt for position $\boldsymbol{R}$ the position of the center of mass of the particle (see the Electromagnetic Radiation chapter).

The particle may be animated by a thermal motion. In solids, this motion is the statistical motion of the phonons. At the moment of emission of a photon the motion of the particle is a quantum-mechanical motion. Consequently, at the moment of emission of the photon we may leave aside the thermal motion and use an initial wavefunction $\frac{1}{\sqrt{V}}e^{i\boldsymbol{K}_i\boldsymbol{R}}\varphi_i(\xi)$ for the particle, where $\hbar\boldsymbol{K}_i$ is the initial momentum; it may be taken equal to zero. Similarly, the final wavefunction is $\frac{1}{\sqrt{V}}e^{i\boldsymbol{K}_f\boldsymbol{R}}\varphi_f(\xi)\cdot\frac{1}{\sqrt{V}}e^{i\boldsymbol{k}\boldsymbol{R}}$, where the wavefunction of the emitted photon is included. The transition rate per unit time is

$$w=\frac{2\pi}{\hbar}\left|H_{fi}\right|^2\delta(E_f-E_i) \ , \tag{7.14}$$

where $E_{i,f}$ are the energies of the initial and final states. It is worth noting that the δ-factor in this equation is obtained by introducing adiabatically the interaction, according to

$$\lim_{\gamma\rightarrow 0}\frac{\partial}{\partial t}\left|\int_{-\infty}^{t}dt'e^{i(E_f-E_i)t'/\hbar+\gamma t'/\hbar}\right|^2=2\pi\hbar\delta(E_f-E_i) \ . \tag{7.15}$$

The time evolution of the wavefunctions is fictitious, until the moment of the sudden transition jump.[5]

The matrix element of the interaction is

$$H_{fi} = -\frac{q\alpha_k}{mc\sqrt{V}}(\boldsymbol{e_k p}_{fi})\delta_{\boldsymbol{K}+\boldsymbol{k},0} \ , \tag{7.16}$$

where $\boldsymbol{K}_i = 0$ and $\boldsymbol{K}_f = \boldsymbol{K}$. We can see that the total momentum is conserved, *i.e.* the particle receives a recoil momentum $\hbar\boldsymbol{K} = -\hbar\boldsymbol{k}$, equal and opposite to the momentum of the photon. Therefore, the energies are $E_i = \hbar\omega_0$ and $E_f = \hbar^2K^2/2M + \hbar\omega = \hbar^2\omega^2/2Mc^2 + \hbar\omega$, where M is the mass of the particle and $\omega = \omega_k$ is the frequency of the photon; since $\hbar\omega/Mc^2 \ll 1$, we may neglect the recoil energy in the δ-function (though, for atomic nuclei it may be within the width of the excited level). We note that this energy conservation is precisely that given by equation (7.2), the initial momentum of the particle included.

Further on, the average over polarizations gives $\overline{(\boldsymbol{e_k p}_{fi})^2} = \frac{1}{3}p_{fi}^2$ and $p_{fi} = -im\omega_0 u_{fi}$, where u_{fi} is the matrix element of the displacement of the charge. For the total transition rate we multiply by the number of states $2V\omega^2 d\omega do/(2\pi)^3c^3$ in the frequency interval $d\omega$ and the solid angle do, such that we get

$$dw = \frac{4q^2}{3\hbar c^3}\omega_0^3 \mid u_{fi} \mid^2 \delta(\omega - \omega_0)d\omega \ , \tag{7.17}$$

or, introducing the width Γ of the excited level,

$$dw = \frac{4q^2}{3\pi\hbar c^3}\omega_0^3 \mid u_{fi} \mid^2 \frac{\Gamma}{(\omega-\omega_0)^2+\Gamma^2}d\omega \ ; \tag{7.18}$$

[5] "It is essential to observe that a statement, according to which a system, independent of a determination by a measurement, contains a well-defined internal energy E_n, or equivalently is in a well-defined stationary state, easily leads to contradictions. This is true particularly where the older quantum theory talks of "transition processes" among the various stationary states of the system". (W. Pauli, *Die allgemeinen Prinzipien der Wellenmechanik*, Springer, Berlin (1958)).

this is the resonance-fluorescence rate.[6] The total rate is

$$w = \frac{4q^2}{3\hbar c^3}\omega_0^3 \mid u_{fi} \mid^2 \quad . \tag{7.19}$$

In an atom the electron which suffers a transition with energy $1eV$ is a peripheral electron; its energy is of the order $1eV$ (the charge of the nucleus is screened by the other electrons), such that its velocity is of the order $v \simeq 10^7 cm/s$ ($v/c \simeq 10^{-3}$). A change $\Delta\varepsilon$ in energy gives a change in momentum $\Delta p \simeq \Delta\varepsilon/v \simeq 10^{-19} g \cdot cm/s$ for $\Delta\varepsilon = 1eV$ and an uncertainty in position $a \simeq \hbar/\Delta p \simeq 10^{-8} cm$ (diameter of an atom). We may take $u_{fi} \simeq a$, such that we get $w \simeq 10^6 s^{-1}$ (for $\omega_0 = 10^{15} s^{-1}$). The decay time (the time between two succesive emissions) is $\tau \simeq 10^{-6} s$.

In an atomic nucleus the single-nucleon low-lying energy levels are of the order $100keV$, which leads to a velocity $v \simeq 10^8 cm/s$ and an uncertainty in position $a \simeq 10^{-12} cm$ for $\Delta\varepsilon = 100keV$ (diameter of a large nucleus). The transition rate is $w \simeq 10^{13} s^{-1}$ (for $\omega_0 = 10^{20} s^{-1}$, energy $100keV$) and the decay time $\tau \simeq 10^{-13} s$ (it is much shorter than the lifetime $\sim \Gamma^{-1}$, which is of the order $10^{-7} s$).

The time between two successive emissions (absorptions) in a sample with N active particles is $\tau = 1/Nw$. For $N = 10^{20}$ we get $\tau = 10^{-16} s$ for atoms and $\tau = 10^{-33} s$ for nuclei. These times are much shorter than the equilibrium times $\Delta t_{eq} = 2 \times 10^{-9} s$ for atoms and $\Delta t_{eq} = 10^{-14} s$ for nuclei. Consequently, the emission (absorption) acts are not correlated, such that the transition rates derived above are applicable. If we view the photon emission/absorption as a scattering process, this is a classical scattering. The emitted flux density of photons is $\Phi_e = N_e w/A = n_e l_e w$, where N_e is the number of active emitters with density n_e in the sample with the cross-section A and length l_e. The number of absorptions per unit time in the target with N_t absorbers and the same cross-section is $N_t w$; the cross-section is $\sigma = (N_t/N_e)A$. A coherence-domain analysis can be applied.

[6] V. Weiskopf, "Zur Theorie der Resonanzfluoreszenz", Ann. Phys. **9** 23 (1931); V. Weisskopf and E. Wigner, "Berechnung der natuerlichen Linienbreite auf Grund der Diracschen Lichttheorie", Z. Phys. **63** 54 (1930); "Ueber die natuerliche Linienbreite in der Strahlung des harmonischen Oszillators", Z. Phys. **65** 18 (1930).

8 Light scattering

8.1 Sound

Molecules in a classical gas collide with each other and against the walls of the container. We define spatial regions in a gas, which are macroscopically small but microscopically large; so, we get a continuum model of a gas. The continuum model allows to define the thermal equilibrium. At thermal equilibrium the collisions in a gas determine the pressure. The pressure p is a function of the gas density n. The function $p(n)$ is the equation of state of the gas. We may have local equilibrium, or a global one. For a perfect, ideal gas the equation of state is $p = nT$, where T is the temperature. In normal conditions, *i.e.* at room temperature $T = 300K$ and atmospheric pressure $p = 10^6 dyn/cm^2$, the gas density is $n = 2.5 \times 10^{19} cm^3$ and the mean inter-molecular distance is $a = 1/n^{1/3} \simeq 30\text{Å}$ ($1K = 1.38 \times 10^{-16} erg$, $1\text{Å} = 10^{-8} cm$). For a molecular mass $m = 10^5 m_e$, where $m_e = 10^{-27} g$ is the electron mass, the thermal velocity of a molecule is of the order $v = \sqrt{T/m} \simeq 2 \times 10^4 cm/s$.

The positions and the velocities of the molecules in a gas are not defined, they are statistical variables. Let us assume that the molecules suffer a displacement $\boldsymbol{u}$; a density imbalance $\delta n = -n div \boldsymbol{u}$ appears in the continuum model, and a force per unit volume $-grad p = -(\partial p/\partial n) grad n$; $\partial p/\partial n$ is related to the coefficient of compressibility $\kappa = -\frac{1}{V}(\partial V/\partial p)$, where V denotes the volume. It follows the equation of motion

$$\frac{\partial^2 \boldsymbol{u}}{\partial t^2} = \frac{1}{m}(\partial p/\partial n) grad\, div \boldsymbol{u} \ . \tag{8.1}$$

This is the equation of motion of a displacement field $\boldsymbol{u}$ which moves along its direction, *i.e.* it is a longitudinal motion with a velocity c

given by $c^2 = \frac{1}{m}(\partial p/\partial n)$. It can be written as a wave equation

$$\frac{\partial^2 \boldsymbol{u}}{\partial t^2} - c^2 grad\, div\boldsymbol{u} = 0 \ , \tag{8.2}$$

or

$$\frac{\partial^2 (div\boldsymbol{u})}{\partial t^2} - c^2 \Delta(div\boldsymbol{u}) = 0 \ , \tag{8.3}$$

where an external force (per mass density) can be added. This is the equation of hydrodynamical sound.

For an ideal gas c is of the order of the thermal velocity v ($p = nT$). Obviously, the periods of the sound waves are limited by $a/v \simeq 10^{-12}s$, at most (frequencies $\simeq 10^{12}s^{-1}$). The thermal motion, *i.e.* the energy and the momentum, may move. In a gas this motion proceeds by collisions. This motion is the thermoconduction and the viscosity. The former is the transport of heat, the latter is the transport of momentum. This motion is controlled by the mean freepath of the molecules. The mean freepath is $l = a^3/\sigma$, where σ is the molecular cross-section; it is of the order of the squared molecular diameter. For a molecular diameter of a few Å's, the mean freepath in normal conditions is of the order $10^2 - 10^3$Å. It follows a mean lifetime (of an elementary excitation, which is a moving molecule) of the order $10^2 - 10^3$Å$/v \simeq 10^{-9} - 10^{-10}s$ (collision time). We can see that the sound periods are much shorter than the molecular lifetime, except for very long wavelengths, which, however, are much less numerous. It follows that the thermal motion has no time to propagate during the sound motion, so the sound velocity should be taken at constant entropy. Sound is an equilibrium process. Sound is a collective excitation, which implies local thermal equilibrium, but not a global equilibrium: the pressure, the density and the temperature vary on a global scale (it is a mechanical motion compatible with the statistical motion; sound is not a transport phenomenon).[1] The adiabatic compressibility ($S = const$) is related to the isothermal compressibiliy ($T = const$) by $(\partial p/\partial n)_S = (c_p/c_v)(\partial p/\partial n)_T$, where $c_{p,v}$ are the specific heats at constant pressure or constant volume.

In estimating the thermodynamical properties of a classical gas we include the statistical motion of individual molecules. This way, for

[1] M. Apostol, "Drag force through gases and plasma", J. Theor. Phys. **301** (2019).

a perfect gas we get the energy $\frac{3}{2}NT$, where N is the number of molecules. As we can see, the thermal motion includes also the collective motion of sound waves. This motion should be included in estimating the thermodynamical properties. The collective motion of hydrodynamical sound takes place in a continuum model. The points in a continuum model are continuity domains with dimension d. This dimension can be estimated as $d = (a/a_0)^3 a$, where a is the mean inter-molecular distance and a_0 is the molecular diameter.[2] In normal conditions d is of the order $10^3 a$. It follows that, in a gas, such a domain includes $\simeq 10^9$ molecules. The sound wavelengths should be larger than the length d. The sound periods should be longer than $d/c \simeq 10^{-8}s$. In comparison to the thermal time $a/v \simeq 10^{-12}s$ and the collision time $10^{-9} - 10^{-10}s$ these periods are much longer; in this respect, sound is a slow process. The perturbations produced by sound waves are adiabatic, so the entropy remains constant, and we find again that the sound waves are a (quasi-) equilibrium process. The thermoconduction and the viscosity are low. The velocity of a continuity domain is of the order $\dot{u} = \sqrt{T/M}$, where M is the domain mass; this velocity is $\simeq 10^{-9/2}v \ll v$ ($\simeq c$), *i.e.*, it is of the order $\simeq 1cm/s$. The displacement of the domains is of the order $ud/\lambda = \dot{u}d/c \simeq 10^{-4}d$, where λ is the sound wavelength. The thermodynamics of the sound waves is the thermodynamics of a classical gas of domains with mass M; it follows that its contribution to the total thermodynamics of the gas is very small, so it may be neglected.

The molecules in liquids are separated by a mean distance a of the order of 10Å. The collisions are rare, but the interaction dominates. The molecules are in quasi-bound states. First, the interaction should be solved and only thereafter the thermodynamics computed. The thermodynamics is controlled by the cohesive energy and the elementary excitations, which are mainly local vibrations.[3] The thermal motion propagates through interaction. The thermal velocity is of the same order as in gases, the compressibility is lower, the sound velocity is higher, the sound is adiabatic. The hydrodynamical sound exists, lim-

[2] M. Apostol, *Statistical Physics*, Cambridge Scholars Publishing, Newcastle upon Tyne (2021).

[3] M. Apostol, "A model for the thermodynamics of simple liquids", Physica **B403** 3946 (2008); *Physical Kinetics*, Cambridge Scholars Publishing, Newcastle upon Tyne (2020) and References therein.

ited by a continuum model and governed by compressibility. The compressibility includes contributions from both the cohesive energy and the thermally activated excitations. The dimension of the domains is smaller than in gases. For shorter wavelengths additional short-range interaction may be effective, such that an anomalous sound may appear (densitons), distinct from the usual, hydrodynamical sound.[4] Both sounds contribute little to thermodynamics.

In solids the mean separation distance between the molecules is less than in liquids. The molecules are in bound states. They may move around their equilibrium positions, and may interact with each other. This interaction gives rise to solid vibrations, which in the harmonic-oscillator limit are phonons. At higher temperatures (room temperature) the anharmonic interactions are important. The cohesive energy and the phonons, the anharmonicities govern the thermodynamics of the solids. The thermal velocity is of the same order as in gases and liquids, the thermoconduction proceeds by interaction, the viscosity is weak,[5] the sound is adiabatic.[6] Apart from a longitudinal sound, the interaction may give rise to a transverse sound in solids. Indeed, apart from the propagation of density perturbations $div\boldsymbol{u}$ governed by a force $\sim grad\,div\boldsymbol{u}$ (equations (8.1) and (8.3)), rotations given by $curl\boldsymbol{u}$ appear, governed by a force $\sim curl\,curl\boldsymbol{u}$ (in isotropic solids). Since $curl\,curl = -\Delta + grad\,div$, the equation of motion of the displacement $\boldsymbol{u}$ is generalized to

$$\frac{\partial^2 \boldsymbol{u}}{\partial t^2} - \left(c_l^2 - c_t^2\right) grad\,div\boldsymbol{u} - c_t^2 \Delta \boldsymbol{u} = 0\ ; \tag{8.4}$$

this is the Navier-Cauchy equation of elasticity. It exhibits the propagation of two kinds of waves, a longitudinal one, propagating with velocity c_l and a transverse one, propagating with velocity c_t. The interaction enters through the elasticity moduli λ (longitudinal modulus) and μ (shear modulus), such that $c_l^2 = (\lambda + 2\mu)/\rho$ and $c_t^2 = \mu/\rho$,

[4] M. Apostol and E. Preoteasa, "Density oscillations in a model of water and other similar liquids", Phys. Chem. Liqs. **46** 653 (2008); M. Apostol, "Dynamics of collective density modes in multi-component molecular mixtures", Phys. Chem. Liqs. **47** 35 (2009); *Physical Kinetics*, Cambridge Scholars Publishing, Newcastle upon Tyne (2020) and References therein.

[5] A high viscosity appears in soft matter.

[6] M. Apostol, *Physical Kinetics*, Cambridge Scholars Publishing, Newcastle upon Tyne (2020).

where ρ is the mass density; an external force (per mass density) can be added. Obviously, $(\partial p/\partial\rho)_S = (\lambda + 2\mu)/\rho$. This is not a hydrodynamical sound, since the main contribution to pressure comes from the cohesive energy (interaction), not from thermal equilibrium. The thermal equilibrium is established by phonons, with anharmonicities. The sound velocity in solids is of the order $c = 10^6 cm/s$.

The displacement can be separated into a longitudinal one, for which $curl\boldsymbol{u} = 0$, and a transverse one, for which $div\boldsymbol{u} = 0$. For a longitudinal displacement the equation of motion becomes

$$\rho\frac{\partial^2\boldsymbol{u}}{\partial t^2} - \rho c^2 grad\, div\boldsymbol{u} = \boldsymbol{F} \ , \tag{8.5}$$

where $c = c_l$ and $\boldsymbol{F}$ is an external force (per unit volume). If we multiply this equation, written for the component u_i, by $\partial u_i/\partial t$, we get

$$\frac{\partial e}{\partial t} + c \cdot div\boldsymbol{s} = \dot{\boldsymbol{u}}\boldsymbol{F} \ , \tag{8.6}$$

where

$$\begin{aligned} e &= \left(\tfrac{1}{2}\rho\dot{\boldsymbol{u}}^2 + \tfrac{1}{2}\rho c^2(div\boldsymbol{u})^2\right) \ , \\ \boldsymbol{s} &= -\rho c(\dot{\boldsymbol{u}} div\boldsymbol{u}) \ . \end{aligned} \tag{8.7}$$

This is the equation of energy conservation. The mechanical work $\dot{\boldsymbol{u}}\boldsymbol{F}$ done by the external force per unit time equals the rate of change of the energy density e and the flow of the energy flux density $c\boldsymbol{s}$. Similarly, for the transverse displacement which satisfies the condition $div\boldsymbol{u}$, we get an energy density

$$e = \frac{1}{2}\rho\dot{\boldsymbol{u}}^2 + \frac{1}{2}\rho c^2(\partial_j u_i)^2 \tag{8.8}$$

and an energy flux density

$$s_i = -\rho c\dot{u}_j\partial_i u_j \tag{8.9}$$

(where the repeated labels are summed up).

A Fourier transform of the displacement leads to an elastic energy

$$E = \sum_{\boldsymbol{k}}\left[\frac{1}{2}\rho\dot{\boldsymbol{u}}(\boldsymbol{k})\dot{\boldsymbol{u}}(-\boldsymbol{k}) + \frac{1}{2}\rho c^2 k^2\boldsymbol{u}(\boldsymbol{k})\boldsymbol{u}(-\boldsymbol{k})\right] \ , \tag{8.10}$$

which indicates three branches of phonons, one longitudinal and two transverse, with frequency given by $\omega^2(k) = c_{l,t}^2 k^2$. We write the Fourier transforms as

$$\boldsymbol{u}(\boldsymbol{r}_i) = \boldsymbol{u}_i = \frac{1}{\sqrt{N}} \sum_{\boldsymbol{k}} \boldsymbol{u}(\boldsymbol{k}) e^{i\boldsymbol{k}\boldsymbol{r}_i} \ , \tag{8.11}$$

where $\boldsymbol{r}_i$ are the positions of the molecules or the continuity domains, in number of N, and $\boldsymbol{u}^*(-\boldsymbol{k}) = \boldsymbol{u}(\boldsymbol{k})$. The total energy E is the summation over i of the energy density given by equation (8.8), where ρ is replaced by mass m of the molecules (or domains). For crystals the wavevectors $\boldsymbol{k}$ are limited to the Brillouin zone. For one branch of phonons $u(\boldsymbol{k}) = \sqrt{\frac{\hbar}{2m\omega(k)}} \left(a_{\boldsymbol{k}} + a_{-\boldsymbol{k}}^*\right)$, with $a_{\boldsymbol{k}}$, $a_{\boldsymbol{k}}^*$ boson operators, leads to

$$E = \sum_{\boldsymbol{k}} \frac{1}{2} \hbar\omega(k) \left(a_{\boldsymbol{k}} a_{\boldsymbol{k}}^* + a_{\boldsymbol{k}}^* a_{\boldsymbol{k}}\right) \tag{8.12}$$

and

$$\boldsymbol{s} = c \sum_{\boldsymbol{k}} \frac{1}{2} \hbar\boldsymbol{k} \left(a_{\boldsymbol{k}} a_{\boldsymbol{k}}^* + a_{\boldsymbol{k}}^* a_{\boldsymbol{k}}\right) \ . \tag{8.13}$$

This is the quantization of the phonons ($\hbar$ is Planck's constant), each phonon mode with an energy $\hbar\omega(k)$ and a momentum $\hbar\boldsymbol{k}$ ($\boldsymbol{s}/c$ is the total momentum). Noteworthy, for each mode we have both a creation (destruction) of a $\boldsymbol{k}$-phonon and a destruction (creation) of a $-\boldsymbol{k}$-phonon. At thermal equilibrium with temperature T the phonon occupation number is

$$n_{\boldsymbol{k}} = \overline{a_{\boldsymbol{k}}^* a_{\boldsymbol{k}}} = \frac{1}{e^{\hbar\omega(k)/T} - 1} \ . \tag{8.14}$$

The number of modes are limited by a Debye frequency (temperature) ω_D defined by

$$N = \frac{V \cdot 4\pi}{(2\pi)^3 c^3} \int_0^{\omega_D} d\omega \cdot \omega^2 \ , \tag{8.15}$$

where N is the number of points with positions $\boldsymbol{r}_i$ in volume V and c is defined by $3/c^3 = 1/c_l^3 + 2/c_t^3$. ω_D is an empirical parameter, which depends on the phonon spectrum. It is approximately of the order c/d, where d is the mean separation distance between the molecules (domains). Typical values are $10-100K$ (with a few exceptions which reach 10^3K).

The mean square displacement is given by

$$\overline{u_i^2} = \frac{\hbar d^3}{4\pi^2 mc^3}\left(\frac{2T}{\hbar}\right)^2 \int_0^{\hbar\omega_D/2T} dx \cdot x \coth x \ . \tag{8.16}$$

At room temperature $x < 1$ and

$$\overline{u_i^2} \simeq \frac{d^3 T\omega_D}{2\pi^2 mc^3} \ ; \tag{8.17}$$

for typical values $d = 4\text{Å}$, $T = 300K$, $\omega_D = 100K$, $m = 10^5 m_e$, $m_e = 10^{-27}g$ and $c = 10^4 cm/s$ we get $\left(\overline{u_i^2}\right)^{1/2}$ of the order $10^{-11}cm$. Also, $T/\hbar\omega$ in $n_{\boldsymbol{k}}$ is at least of the order 10, which indicates that the occupation number of the phonon states is very large, such that, at room temperature, the motion of the molecules is classical. The anharmonic interactions in solids are large in this case. This motion is quantum-mechanical only at low temperatures.

A similar treatment can be used for a more general interaction between the molecules, as well as for optical phonons, associated with the motion of the molecules inside of the unit cells of a crystal.

8.2 Scattering by sound

Let us consider a polarizable body. The Maxwell equations

$$\begin{aligned} curl\boldsymbol{E} &= -\tfrac{1}{c}\tfrac{\partial \boldsymbol{H}}{\partial t} \ , \\ curl\boldsymbol{H} &= \tfrac{1}{c}\tfrac{\partial \boldsymbol{E}}{\partial t} + \tfrac{4\pi}{c}\boldsymbol{j} \end{aligned} \tag{8.18}$$

lead to

$$curl\, curl\boldsymbol{E} = -\frac{1}{c^2}\frac{\partial}{\partial t}\left(\frac{\partial E}{\partial t} + 4\pi\boldsymbol{j}\right) \ , \tag{8.19}$$

where $\boldsymbol{E}$ is the electric field, $\boldsymbol{H}$ is the magnetic field, $\boldsymbol{j}$ is the induced current density and c is the speed of light in vacuum. The electric field $\boldsymbol{E}$ is the total electric field, consisting of the external field $\boldsymbol{E}_0 = \boldsymbol{D}$

(electric displacement, induction) and the internal electric field $\boldsymbol{E}_i = -4\pi\boldsymbol{P}$, where $\boldsymbol{P}$ is the polarization, *i.e.* $\boldsymbol{E} = \boldsymbol{D} - 4\pi\boldsymbol{P}$. From the Gauss equation $div\boldsymbol{E} = 4\pi\delta\rho$, where $\delta\rho$ is the induced charge density, we get $\delta\rho = -div\boldsymbol{P}$ (since $div\boldsymbol{D} = 0$). Therefore, from the continuity equation, $\boldsymbol{j} = \frac{\partial\boldsymbol{P}}{\partial t}$. Consequently, equation (8.19) becomes

$$curl\, curl\boldsymbol{E} = -\frac{1}{c^2}\frac{\partial^2\boldsymbol{D}}{\partial t^2} \ , \tag{8.20}$$

or, by using $\boldsymbol{E} = \boldsymbol{D} - 4\pi\boldsymbol{P}$,

$$curl\, curl\boldsymbol{D} - 4\pi curl\, curl\boldsymbol{P} = -\frac{1}{c^2}\frac{\partial^2\boldsymbol{D}}{\partial t^2} \ ; \tag{8.21}$$

this equation can be rewritten as

$$\frac{1}{c^2}\frac{\partial^2\boldsymbol{D}}{\partial t^2} - \Delta\,\boldsymbol{D} = 4\pi curl\, curl\boldsymbol{P} \ . \tag{8.22}$$

We solve this equation for $\boldsymbol{D} = \boldsymbol{E}$ at large distance from the body, with the source term $4\pi curl\, curl\boldsymbol{P}$, where $\boldsymbol{P}$ is determined by an incident electric field $\boldsymbol{E}_0$. This defines the problem of light scattering.

In the first approximation $\boldsymbol{P}$ should depend linearly on $\boldsymbol{E}_0$; for instance, $\boldsymbol{P} = \alpha\boldsymbol{E}_0$, where α is the polarizability (always positive). We have $\chi = \alpha\varepsilon$, or $\frac{\varepsilon-1}{4\pi} = \alpha\varepsilon$, where χ is the electrical susceptibility ($\boldsymbol{P} = \chi\boldsymbol{E}_t$) and ε is the permittivity ($\boldsymbol{E}_0 = \varepsilon\boldsymbol{E}_t$); $\boldsymbol{E}_t = \boldsymbol{E}_0 + \boldsymbol{E}_i = \boldsymbol{E}_0 - 4\pi\boldsymbol{P}$ is the total electric field. In general, the displacement of the charges should follow the deformation of the body, *i.e.* the polarizability should be a tensor proportional to the strain tensor u_{ij}. For an isotropic body it contains $const{\cdot}\delta_{ij}$, $div\boldsymbol{u}\cdot\delta_{ij}$ and u_{ij}, where $\boldsymbol{u}$ is the displacement of the deformed body. This observation was made by Brillouin and Mandelstam.[7]

Let us assume a density n of charges q with mass m which are displaced by the external field $\boldsymbol{E}_0$; a density imbalance $-ndiv\boldsymbol{u}$ appears, where $\boldsymbol{u}$ is the displacement, and an internal field $\boldsymbol{E}_i = -4\pi nq\boldsymbol{u}$ ($\boldsymbol{P} = nq\boldsymbol{u}$). From the equation of motion

$$m\ddot{\boldsymbol{u}} + \omega_c^2\boldsymbol{u} = q\boldsymbol{E}_0 - 4\pi nq^2\boldsymbol{u} \ , \tag{8.23}$$

[7] L. Brillouin, "Diffusion de la lumiere et des rayons X par un corps transparent homogene. Influence de l'agitation thermique", Ann. Physique **9** 88 (1922); L. Mandelstam, "Light scattering by inhomogeneous media", Zh. Russ. Fiz.-Khim. Obsc. Fiz. **58** 381 (1926) (in Russian).

where ω_c is a characteristic frequency related to internal elastic forces, we get

$$\boldsymbol{u} = -\frac{q}{m}\boldsymbol{E}_0\frac{1}{\omega^2 - \omega_c^2 - \omega_p^2} , \tag{8.24}$$

where $\omega_p = (4\pi nq^2/m)^{1/2}$ is the plasma frequency and ω is the frequency of the external field ($\omega_p > \omega_c$). We neglect the spatial dispersion, since the dimension of the body l is much smaller than ct, where t is the time of interest; this is in agreement with the non-relativistic equation of motion that we use, and with the continuum model of matter, for which we write the Maxwell equations. It amounts to considering well-defined photons in the body. The total field is

$$\boldsymbol{E}_t = \boldsymbol{E}_0 + \boldsymbol{E}_i = \frac{\omega^2 - \omega_c^2}{\omega^2 - \omega_c^2 - \omega_p^2}\boldsymbol{E}_0 , \tag{8.25}$$

the permittivity (dielectric function) is

$$\varepsilon = 1 - \frac{\omega_p^2}{\omega^2 - \omega_c^2} \tag{8.26}$$

and the polarizability is

$$\alpha = -\frac{nq^2}{m}\frac{1}{\omega^2 - \omega_c^2 - \omega_p^2} \tag{8.27}$$

(for $\omega < \omega_c$, since $\alpha > 0$ and $\varepsilon > 1$).[8] This is the Lorentz-Drude model of polarization.[9] The density n may include a variation $\sim div\boldsymbol{u}$, besides a constant, where $\boldsymbol{u}$ is distinct from the displacement given by equation (8.24); in addition, ω_c^2 (elastic force) should be proportional to the strain tensor u_{ij}.[10] This is the origin of the dependence of the polarizability on the deformation of the body. In the following we assume $\alpha = \alpha_0 div\boldsymbol{u}$, where α_0 is $-\alpha$ given by equation (8.27).

[8] L. Landau and E. Lifshitz, *Course of Theoretical Physics*, vol. 8, *Electrodynamics of Continuous Media*, Pergamon, Oxford (1984).

[9] M. Apostol, *Essays in Electromagnetism and Matter*, Lambert, Saarbruecken (2013) and References therein.

[10] L. Landau and E. Lifshitz, *Course of Theoretical Physics*, vol. 7, *Elasticity*, Pergamon, Oxford (1986).

Let $\boldsymbol{E}_0 e^{i\boldsymbol{kr}-i\omega t}$ be an incident electric field, with $\omega = ck$, and $\boldsymbol{u} = \boldsymbol{b}e^{i\boldsymbol{qr}-i\Omega t}$ a displacement, with $\Omega = vq$, where v is the sound velocity. The polarization is

$$\boldsymbol{P} = \boldsymbol{P}_0 e^{i(\boldsymbol{k}+\boldsymbol{q})\boldsymbol{r}-i(\omega+\Omega)t} \ , \ \boldsymbol{P}_0 = i\alpha_0(\boldsymbol{qb})\boldsymbol{E}_0 \ . \tag{8.28}$$

The solution of the equation (8.22) is

$$\begin{aligned} &\boldsymbol{E}(\boldsymbol{r},t) = \\ &= \int d\boldsymbol{r}' \frac{curl'\, curl' \boldsymbol{P}_0 e^{i(\boldsymbol{k}+\boldsymbol{q})\boldsymbol{r}'-i(\omega+\Omega)t'}}{|\boldsymbol{r}-\boldsymbol{r}'|} \Big|_{t'=t-|\boldsymbol{r}-\boldsymbol{r}'|/c} \ , \end{aligned} \tag{8.29}$$

which, for $| \boldsymbol{r} - \boldsymbol{r}' | \simeq r - \boldsymbol{rr}'/r$ becomes

$$\boldsymbol{E}(\boldsymbol{r},t) = -V\boldsymbol{k}' \times (\boldsymbol{k}' \times \boldsymbol{P}_0) \frac{e^{ik'r-(\omega+\Omega)t}}{r} \delta_{\boldsymbol{k}+\boldsymbol{q},\boldsymbol{k}'} \ , \tag{8.30}$$

where V is the volume of the body and $\boldsymbol{k}' = (\omega+\Omega)\boldsymbol{r}/cr$. We can see that the field is scattered with frequency $\omega' = \omega + \Omega$ and the wavevector $\boldsymbol{k}' = \boldsymbol{k} + \boldsymbol{q}$.

By dividing the energy flux $c \mid \boldsymbol{E} \mid^2 /4\pi \times r^2 do$ scattered in the solid angle do by the incident flux density $c \mid \boldsymbol{E}_0 \mid^2 /4\pi$, we get the cross-section

$$d\sigma = \frac{1}{2} k'^4 (1+\cos^2\theta)\alpha_0^2 \mid \boldsymbol{qb} \mid^2 V^2 do \ , \tag{8.31}$$

where the scattering angle θ is the angle made by $\boldsymbol{k}$ with $\boldsymbol{k}'$ (for unpolarized light). This cross-section corresponds to the peak given by $\boldsymbol{k}' = \boldsymbol{k}+\boldsymbol{q}$. The total cross-section is obtained by integrating over the solid angle $\Delta o = (2\pi/L)^2/k'^2$, where $L = V^{1/3}$; we get

$$\sigma = 2\pi^2 k'^2 (1+\cos^2\theta)\alpha_0^2 \mid \boldsymbol{qb} \mid^2 V^{4/3} \ . \tag{8.32}$$

By measuring the scattering for every $\boldsymbol{k}'$, we get the sound wavevector $\boldsymbol{q} = \boldsymbol{k}' - \boldsymbol{k}$ and the sound frequency $\Omega = \omega' - \omega$ (momentum and energy conservation). This way we can derive the phonon spectrum.

From equation (8.11) we get $b = \sqrt{\frac{\hbar}{2mN\Omega_q}} a_{\boldsymbol{q}}$, such that

$$\mid \boldsymbol{qb} \mid^2 = \frac{\hbar q^2}{2mN\Omega_q} n_{\boldsymbol{q}} \simeq \frac{T}{2mNv^2} \tag{8.33}$$

at room temperature.

A coherence-domain analysis can be done for light scattering by sound. The energy transfer is the energy $\hbar\Omega$ of the sound quanta, such that the equilibrium time is the sound period. A coherent diffraction appears, which may imply coherence domains.

Finally, we note that the change $\alpha \sim div\boldsymbol{u}$ in the polarizability indicates also a change $\alpha \sim \delta n$ produced by the density fluctuations δn.[11] Close to a critical point, the scattering of light in this case exhibits opalescence. The cross-section is proportional to $\delta n \delta n$, which involves density correlation functions.[12] Also, non-linear effects in matter lead to stimulated Brillouin and Raman scattering.[13]

[11] M. Smoluchowski, "Molekular-kinetische Theorie der Opaleszenz von Gasen im kritischen Zustande, sowie einiger verwandter Erscheinungen", Ann. Physik **25** 205 (1908); A. Einstein, "Theorie der Opaleszenz von homogenen Fluessigkeiten und Fluessigkeitsgemischen in der Naehe des kritischen Zustandes", Ann. Physik **33** 1275 (1910).

[12] L. van Hove, "Correlations in space and time and Born approximation scattering in systems of interacting particles", Phys. Rev. **95** 249 (1954).

[13] R. Pike and P. Sabatier, eds., *Scattering*, Academic, NY (2002); see also L. Landau and E. Lifshitz, *Course of Theoretical Physics*, vol. 8, *Electrodynamics of Continuous Media*, Elsevier, Oxford (1984) and vol. 5, *Statistical Physics*, Elsevier, Oxford (1980).

9 Scattering by laser pulses

9.1 Charge scattering by electromagnetic radiation

The cross-section is computed herein for non-relativistic charged particles (like electrons, ions) scattered by electromagnetic radiation confined to a finite region (like the focal region of optical laser beams). The cross-section exhibits maxima at scattering angles given by the energy and momentum conservation in multi-photon absorption or emission processes. For convenience, a potential scattering is included and a comparison is made with the well-known Kroll-Watson scattering formula. The scattering process addressed here is distinct from the process dealt with in previous studies, where the scattering is immersed in the radiation field.[1]

The potential scattering of charged particles in the presence of electromagnetic radiation enjoyed, and it is still enjoying, much interest.[2] In the context of lasers development, these studies revealed the potential scattering assisted by multiple-photon exchange, as shown by the well-known Kroll-Watson cross-section. In the formulation of this problem the scattering is immersed in the radiation field, *i.e.* the asymptotic incoming and outgoing particle states and the scattering potential are included in the region containing the radiation. The duration of the laser pulse is much longer than the scattering time. The starting point of these approaches is the standard non-relativistic hamiltonian where the particle momentum $\mathbf{p} = m\mathbf{v} + e\mathbf{A}/c$ includes the electromagnetic contribution $e\mathbf{A}/c$ besides the purely mechanical

[1] M. Apostol, "Scattering of non-relativistic charged particles by electromagnetic radiation", Z. Naturforsch. **72a** 1173 (2017).

[2] N. M. Kroll and K. M. Watson, "Charged-particle scattering in the presence of a strong electromagnetic wave", Phys. Rev. **A8** 804 (1973).

contribution $m\mathbf{v}$ (the notations are the usual ones, *i.e.* m and e denote the particle mass and the charge, $\mathbf{v}$ is the particle velocity, $\mathbf{A}$ is the vector potential of the radiation field and c denotes the speed of light in vacuum). The Kroll-Watson cross-section corresponds to radiation-assisted potential scattering, *i.e.* it shows how the potential cross-section is modified by the presence of the radiation; it becomes zero when the potential is removed.

With the advent of high-intensity lasers and strongly focused laser beams,[3] it appears the possibility of scattering charged particles by the radiation field confined to the focal region of the beam (the radiation is vanishing smoothly outside the focal region). In this case the asymptotic scattering states are radiation free; they are eigenstate of the quantum-mechanical momentum corresponding to the purely mechanical momentum, without including the electromagnetic contribution. This is the scattering problem addressed here. It resembles to some extent the electron diffraction from standing light waves, where the scattering proceeds by spontaneous emission of Compton photons (Kapitsa-Dirac effect).[4] We envisage charged particles like electrons or ions, with non-relativistic energies, scattered off electromagnetic radiation confined to the focal region (in vacuum) of an optical laser beam. Usually, the order of magnitude of the dimension of the focal region is a few tens of radiation wavelengths, and the radiation field has a reasonably high intensity, such that the non-relativistic character of the particle motion can be preserved. For convenience, we include in the focal region a static potential. We assume first a laser pulse much longer than the radiation period.

We consider a non-relativistic particle with mass m and charge e scattered by an electromagnetic radiation field with the vector potential

[3] Y. I. Salamin, S. X. Hu, K. Z. Hatsagortsyan and C. H. Keitel, "Relativistic high-power laser-matter interaction", Phys. Reps. **427** 41 (2006); M. Marklund and P. K. Shukla, "Nonlinear collective effects in photon-photon and photon-plasma interactions", Revs. Mod. Phys. **78** 591 (2006); H. K. Avetissian, *Relativistic Nonlinear Electrodynamics. Interaction of Charged Particles with Strong and Super Strong Laser Fields*, Springer Series in Optical Sciences **120**, ed. W. T. Rhodes, Springer, New York (2006); A. Di Piazza, C. Muller, K. Z. Hatsagortsyan and C. H. Keitel, "Extremely high-intensity laser interactions with fundamental quantum systems", Revs. Mod. Phys. **84** 1177 (2012).

[4] P. L. Kapitza and P. A. M. Dirac, "The reflection of electrons from standing light waves", Proc. Cambridge Phil. Soc. **29** 297 (1933).

$\mathbf{A} = \mathbf{A}_0 \cos(\omega t - \mathbf{kr})$ (linear polarization), where $\mathbf{A}_0$ is the amplitude and ω and $\mathbf{k}$ are the radiation frequency and the wavevector; t and $\mathbf{r}$ denote time and position. Since the phase velocity of the non-relativistic charge is much smaller than the speed of light c in vacuum ($\omega = ck$), we may neglect the spatial phase $\mathbf{kr}$ in comparison with the temporal phase ωt; consequently, the vector potential may be approximated by $\mathbf{A} \simeq \mathbf{A}_0 \cos \omega t$. The hamiltonian of the particle in the radiation field becomes the well-known dipole hamiltonian[5]

$$H = \frac{1}{2m} p^2 - e\mathbf{r}\mathbf{E}(t) + V(\mathbf{r}) \quad , \tag{9.1}$$

where $\mathbf{E}(t) = \mathbf{E}_0 \sin \omega t$, $\mathbf{E}_0 = \omega \mathbf{A}_0 / c$, is the electric field and $V(\mathbf{r})$ is a potential which does not depend on time. The non-relativistic character of the motion is preserved in the radiation field provided $eA_0/mc^2 \ll 1$. Making use of the notations $H_0 = p^2/2m$ and $U = -e\mathbf{r}\mathbf{E}$, the wavefunction ψ_i of the incident particle satisfies the Schroedinger equation

$$i\hbar \frac{\partial \psi_i}{\partial t} = (H_0 + U + V)\psi_i \tag{9.2}$$

with the initial condition (incoming state)

$$\psi_i^0 = \frac{1}{\sqrt{v}} e^{-\frac{i}{\hbar} E_i t + \frac{i}{\hbar} \mathbf{p}_i \mathbf{r}} \quad , \tag{9.3}$$

where $\mathbf{p}_i$ is the initial momentum, E_i is the initial energy and v denotes the volume ($\hbar$ is Planck's constant); the solution of equation (9.2) is given by

$$\psi_i = \psi_i^0 - \frac{i}{\hbar} e^{-\frac{i}{\hbar} H_0 t} \int_{-\infty}^{t} dt' e^{\frac{i}{\hbar} H_0 t'} (U + V) \psi_i \ . \tag{9.4}$$

The wavefunction of the final scattering state (outgoing state) is

$$\psi_f = \frac{1}{\sqrt{v}} e^{-\frac{i}{\hbar} E_f t + \frac{i}{\hbar} \mathbf{p}_f \mathbf{r}} \quad , \tag{9.5}$$

---

[5] H. A. Bethe and E. E. Salpeter, *Quantum Mechanics of One- and Two-Electron Atoms*, Springer, Berlin (1957); L. D. Landau and E. M. Lifshitz, *Course of Theoretical Physics*, vol. 3, *Quantum Mechanics*, 3rd ed., Elsevier Butterworth-Heinemann, Oxford (2005), p. 297.

where $\mathbf{p}_f$ is the final momentum and E_f is the final energy. The transition amplitude (S-matrix) reads

$$a_{fi} = -\frac{i}{\hbar}\int_{-\infty}^{+\infty} dt(\psi_f, (U+V)\psi_i) \ . \tag{9.6}$$

In equation (9.2) we insert successively the Goeppert-Mayer transform[6]

$$\begin{gathered} \psi_i = e^{iS_1}\phi_i \ , \ S_i = -\tfrac{1}{\hbar}\int_{-\infty}^t dt' U(t') = -\tfrac{e}{\hbar c}\mathbf{rA} \ , \\ i\hbar\tfrac{\partial\phi_i}{\partial t} = \left[\tfrac{1}{2m}(\mathbf{p}-\tfrac{e}{c}\mathbf{A})^2 + V\right]\phi_i \end{gathered} \tag{9.7}$$

and the Kramers-Henneberger transform[7]

$$\begin{aligned} \phi_i = e^{iS_2}\chi_i \ , \ S_2 &= \tfrac{e}{\hbar mc}\int_{-\infty}^t dt'\mathbf{pA} - \tfrac{e^2}{2\hbar mc^2}\int_{-\infty}^t dt' A^2 = \\ &= \tfrac{e}{\hbar mc\omega}\mathbf{pA}_0\sin\omega t - \tfrac{e^2}{8\hbar mc^2\omega}A_0^2(\sin 2\omega t + 2\omega t) \ , \\ i\hbar\tfrac{\partial\chi_i}{\partial t} &= \left(H_0 + \widetilde{V}\right)\chi_i \ , \ \widetilde{V}(\mathbf{r}) = V(\mathbf{r} - e\mathbf{E}/m\omega^2) \ . \end{aligned} \tag{9.8}$$

We recognize in equations (9.7) the standard non-relativistic hamiltonian of the particle in the radiation field. As regards the potential V we limit ourselves to the Born approximation; consequently, the wavefunction χ_i can be written as

$$\chi_i = \psi_i^0 - \frac{i}{\hbar}e^{-\frac{i}{\hbar}H_0 t}\int_{-\infty}^t dt' e^{\frac{i}{\hbar}H_0 t'}\widetilde{V}\psi_i^0 \tag{9.9}$$

[6] M. Goeppert-Mayer, "Uber Elementarakte mit zwei Quantensprungen", Ann. Physik **401** 273 (1931).

[7] W. C. Henneberger, "Perturbation method for atoms in intense light beams", Phys. Rev. Lett. **21** 838 (1968); H. A. Kramers, "Non-relativistic quantum-electrodynamics and the correspondence principle", Solvay Congress 1948, in *Collected Scientific Papers*, North-Holland, Amsterdam (1956), p. 865; N. G. van Kampen, "Contribution to the quantum theory of light scattering", Kgl. Danske Videnskab. Selskab, Mat.-fys. Medd. **26** (15) (1951); W. Pauli and M. Fierz, "Zur Theorie der Emission langwelliger Lichtquanten", Nuovo Cimento **15** 167 (1938).

and the transition amplitude becomes

$$a_{fi} = -\frac{i}{\hbar}\int_{-\infty}^{+\infty} dt(\psi_f, (U+V)e^{iS_1}e^{iS_2}\psi_i^0)+$$
$$+\left(-\frac{i}{\hbar}\right)^2\int_{-\infty}^{+\infty} dt(\psi_f, Ue^{iS_1}e^{iS_2}e^{-\frac{i}{\hbar}H_0t}\cdot \qquad (9.10)$$
$$\cdot\int_{-\infty}^{t} dt' e^{\frac{i}{\hbar}H_0t'}\widetilde{V}\psi_i^0) \ .$$

The advantage of the canonical transformations carried out above consists in the possibility of applying the Born approximation. The U-term in equation (9.10) corresponds to the scattering by the radiation field; we can see in equation (9.10) that the scattering by the potential V is dressed with radiation (terms $\sim Ve^{iS_1}e^{iS_2}$, $\sim \widetilde{V}$); in addition, there appears an interference term, which includes the product $U\widetilde{V}$.

We may make certain simplifications in equation (9.10). It is easy to see that the phase S_1 (equations (9.7)) is of the order $(r/\lambda)(eA_0/\hbar\omega) \gg eA_0/\hbar\omega$, where λ is the radiation wavelength; while the phase S_2 (equations (9.8)) is of the order $(p/mc\,,\, eA_0/mc^2)(eA_0/\hbar\omega) \ll eA_0/\hbar\omega$; consequently, we may neglect the phase S_2 in equation (9.10). In addition, the position given by the argument of $\widetilde{V}$ (equations (9.8)) is of the order $r - \lambda_c(eA_0/\hbar\omega)$, where $\lambda_c = \hbar/mc$ is the Compton wavelength of the particle; it takes the potential $\widetilde{V}$ far away from its short range in very short times, especially for (reasonably) high-intensity radiation; similarly, the Coulomb potential is rapidly reduced to an appreciable extent by the radiation, such that we may neglect the potential $\widetilde{V}$ in equation (9.10).

Making use of these simplifications the transition amplitude given by equation (9.10) becomes

$$a_{fi} \simeq -\frac{i}{\hbar}\int_{-\infty}^{+\infty} dt(\psi_f, Ue^{iS_1}\psi_i^0)-$$
$$-\frac{i}{\hbar}\int_{-\infty}^{+\infty} dt(\psi_f, Ve^{iS_1}\psi_i^0) =$$
$$= \int_{-\infty}^{+\infty} dt(\psi_f, \left(\frac{\partial}{\partial t}e^{iS_1}\right)\psi_i^0)- \qquad (9.11)$$
$$-\frac{i}{\hbar}\int_{-\infty}^{+\infty} dt(\psi_f, Ve^{iS_1}\psi_i^0) \ ,$$

or

$$a_{fi} = \frac{1}{v}\int_{-\infty}^{+\infty} dt d\mathbf{r} e^{\frac{i}{\hbar}(E_f-E_i)t}\frac{\partial}{\partial t}e^{-\frac{ie}{\hbar\omega}\mathbf{r}\mathbf{E}_0\cos\omega t}e^{\frac{i}{\hbar}\mathbf{p}\mathbf{r}}-$$
$$-\frac{i}{\hbar v}\int_{-\infty}^{+\infty} dt d\mathbf{r} e^{\frac{i}{\hbar}(E_f-E_i)t}Ve^{-\frac{ie}{\hbar\omega}\mathbf{r}\mathbf{E}_0\cos\omega t}e^{\frac{i}{\hbar}\mathbf{p}\mathbf{r}} \quad , \tag{9.12}$$

where $\mathbf{p} = \mathbf{p}_i - \mathbf{p}_f$ is the momentum transfer. In equation (9.12) we use the decomposition

$$e^{-\frac{ie}{\hbar\omega}\mathbf{r}\mathbf{E}_0\cos\omega t} = \sum_{n=-\infty}^{+\infty}(-i)^n J_n(e\mathbf{r}\mathbf{E}_0/\hbar\omega)e^{-in\omega t} \quad , \tag{9.13}$$

where J_n are Bessel functions (n being any integer), and get

$$a_{fi} = -\frac{2\pi i}{v}\sum_{n=-\infty}^{+\infty}(-i)^n\delta(E_f - E_i - n\hbar\omega)\cdot$$
$$\cdot\int d\mathbf{r}\left[n\hbar\omega + V(\mathbf{r})\right]J_n(e\mathbf{r}\mathbf{E}_0/\hbar\omega)e^{\frac{i}{\hbar}\mathbf{p}\mathbf{r}} \quad ; \tag{9.14}$$

we can see the occurrence of multiple-photon scattering processes with energy conservation $E_f = E_i + n\hbar\omega$. The transition probability per unit time is given by

$$w_{fi} = \frac{2\pi}{\hbar v^2}\sum_{n=-\infty}^{+\infty}\delta(E_f - E_i - n\hbar\omega)\cdot$$
$$\cdot\left|\int d\mathbf{r}\left[n\hbar\omega + V(\mathbf{r})\right]J_n(e\mathbf{r}\mathbf{E}_0/\hbar\omega)e^{\frac{i}{\hbar}\mathbf{p}\mathbf{r}}\right|^2 \quad ; \tag{9.15}$$

we multiply w_{fi} by the density of final states $vp_f^2 dp_f d\Omega/(2\pi\hbar)^3$, where $d\Omega$ is the element of solid angle, divide by the current density v_i/v, where v_i is the initial velocity, and integrate over the final momentum p_f in order to get the differential cross-section for the n-process

$$d\sigma_n = \frac{p_{fn}}{p_i}\left|\frac{m}{2\pi\hbar^2}\int d\mathbf{r}\left[n\hbar\omega + V(\mathbf{r})\right]J_n(e\mathbf{r}\mathbf{E}_0/\hbar\omega)e^{\frac{i}{\hbar}\mathbf{p}\mathbf{r}}\right|^2 d\Omega \quad ; \tag{9.16}$$

the momentum p_{fn} is given by the energy conservation $p_{fn}^2/2m = p_i^2/2m + n\hbar\omega$.

We can see in equation (9.16) the cross-section of the scattering by radiation and the cross-section of the scattering from the potential V;

in addition, there appear mixed radiation-potential scattering terms; for $n = 0$ we get from equation (9.16) the elastic Born scattering in the field of the potential V affected by radiation, due to the presence of the function J_0. The argument of the Bessel function in equation (9.16) varies rapidly over the integration domain; therefore, we may use the asymptotic expression for the Bessel function; by doing so, we can see easily that the **r**-integration in the contribution of the radiation scattering is non-vanishing for $\mathbf{p} \simeq e\mathbf{E}_0/\omega$, *i.e.* the momentum is transferred along the direction of the electric field, as expected; this is the momentum conservation. Making use of this relation and the energy conservation, we get $p_{fn} \simeq p_i(1 + n\hbar\omega/2E_i)$ (for $n\hbar\omega \ll E_i$) and

$$-1 < \cos\theta_n = \frac{1 - e\mathbf{p}_i\mathbf{E}_0/p_i^2\omega}{1 + n\hbar\omega/2E_i} < 1 \ , \tag{9.17}$$

where θ_n is the scattering angle. For $n > 0$ (photon absorption) the scattering angle θ_n increases, while for $n < 0$ (photon emission) the scattering angle decreases, with respect to the elastic scattering angle θ_0. Equation (9.17) indicates that there exists a limitation which can be written as $-n\hbar\omega/2E_i < (eA_0\cos\alpha/2E_i)(v/c) < 2 + n\hbar\omega/2E_i$, where α is the angle the initial momentum makes with the electric field; for very low incident energies the scattering occurs only for angles close to the right angle made by the incident momentum with the electric field. Finally, we note that the order of magnitude of the cross-section of the scattering due to radiation is $\simeq d^2[n\hbar\omega/(\hbar^2/md^2)]^2(\hbar c/eA_0d)$, where $(\hbar c/eA_0d) \ll 1$ and d is the dimension of the region where the radiation is confined to. We can see from equation (9.16) that the cross-section due to radiation may acquire large values, as a consequence of the large dimension of the region containing radiation; the cross-section increases with increasing n, *i.e.* for large scattering angles, where, however, the scattering maxima coalesce.

As regards the modification brought about by the presence of the radiation in the scattering produced by the potential $V(\boldsymbol{r})$, there are two differences in the present case, where the radiation is limited to a finite region, in comparison with the case where the scattering is immersed in radiation. If, in the latter case, we neglect the interaction U, leave aside the corresponding transformation given by U_1 and start directly with the standard non-relativistic hamiltonian given by equation (9.7), then it is easy to see that we get from equations (9.11) and (9.16) the Kroll-Watson formula $d\sigma_n = (p_{fn}/p_i)J_n^2(e\boldsymbol{p}\boldsymbol{E}_0/m\hbar\omega^2)d\sigma_B$, where $d\sigma_B$ is the elastic Born cross-section.[8] This formula shows that the Born cross-section is modulated by the function J_n^2, whose argument includes the energy conservation with exchange of n photons. In our case the first difference consists in the presence of the interference term $\sim n\hbar\omega V(\boldsymbol{r})J_n(e\boldsymbol{r}\boldsymbol{E}_0/\hbar\omega)$ in equation (9.16), arising from both potentials U and V, which is absent in the Kroll-Watson formula. The second difference is related to the Bessel function $J_n(e\boldsymbol{r}\boldsymbol{E}_0/\hbar\omega)$, which is included in the integral in equation (9.16) and depends on the range of the interaction $V(\boldsymbol{r})$ through its argument $e\boldsymbol{r}\boldsymbol{E}_0/\hbar\omega$, while in the Kroll-Watson formula the Bessel function depends on $e\boldsymbol{p}\boldsymbol{E}_0/m\hbar\omega^2$ and is placed outside the integral. These differences are important, especially for Coulomb potentials. In addition, while in the Kroll-Watson formula the momentum transfer $\boldsymbol{p}$ is given by the potential $V(\boldsymbol{r})$, in our case it is given both by the potential $V(\boldsymbol{r})$ and the radiation potential $U(\boldsymbol{r})$ through the function $J_n(e\boldsymbol{r}\boldsymbol{E}_0/\hbar\omega)$.

The above calculations are done for a sufficiently long laser pulse. In practice, the pulse has a finite duration τ and a repetition time Δt. In these conditions the expansion given by equation (9.13) remains valid, but the function $\delta(\Delta E)$ in the scattering amplitude a_{fi} (equation (9.14)), where $\Delta E = E_f - E_i - n\hbar\omega$), is replaced by the function $\zeta(\Delta E) = e^{\frac{i}{\hbar}\Delta E\cdot t_i}\sin\alpha\Delta E/\pi\Delta E$, where $\alpha = \tau/2\hbar$ and t_i denotes the time moment of the pulse (the pulse lasts from $t_i - \tau/2$ to $t_i + \tau/2$). For large α the function $\zeta(\Delta E)$ has a maximum for $\Delta E = 0$ and extends approximately over a bandwidth $\delta E \simeq \pi/\alpha = 2\pi\hbar/\tau$. Since, usually τ is much longer than the radiation period $T = 2\pi/\omega$, the energy separation δE is much smaller than the radiation quanta of energy $\hbar\omega$. It follows that the functions $\zeta(\Delta E)$ for different n can be viewed as being well separated. Under these conditions the cross-section $d\sigma_n$ (equation (9.16)) preserves its form, except that it is multiplied by the reduction factor $\tau/(\tau + \Delta t)$.

[8] N. M. Kroll and K. M. Watson, "Charged-particle scattering in the presence of a strong electromagnetic wave", Phys. Rev. **A8** 804 (1973); N. K. Rahman, "Effect of intense electromagnetic wave in charged-particle scattering", Phys. Rev. **A10** 440 (1974).

In conclusion, it is shown here that non-relativistic charged particles may suffer scattering as a result of their interaction with the electromagnetic radiation in the focal region of laser beams. The cross-section of this scattering process is computed here for a single-mode radiation with linear polarization. As expected, the cross-section exhibits maxima at certain scattering (diffraction) angles θ_n, as given by equation (9.17), determined by the energy and momentum conservation in multiple-photon exchange processes. The calculations can be easily extended to any polarization; for realistic laser beams, or for multi-mode radiation, we should take into account the particular beam shape and the amplitude and frequency fluctuations.[9] For convenience, we included also the scattering from a potential placed in the radiation field, in the Born approximation. The cross-section of the potential scattering is modified by the presence of the radiation, because the scattering states are dressed by radiation. The cross-section is reduced for high-intensity radiation (preserving the non-relativistic character of the particle motion) and, similarly, the multi-photon scattering is diminished for high energy of the particle, when the process reduces to elastic scattering in the forward direction. In contrast with previous studies, where the scattering process is immersed in the radiation field, in the scattering process addressed here the radiation field is confined to a finite region. In these circumstances, the modifications brought by the radiation to the potential scattering are different from the well-known Kroll-Watson formula.

9.2 Cross-section

The charges envisaged above are electrons or ions, and the electromagnetic radiation is confined to the focal region of the laser beam, which has a finite spatial extension d. This scattering process is distinct from the Kroll-Watson scattering, where the charges are immersed in the laser radiation and are scattered by atomic or nuclear targets. Further details are given below, which may be useful for experiments.[10]

[9] P. Zoller, "Laser photon correlation effects in electron scattering", J. Phys. B: At. Mol. Opt. Phys. **13** L249 (1980).

[10] M. Apostol and L. C. Cune, "Analysis of the cross-section of charge scattering by electromagnetic radiation", Roum. J. Phys. **63** 110 (2018).

We assume optical lasers with photon frequency of the order $\omega = 10^{15}s^{-1}$ (wavelength λ of the order $1\mu m$) and the dimension d of the focal region of the order of a few tens of photon wavelengths (*e.g.*, $d \simeq 3{\cdot}10^{-3}cm$). In the non-relativistic approximation the electric field $\mathbf{E} = \mathbf{E}_0 \sin\omega t$ is purely oscillating in time (t), where $\mathbf{E}_0$ is its amplitude, for linear polarization (its spatial dependence may be neglected, as well as the radiation magnetic field). The non-relativistic approximation requires $\mid e \mid A_0/mc^2 \ll 1$, where e is the particle charge, m is the particle mass, $A_0 = cE_0/\omega$ is the amplitude of the vector potential and c ($= 3\times 10^{10}cm/s$) is the speed of light in vacuum. This condition imposes $E_0 \ll 10^8 esu$ for electrons and $E_0 \ll 10^{11}A(esu)$ for ions, where A is the mass number of the ion. The laser intensity ($I = cE_0^2/8\pi$) is limited to $10^{18}w/cm^2$ for electrons and $10^{24}w/cm^2$ for ions.

The cross-section of charge scattering by electromagnetic radiation given above (equation (9.16)) is

$$d\sigma_n = \tfrac{p_{fn}}{p_i} \mid \tfrac{m}{2\pi\hbar^2}\int d\mathbf{r}\,[n\hbar\omega + V(\mathbf{r})]\cdot$$
$$\cdot J_n(e\mathbf{r}\mathbf{E}_0/\hbar\omega)e^{-\frac{i}{\hbar}\mathbf{p}_n\mathbf{r}} \mid^2 d\Omega \quad , \tag{9.18}$$

where J_n is the Bessel function of (integral) order n, $\mathbf{p}_i$ and $\mathbf{p}_{fn}$ are the initial and final momenta of the projectile, $\mathbf{p}_n = \mathbf{p}_{fn} - \mathbf{p}_i$ is the momentum transfer, $V(\mathbf{r})$ denotes an external potential, $d\Omega$ is the scattering solid angle and $\hbar$ is the Planck constant. For brevity, we omit the suffix n of the cross-section ($d\sigma$), the final momentum ($\mathbf{p}_f$) and the momentum transfer ($\mathbf{p}$). The integral in equation (9.18) is performed over the spatial region of dimension d. The initial and final momenta are related by the law of energy conservation

$$p_f^2/2m = p_i^2/2m + n\hbar\omega \ ; \tag{9.19}$$

this relation indicates scattering processes associated with emission and absorbtion of multiple photons.

We neglect for the moment the external potential $V(\mathbf{r})$ in equation (9.18). The cross-section due solely to the radiation scattering can be written as

$$d\sigma = \frac{p_f}{p_i}\left(\frac{mn\omega}{2\pi\hbar}\right)^2 \mid K_n \mid^2 d\Omega \quad , \tag{9.20}$$

where

$$K_n = \int d\mathbf{r} J_n(\mathbf{ar})e^{-i\mathbf{qr}} \ , \ \ \mathbf{a} = e\mathbf{E}_0/\hbar\omega \ , \ \ \hbar\mathbf{q} = \mathbf{p} \ , \tag{9.21}$$

or

$$K_n = \int dx J_n(ax)e^{-iq_{\parallel}x} \int d\mathbf{r}_{\perp} e^{-i\mathbf{q}_{\perp}\mathbf{r}_{\perp}} \ , \tag{9.22}$$

where the vectors $\mathbf{r} = (x, \mathbf{r}_{\perp})$, $\mathbf{q} = (q_{\parallel}, \mathbf{q}_{\perp})$ have been decomposed along two directions, one parallel with $\mathbf{a}$ (components x and $q_{\parallel}$) and the other perpendicular to $\mathbf{a}$ (components $\mathbf{r}_{\perp}$ and $\mathbf{q}_{\perp}$). The integral with respect to $\mathbf{r}_{\perp}$ shows that $q_{\perp}$ is approximately of the order $1/d$ and the integral can be approximated by d^2. Within this approximation we can see that the momentum transfer reduces to $p = \hbar q_{\parallel}$, *i.e.* the momentum is transferred along the electric field $\mathbf{E}_0$ ($\mathbf{a}$). We omit the suffix $\parallel$ in $q_{\parallel}$ and write $K_n d^2 = I_n$, where

$$I_n = \int dx J_n(ax)e^{-iqx} \ , \ \ \hbar q = p \ ; \tag{9.23}$$

since $\mid a \mid = \mid e \mid E_0/\hbar\omega \ll 1/d$ and $q = p/\hbar \ll 1/d$, we may extend the integration in this integral to the whole space. The restriction of the momentum transfer to the direction $\mathbf{E}_0$ implies a cylindrical geometry for the scattering; the integration over angle φ in equation (9.20) leads to formally replacing $p_f \sin\theta d\varphi$ by $\hbar/d$; therefore, the cross-section given by equation (9.20) becomes

$$d\sigma = \frac{\hbar d^3}{p_i}\left(\frac{mn\omega}{2\pi\hbar}\right)^2 \mid I_n \mid^2 d\theta \ . \tag{9.24}$$

We pass now to the momentum conservation law

$$\mathbf{p} = \mathbf{p}_f - \mathbf{p}_i \ , \tag{9.25}$$

where $\mathbf{p}$ is directed along the electric field $\mathbf{E}_0$ and $p_f = R_n p_i$, $R_n = \sqrt{1 + n\hbar\omega/E_i}$ from the energy conservation law given by equation (9.19), $E_i = p_i^2/2m$ being the initial energy. Let us assume that the electric field $\mathbf{E}_0$ makes an angle α with the initial direction of the projectile; it is sufficient to assume $0 < \alpha < \pi/2$. We parametrize the vectors occurring in equation (9.25) by $\mathbf{p} = p(\cos\alpha, \sin\alpha)$, $\mathbf{p}_f = p_f(\cos\theta, \sin\theta)$ and $\mathbf{p}_i = (1, 0)$ and get from equation (9.25)

$$p\cos\alpha = p_f\cos\theta - p_i \ , \ \ p\sin\alpha = p_f\sin\theta \tag{9.26}$$

and

$$\sin(\alpha - \theta) = \frac{\sin\alpha}{R_n} \ . \tag{9.27}$$

The solutions of this latter equation are

$$\begin{aligned} \theta_n &= \alpha - \arcsin\left(\tfrac{\sin\alpha}{R_n}\right) \ , \\ \theta_n &= \alpha - \pi + \arcsin\left(\tfrac{\sin\alpha}{R_n}\right) \ ; \end{aligned} \tag{9.28}$$

we can see that the scattering angles take discrete values, like for a diffraction scattering; the angles in the first row of equation (9.28) range from 0 to α, while those in the second row are comprised between $\alpha - \pi$ and 0. The momentum transfer given by equations (9.26) is

$$\begin{aligned} p &= p_i\tfrac{n\hbar\omega/E_i}{\sqrt{\cos^2\alpha + n\hbar\omega/E_i} + \cos\alpha} \ , \ \ 0 < \theta_n < \alpha \ , \\ p &= -p_i\left(\sqrt{\cos^2\alpha + n\hbar\omega/E_i} + \cos\alpha\right) \ , \ \ \alpha - \pi < \theta_n < 0 \ . \end{aligned} \tag{9.29}$$

There exist a few special scattering angles, according to these equations. For $n \to \infty$ the scattering angles are α and $\alpha - \pi$, *i.e.* the scattering takes place along the electric field (in both directions), with an infinite momentum and energy transfer. For $n = 0$ the scattering is elastic and it takes place at angles 0 and $2\alpha - \pi$; the momentum transfer is zero for $\theta_0 = 0$ and $-2p_i\cos\alpha$ for $\theta_0 = 2\alpha - \pi$. For $\theta_{n_0} = \alpha - \pi/2$, where n_0 is given by $R_{n_0} = \sin\alpha$ ($n_0 = -(E_i/\hbar\omega)\cos^2\alpha$) the scattering takes place along the direction perpendicular to the electric field, with the momentum transfer $-p_i\cos\alpha$. For $0 < \theta_n < \alpha$ and $\alpha - \pi < \theta_n < 2\alpha - \pi$ the scattering is associated with photon absorption ($n > 0$), while for $2\alpha - \pi < \theta_n < 0$ the scattering implies photon emission. All this information is shown in Fig. 9.1. The final momenta $\mathbf{p}_f$ can be obtained by a graphical method. In the plane made by the vectors $\mathbf{p}_i$ and $\mathbf{E}_0$ we draw a succession of concentric

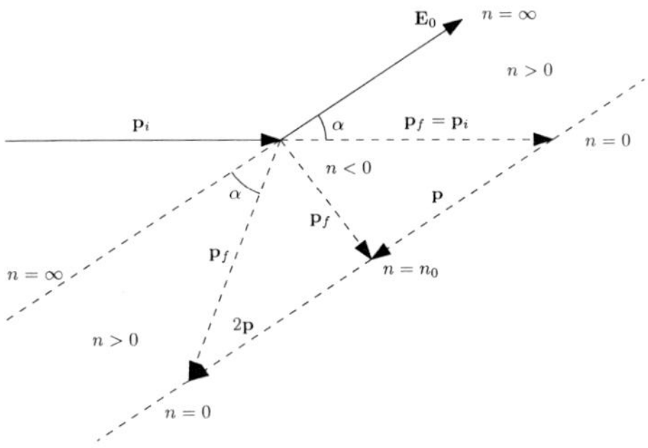

Figure 9.1: Special scattering angles for electric field $\mathbf{E}_0$, initial momentum $\mathbf{p}_i$, final momentum $\mathbf{p}_f$ and momentum transfer $\mathbf{p}$ and $2\mathbf{p}$; the regions with $n > 0$ correspond to photon absorption, while the region with $n < 0$ corresponds to photon emission. The line $n = n_0 = -(E_i/\hbar\omega)\cos^2\alpha$, where E_i is the projectile energy and $\hbar\omega$ is the photon energy indicates the scattering angle $\theta = \alpha - \pi/2$.

spheres with radii $p_{fn} = p_i\sqrt{1 + n\hbar\omega/E_i}$, n integer; the points of the intersection of these spheres with the line parallel with $\mathbf{E}_0$ and passing through the point indicated by the vector $\mathbf{p}_i$ give the scattering momenta $\mathbf{p}_{fn}$.

In order to get an observable diffraction pattern the integer n should take many values in the vicinity of $n = 0$ and n_0. (On the other hand, n must be limited, of course, to the maximum number of photons $N \simeq E_0^2 d^3/8\pi\hbar\omega \simeq 4 \times 10^5 E_0^2$ comprised in the scattering region). Therefore, we should impose the condition $\hbar\omega/E_i \ll 1$. From equations (9.28) the angle separation is given by

$$\Delta\theta_n = \frac{\sin\alpha}{2R_n^2\sqrt{R_n^2 - \sin^2\alpha}}\frac{\hbar\omega}{E_i} \; ; \tag{9.30}$$

we can see that for large n the diffraction maxima coalesce. In the vicinity of $n = 0$ the separation angle is given by

$$\Delta\theta_0 \simeq \tan\alpha \cdot \frac{\hbar\omega}{2E_i} \; , \tag{9.31}$$

while the separation angle near n_0 it is

$$\Delta\theta_n \simeq \frac{\sqrt{\hbar\omega/E_i\Delta n}}{2\sin\alpha(1+\Delta n\hbar\omega/E_i\sin^2\alpha)} \ , \tag{9.32}$$

where $\Delta n = n - n_0 > 0$; we can see that the diffraction maxima are more separated in this region in comparison with region $n \simeq 0$. In order to have a good resolution, the separation angle $\Delta\theta_0$ should not be too small; for instance, a good resolution would be $\Delta\theta_0 \simeq 10^{-3}$, which imposes E_i of the order $1keV$ (for $\hbar\omega = 1eV$).

We turn now to the estimation of the cross-section given by equation (9.24).

By using the generating function of the Bessel functions

$$e^{iz\cos\alpha} = \sum_n i^n J_n(z)e^{in\alpha} , \ J_n(z) = \frac{(-i)^n}{2\pi}\int_{-\pi}^{\pi} d\alpha e^{iz\cos\alpha - in\alpha} , \tag{9.33}$$

we get

$$I_n = (-i)^n \int_{-\pi}^{\pi} d\alpha e^{-in\alpha}\delta(a\cos\alpha - q) \ . \tag{9.34}$$

The integral in equation (9.23) can be computed immediately, with the result

$$I_n = \frac{2(-i)^n}{\sqrt{a^2-q^2}}\cos n\alpha_0 \ , \tag{9.35}$$

where $0 < \alpha_0 = \arccos(q/a) < \pi$ and $\mid q \mid < a$. For $q \ll a$, $\alpha_0 \simeq \pi/2$ and $I_{2k} \simeq 2/a$ does not depend on k. The cross-section in this case is given by

$$d\sigma_{2k} \simeq \frac{\hbar d}{\pi^2 p_i}\left(\frac{\hbar\omega}{eE_0 d}\right)^2\left(\frac{\hbar\omega}{\hbar^2/2md^2}\right)^2 k^2 d\theta \ . \tag{9.36}$$

This result is valid for scattering angles $\theta \simeq 0$ (forward scattering), where the momentum transfer is small. With the numerical data used here and $E_i = 1keV$, $E_0 = 10^8 esu$ the differential cross-section given by equation (9.36) is of the order $\simeq 10^{-8}k^2$. For larger scattering angles odd-order diffraction maxima appear and the diffraction spots increase in intensity (the projectile acquires more momentum transfer along directions parallel with the electric field). In the emission region

$n < 0$ the momentum transfer is of the order p_i. For large momentum transfer the denominator in equation (9.35) may vanish; then, it should be replaced by $\sqrt{a/d}$, and the cross-section given by equation (9.36) acquires an additional factor of the order $(\mid e \mid E_0 d/\hbar\omega)$.

For $\mid q \mid > \mid a \mid$ the integrals given by equation (9.23) are zero. It follows that the momentum transfer given by equations (9.29) is limited to $\mid p \mid < \mid e \mid E_0/\omega$ (the diffraction spots disappear for $\mid p \mid < \mid e \mid E_0/\omega$, in particular they are absent in the limit $n \to \infty$). Indeed, in equation (9.23) we may limit ourselves to $n \geq 0$ and a, $q > 0$. The integrals I_n reduce to

$$L_n = [1 + (-1)^n] L_n^c + [1 - (-1)^n] L_n^s \ , \tag{9.37}$$

where

$$L_n^c = \int_0^\infty dx J_n(x) \cos\lambda x \ , \ L_n^s = \int_0^\infty dx J_n(x) \sin\lambda x \tag{9.38}$$

and $\lambda = q/a > 1$; further on, integrating by parts,

$$L_n^c = \frac{1}{2\lambda}\left(L_{n+1}^s - L_{n-1}^s\right) \tag{9.39}$$

for $n \geq 1$. The integral L_n^s is known:[11]

$$L_n^s = \frac{\cos(n\pi/2)}{\sqrt{\lambda^2 - 1}\left[\lambda + \sqrt{\lambda^2 - 1}\right]^n} \ , \ \lambda > 1 \ . \tag{9.40}$$

We can see that the integrals L_n, $n \geq 0$ are zero for $\lambda > 1$.

Two particular cases may be worth discussing. First, let us assume that the initial momentum $\mathbf{p}_i$ is perpendicular to the electric field $\mathbf{E}_0$ (transverse scattering, $\alpha = \pi/2$). The scattering angles are given in this case by $\cos\theta_n = 1/R_n$, the momentum transfer is $p_n = p_i \tan\theta_n$ and the separation angles are $\Delta\theta_n = \sqrt{\hbar\omega/nE_i}/2R_n^2$ ($R_n = \sqrt{1 + n\hbar\omega/E_i}$) for $n \geq 0$ (there exists only absorption). For $\alpha = 0$ (longitudinal "scattering") there is no scattering, the motion takes place along the direction of the electric field $\mathbf{E}_0$; the final momenta

[11] I. S. Gradshteyn and I. M. Ryzhik, *Table of Integrals, Series and Products*, eds. A. Jeffrey and D. Zwillinger, 6*th* ed., Academic Press (2000), p. 1117, 17.33(38).

are given by $p_{fn} = p_i\sqrt{1 + n\hbar\omega/E_i}$, with absorption for $n > 0$ and emission for $0 < n < -E_i/\hbar\omega$ (providing $\hbar\omega < E_i$).

The contribution of the external potential $V(\mathbf{r})$ to the cross-section given by equation (9.18) can be estimated by using the Fourier transform of the potential. The momentum transfer is not confined to the direction of $\mathbf{E}_0$ anymore. The diffraction maxima are placed on the circles resulting from the intersection of the spheres with radii p_{fn} with the plane perpendicular to the plane made by the vectors $\mathbf{p}_i$ and $\mathbf{E}_0$, parallel with the vector $\mathbf{E}_0$ and passing through the end point of $\mathbf{p}_i$.

The electric field with elliptic polarization has two components $E_1 \sin\omega t$ and $E_2 \cos\omega t$ in the plane perpendicular to the propagation vector. The energy conservation reads $p_f^2/2m = p_i^2/2m + (n_1+n_2)\hbar\omega$ and the cross-section is given by

$$\begin{aligned} d\sigma = \frac{p_f}{p_i} \mid \frac{m}{2\pi\hbar^2} \int d\mathbf{r}[(n_1+n_2)\hbar\omega + V(\mathbf{r})]\cdot \\ \cdot J_{n_1}(eE_1x/\hbar\omega)J_{n_2}(eE_2y/\hbar\omega)e^{-\frac{i}{\hbar}\mathbf{pr}} \mid d\Omega \ , \end{aligned} \tag{9.41}$$

where $n_{1,2}$ are integers. The diffraction maxima are placed on ellipses, which depend on the sum $n_1 + n_2$, and have a variable intensity, depending on $n_{1,2}$.

9.3 Quasi-classical approximation

It is well known that a quantum-mechanical charge in low-intensity electromagnetic radiation may suffer transitions by absorbing or emitting photons. However, very often, in typical scattering experiments the projectile momentum (energy) is sufficiently high to make the projectile motion a (non-relativistic) quasi-classical motion. Then, if the energy of the electromagnetic field is sufficiently low, the quasi-classical character of the motion is preserved in scattering and the charge may be diffracted (elastically scattered) by the radiation field. For higher radiation intensity the projectile motion in the radiation field becomes quantum-mechanical; then, inelastic scattering may appear, by absorption and emission of multiple "photons" (a mechanism distinct from the quantum-mechanical photon transitions). For very high-intensity radiation the charge may be rapidly accelerated by radiation up to relativistic velocities.

Strongly focused optical laser beams are able to generate high-intensity electromagnetic radiation in the focal region;[12] the electrons accelerated in such fields, up to energies in the GeV range, can be used as projectiles in nuclear physics experiments.[13] Via non-linear Compton scattering promising ways are opened up to investigating fundamental phenomena like electron-positron pairs generation, vacuum polarization, vacuum breakdown or non-linear quantum electrodynamics effects.[14] In this context, special efforts are devoted to connect typical nuclear physics with laser radiation. We present below a de-

---

[12] V. Yanovski, V. Chvykov, G. Kalinchenko, P. Rousseau, T. Planchon, T. Matsuoka, A. Maksimchuk, J. Nees, G. Cheriaux, G. Mourou and K. Krushelnik, "Ultra-high intensity-300-Tw laser at $0.1Hz$ repetition rate", Opt. Express **16** 2109 (2008); A. V. Korzhimanov, A. A. Gonoskov, E. A. Khazanov and A. M. Sergeev, "Horizons of petawatt laser technology", Phys. Usp. **54** 9 (2011).

[13] A. di Piazza, C. Muller, K. Z. Hatsagortsyan and C. H. Keitel, "Extremely high-intensity laser interactions with fundamental quantum systems", Revs. Mod. Phys. **84** 1177 (2012); D. Haberberger, S. Tochitsky, F. Fiuza, C. Gong, R. A. Fonseca, L. O. Silva, W. B. Mori and C. Joshi, "Collisionless shocks in laser-produced plasma generate monoenergetic high-energy proton beams", Nature Phys. **8** 95 (2012); V. Malka, "Laser plasma accelerators", Phys. Plasmas **19** 055501 (2012); D. L. Burke, R. C. Field, G. Horton-Smith, J. E. Spencer, D. Walz, S. C. Berridge, W. M. Bugg, K. Shmakov, A. W. Weidemann, C. Bula, K. T. McDonald, E. J. Prebys, C. Bamber, S. G. Boege, T. Koffas, T. Kotseroglou, A. C. Melissinos, D. D. Meyerhofer, D. A. Reiss and W. Ragg, "Positron production in multiphoton light-by-light scattering", Phys. Rev. Lett. **79** 1626 (1997); C. Bamber, S. J. Boege, T. Koffas, T. Kotseroglou, A. C. Melissinos, D. D. Meyerhofer, D. A. Reiss and W. Ragg, C. Bula, K. T. McDonald, E. J. Prebys, D. L. Burke, R. C. Field, G. Horton-Smith, J. E. Spencer, D. Walz, S. C. Berridge, W. M. Bugg, K. Shmakov, A. W. Weidemann, "Studies of nonlinear QED in collisions of $46.6GeV$ electrons with intense laser pulses", Phys. Rev. **D60** 092004 (1999).

[14] Y. I. Salamin, S. X. Hu, K. Z. Hatsagortsyan and C. H. Keitel, "Relativistic high-power laser-matter interaction", Phys. Reps. **427** 41 (2006); M. Marklund and P. K. Shukla, "Nonlinear collective effects in photon-photon and photon-plasma interactions", Revs. Mod. Phys. **78** 591 (2006); M. Marklund, "Fundamental optical physics: probing the quantum vacuum", Nature Photon. **4** 72 (2010); F. Ehlotzky, K. Krajewska and J. Z. Kaminski, "Fundamental processes of quantum electrodynamics in laser fields of relativistic power", Reps. Progr. Phys. **72b** 046401 (1-32) (2009); H. K. Avetissian, *Relativistic Nonlinear Electrodynamics. Interaction of Charged Particles with Strong and Super Strong Laser Fields*, Springer Series in Optical Sciences **120,** ed. W. T. Rhodes, Springer, New York (2006).

scription of quasi-classical charges scattered by radiation field. It is shown that a new diffraction phenomenon may appear in the scattering by low- and moderate-intensity laser radiation. It resembles the well known Kapitza-Dirac electron diffraction in an electromagnetic standing wave.[15] Also, the inelastic scattering cross-section is estimated for high-intensity radiation.

The scattering in the presence of the electromagnetic radiation was previously approached for projectiles dressed with radiation;[16] the interest in these investigations was to estimate the radiation effect on the scattering by a potential. In these approaches the incoming and outgoing wavefunctions of the projectile include the effects of the radiation (the scattering experiment is embedded in the radiation field). In the treatment below the asymptotic wavefunctions of the projectile are free wavefunctions, the scatterer being the radiation field or, possibly, a potential similar to the nuclear potential.

First, we consider the scattering of a non-relativistic ion with mass M by an electromagnetic radiation field; for convenience, we include also a target placed at the origin, which gives rise to a potential $V(\mathbf{r})$; the scattering potential $V(\mathbf{r})$ may be viewed as a typical nuclear potential. The electromagnetic radiation is described by a vector potential $\mathbf{a} = \mathbf{a}_0 \cos(\omega t - \mathbf{kr})$, $\mathbf{ka}_0 = 0$ (linear polarization), confined to a finite region F; the region F corresponds to the focal region of a laser beam; ω is the radiation frequency and $\mathbf{k}$ is the wavevector, $\omega = ck$, where

[15] P. L. Kapitza and P. A. M. Dirac, "The reflection of electrons from standing light waves", Proc. Cambridge Phil. Soc. **29** 297 (1933).

[16] N. M. Kroll and K. M. Watson, "Charged-particle scattering in the presence of a strong electromagnetic wave", Phys. Rev. **A8** 804 (1973); N. K. Rahman, "Effect of intense electromagnetic wave in charged-particle scattering", Phys. Rev. **A10** 440 (1974); N. J. Mason, "Laser-assisted electron-atom collisions", Reps. Progr. Phys. **56** 1275 (1993); B. Wallbark and J. K. Holmes, "Laser-assisted elastic electron-atom collisions: low electron energy and small scattering angle", J. Phys. B: At. Mol. Opt. Phys. **27** 1221 (1994); "Differential cross sections for laser-assisted elastic electron scattering from argon", J. Phys. B: At. Mol. Opt. Phys. **27** 5405 (1994); L. B. Madsen and K. Taulbjerg, "Theory of laser-assisted electron-atom scattering: beyond the Kroll-Watson approximation, J. Phys. B: At. Mol. Opt. Phys. **28** 5327 (1995); A. F. Klinskikh and L. P. Rapoport, "Quasi-classical theory of potential scattering of an electron in the field of a laser wave", Sov. Phys.-JETP **66** 53 (1987) (ZhETF **93** 95 (1987)); I. F. Barna and S. Varro, "Laser-assisted proton collision on light nuclei at moderate energies", Laser&Part. Beams **33** 299 (2015).

c is the speed of light in vacuum; t and $\mathbf{r}$ denote the time and the position vector. Usually, the size d of the focal region F may be of the order of a few tens of radiation wavelengths $\lambda = 2\pi/k$.

We consider the energy of the projectile in the range $1MeV$ to $\simeq 100MeV$, with a typical value $\mathcal{E} = 10MeV$, used for illustrating numerical estimations. For this energy the velocity of the projectile is $v = \sqrt{2\mathcal{E}/M} \simeq 3 \times 10^9/\sqrt{A}cm/s$, where A is the mass number of the projectile; we assume $v/c \ll 1$, *i.e.* non-relativistic ions (electron mass $\simeq 10^{-27}g$). The potential $V(\mathbf{r})$ is sufficiently small and localized, such that the scattering on it may be treated within the Born approximation; we shall show that the Born approximation is warranted also for the scattering by low- and moderate-intensity radiation field. For instance, for a laser intensity $I = 10^{18}w/cm^2$ in the focus, we get an electric field of the order $E_0 = \sqrt{8\pi I/c} \simeq 10^8 esu$ ($\simeq 3 \times 10^3 V/m$) (and a magnetic field with the same strength). For an optical radiation with frequency $\omega = 10^{15}s^{-1}$ the vector potential is of the order $a_0 \simeq (c/\omega)E_0 \simeq 3 \times 10^3 statvolt$, which implies an energy $ea_0 \simeq 10^{-6}erg \simeq 1MeV$ ($e = 4.8 \times 10^{-10}esu$ being the electron charge). The wavelength of the projectile is of the order $\Lambda = 2\pi\hbar/\sqrt{2M\mathcal{E}} \simeq 10^{-12}/\sqrt{A}cm$ (for $\mathcal{E} = 10MeV$), which is much shorter than the radiation wavelength $\lambda = 2\pi c/\omega \simeq 1.8 \times 10^{-4}cm$ (for $\omega = 10^{15}s^{-1}$) and the size d of the focal region F. It follows that the projectile motion may be viewed as quasi-classical.

The Schroedinger equation of the projectile in the region F reads

$$i\hbar\frac{\partial\psi}{\partial t} = \left[\frac{1}{2M}\mathbf{P}^2 - Ze\mathbf{r}\mathbf{E} + V(\mathbf{r})\right]\psi \ , \tag{9.42}$$

where $\mathbf{P}$ denotes momentum, Z is the charge number of the projectile and $\mathbf{E} = -(1/c)\partial\mathbf{a}/\partial t = \mathbf{E}_0\sin(\omega t - \mathbf{kr})$, $\mathbf{E}_0 = \omega\mathbf{a}_0/c$, is the electric field. First, we note that the effect of the radiation on the projectile motion is governed by the parameter $\xi = ZeE_0 d/\mathcal{E}$; this is also the parameter which controls the validity of the Born approximation. With our numerical data we get ξ of the order unity (for $\mathcal{E} = 10MeV$ and d a few tens of λ). We assume that the radiation has a sufficiently low intensity and the projectile energy is sufficiently high such that $\xi \ll 1$; under these conditions the quasi-classical character of the projectile motion in the radiation field is preserved and the Born approximation is meaningful. Second, we note that the spatial phase $\mathbf{kr}$ in the potential $\mathbf{a} = \mathbf{a}_0 \cos(\omega t - \mathbf{kr})$ is of the order kd at most, while the temporal phase ωt is of the order $(c/v)kd$; since $c/v \gg 1$, it follows that we may retain only the temporal phase ωt in the vector potential and may neglect the spatial phase $\mathbf{kr}$, a typical non-relativistic approximation (this amounts to recognizing that the effect of the magnetic field is much smaller than the effect of the electric field, due to the relativistic factor v/c). Therefore, we may use $\mathbf{a} = \mathbf{a}_0 \cos \omega t$ for the vector potential and $\mathbf{E} = \mathbf{E}_0 \sin \omega t$ for the electric field ($\mathbf{E}_0 = \omega \mathbf{a}_0/c$). In addition, by virtue of the quasi-classical motion of the projectile, we may write $\omega t = \omega l/v$, where l is the length of the projectile path in the radiation region; since v is not changed appreciably by the radiation, we may use the initial velocity $v = \sqrt{2\mathcal{E}/M}$; further on, by the same token (and in accordance with the assumptions of the perturbation theory), we may neglect the transverse distances x, y in the path length, and write $l = z$ (we take the origin of the time at the spatial origin). Finally, we get for the Schroedinger equation

$$i\hbar\frac{\partial\psi}{\partial t} \simeq \left[\frac{1}{2M}\mathbf{P}^2 - ZezE_{0z}\sin\omega z/v + V(\mathbf{r})\right]\psi \ , \tag{9.43}$$

where

$$U(\mathbf{r}) = -ZezE_{0z}\sin\omega z/v + V(\mathbf{r}) \tag{9.44}$$

is viewed as a perturbation; the coordinate z extends to a range of the order of the size d of the focal region about the origin. The approximation caused by neglecting the transverse coordinates in equation (9.44) is valid for small scattering angles (forward scattering).

Recently, laser-assisted proton collisions on light nuclei have been discussed, where the spatial phase $\mathbf{kr}$ is neglected in comparison with the temporal phase ωt for independent t and $\mathbf{r}$ (the so-called dipole approximation).[17] This approximation is valid as long as the energy is changed appreciably by the radiation and the quasi-classical approximation does not hold anymore; it applies to a quantum particle in the laser focal region, where it is dressed with radiation. This assumption does not apply to the problem of ion scattering discussed here, where the ion motion is quasi-classical.

[17] I. F. Barna and S. Varro, "Laser-assisted proton collision on light nuclei at moderate energies", Laser&Part. Beams **33** 299 (2015).

In the Born approximation for the scattering potential $U(\mathbf{r})$ in equation (9.43) we get the scattering amplitude

$$f = f_{rad} + f_V \quad , \tag{9.45}$$

where

$$f_{rad} = \frac{MZeE_{0z}}{2\pi\hbar^2} \int d\mathbf{r} \cdot z \sin \omega z/v \cdot e^{i\mathbf{qr}} \tag{9.46}$$

is the scattering amplitude due to the radiation and

$$f_V = \frac{M}{2\pi\hbar^2} \int d\mathbf{r} V(\mathbf{r}) e^{i\mathbf{qr}} \tag{9.47}$$

is the well-known Born amplitude of the scattering potential $V(\mathbf{r})$; $\mathbf{q} = \mathbf{K} - \mathbf{K}^{'}$ is the wavevector exchanged by the projectile during collision. We can see that apart from the regular scattering Born amplitude f_V due to the potential $V(\mathbf{r})$, there exists, in the presence of the radiation field, an additional scattering amplitude f_{rad} due to the radiation.

The evaluation of the integral in equation (9.46) requires special attention. We can write

$$f_{rad} = \frac{MZeE_{0z}}{2\pi\hbar^2} \int dz \cdot z \sin \omega z/v \cdot e^{iq_z z} \int d\mathbf{r}_\perp e^{i\mathbf{q}_\perp \mathbf{r}_\perp} \quad , \tag{9.48}$$

where q_z is the z-component of $\mathbf{q}$ and $\mathbf{q}_\perp$, $\mathbf{r}_\perp$ are the transverse components of $\mathbf{q}$ and $\mathbf{r}$. From $\mathbf{q} = \mathbf{K} - \mathbf{K}^{'}$ and $K = K^{'}$ we have $q_z = 2K \sin^2 \theta/2$ and $q_\perp = K \sin \theta$ ($q = 2K \sin \theta/2$), where θ is the scattering angle. The integration over z gives a longitudinal factor

$$F_l = -\frac{\partial}{\partial q_z} \left[\frac{\sin(q_z + \omega/v)d/2}{q_z + \omega/v} - \frac{\sin(q_z - \omega/v)d/2}{q_z - \omega/v} \right] \quad , \tag{9.49}$$

which, for large d, may be approximated by

$$F_l \simeq -\pi \left[\delta^{'}(q_z + \omega/v) - \delta^{'}(q_z - \omega/v) \right] \quad . \tag{9.50}$$

Due to the quasi-classical character of the ion motion, the transverse integration over $\mathbf{r}_\perp$ for each ion extends over a very small region of cross-section $d^{'2}$, where $d^{'}$ is of the order of the ion wavelength Λ, placed at the impact parameter $\mathbf{r}_{\perp i}$ for the $i-th$ ion; the transverse factor in equation (9.48) is

$$F_t = \int d\mathbf{r}_\perp e^{i\mathbf{q}_\perp \mathbf{r}_\perp} \simeq d'^2 e^{i\mathbf{q}_\perp \mathbf{r}_{\perp i}} \simeq \Lambda^2 e^{i\mathbf{q}_\perp \mathbf{r}_{\perp i}} \ ; \tag{9.51}$$

we get

$$f_{rad} \simeq \frac{MZeE_{0z}}{2\hbar^2}\Lambda^2 \left[\delta'(q_z - \omega/v) - \delta'(q_z + \omega/v)\right] e^{i\mathbf{q}_\perp \mathbf{r}_{\perp i}} \ . \tag{9.52}$$

The summation over i can be transformed into an integral which includes the Bessel function $J_0(q_\perp d)$; making use of the integral[18]

$$\int_0^1 dx \cdot x J_0(q_\perp d x) = \frac{1}{q_\perp d} J_1(q_\perp d) \tag{9.53}$$

and using the asymptotic expression of the Bessel function J_1, we get

$$\begin{gathered} f_{rad} \simeq \sqrt{\pi/2}\xi \cos\alpha \tfrac{\Lambda^{3/2} n^{2/3}}{d^{1/2}\theta^{3/2}} \cdot \\ \cdot \left[\delta'(q_z - \omega/v) - \delta'(q_z + \omega/v)\right] \cos(\theta d/\Lambda - 3\pi/4) \ , \end{gathered} \tag{9.54}$$

where α is the angle between the electric field and the direction of ion motion and n is the ion density.

We can see that the scattering amplitude f_{rad} in the radiation field (Born approximation) exhibits a "double" maximum at $q_z = \omega/v$, *i.e.* at scattering angles $\theta = \pm\sqrt{2\omega/vK} = \pm\sqrt{\hbar\omega/\mathcal{E}}$, corresponding to the emission of one "photon" with momentum $\hbar\omega/v$ (only the $\delta'(q_z - \omega/v)$ must be preserved in equation (9.54)); in higher-orders of the perturbation theory similar diffraction maxima (with richer internal structure) appear at angles given by $\sin^2\theta/2 = n\hbar\omega/4\mathcal{E}$, n any integer, corresponding to multiple-"photon" emission with momenta $n\hbar\omega/v$, where n is any integer; these maxima affect the regular scattering amplitude generated by the potential $V(\mathbf{r})$. It is easy to see that the order of magnitude of the scattering amplitude f_{rad} is given

[18] I. S. Gradshteyn and I. M. Ryzhik, *Table of Integrals, Series and Products*, eds. A. Jeffrey and D. Zwillinger, 6th edition, Academic Press, NY (2000), p.668, 6.561.5.

by $\xi \cos\alpha(\Lambda d/\theta)^{3/2}n^{2/3}$; in the scattering cross-section δ'^2 may be replaced by $d\left|\delta'\right|$ and the rapidly varying factor $\cos^2(\theta d/\Lambda - 3\pi/4)$ must be averaged. We emphasize that the radiation part of the scattering exists independently of the scattering potential $V(\mathbf{r})$; it is a diffraction effect suffered by ions in the radiation field, where the role of the radiation is played by ions, while the role of the diffraction-grating distance is played by the radiation wavelength; in contrast with the usual Bragg diffraction condition $\theta \simeq n\Lambda/\lambda = n\hbar\omega/cP$, the condition of ion diffraction given here reads $\theta \simeq \pm\sqrt{n\hbar\omega/\mathcal{E}}$. The effect resembles the electron diffraction in a standing electromagnetic wave.[19] We note that the "photons" involved in this diffraction effect can be viewed as photons dressed with interaction, due to the ion velocity v which appears in their momenta $n\hbar\omega/v$, instead of the light speed c. Similar results can be obtained for a general polarization of the radiation field.

If the parameter ξ is much larger than unity the quasi-classical approximation and the Born approximation are not valid anymore; the ion suffers a quantum-mechanical motion in the radiation field and the time t is disentangled from the position $\mathbf{r}$; this happens for high-intensity radiation or low-energy projectiles. In order to account for the effect of the radiation field in this case we may use the Goeppert-Mayer unitary transform[20]

$$\psi = e^{-\frac{iZe}{\hbar\omega}\mathbf{E}_0\mathbf{r}(\cos\omega t - 1)}\phi \quad , \tag{9.55}$$

where ψ is the wavefunction in equation (9.42) and ϕ satisfies the Schroedinger equation

$$i\hbar\frac{\partial\phi}{\partial t} = \frac{1}{2M}\left[\mathbf{P} - \frac{Ze}{c}(\mathbf{a} - \mathbf{a}_0)\right]^2 \phi + V(\mathbf{r})\phi \quad , \tag{9.56}$$

[19] The diffraction of electrons in standing electromagnetic wave (Kapitsa-Dirac effect, P. L. Kapitza and P. A. M. Dirac, "The reflection of electrons from standing light waves", Proc. Cambridge Phil. Soc. **29** 297 (1933)) is a Bragg diffraction due to stimulated Compton scattering; in Bragg diffraction the change in the momentum is transverse with respect to the propagtion direction; in the diffraction presented here the momentum change is longitudinal.

[20] M. Goeppert-Mayer, "Uber Elementarakte mit zwei Quantensprungen", Ann. Physik **401** 273 (1931).

with $\psi(t=0)=\phi(t=0)$ (we note that $\mathbf{E}(t=0)=0$); we recognize in equation (9.56) the standard form of the non-relativistic hamiltonian of the charge Ze in the radiation field with the vector potential $\mathbf{a}$ (the constant term $\mathbf{a}_0$ appears as a consequence of the initial conditions).[21] At this point we note that the hamiltonian given by equation (9.56) is not suitable for the scattering problem discussed here with quasi-classical projectiles since it includes the radiation contribution in the asymptotic momentum $\mathbf{P}$; this would correspond to a scattering process embedded in the radiation region, in contrast with the scattering process presented here, where we assume that the radiation field is confined to a finite region F. Moreover, it is worth noting that the radiation effects in the hamiltonian given by equation (9.56) are controlled by the ratio $Zea_0/cP=(Zea_0/Mc^2)(c/v)$, which includes the well-known "relativistic" parameter $\eta=Zea_0/Mc^2$; usually, this parameter is very small for ions, even for high-intensity radiation; *e.g.*, with our numerical data we get $\eta\simeq 10^{-4}/A$. It follows that for reasonably high velocities the ratio Zea_0/cP is small, and we may neglect the radiation interaction in equation (9.55); it follows that the effect of the radiation is, practically, included in the unitary transform given by equation (9.55).

As a matter of fact, if we still maintain the radiation interaction in equation (9.56) (for low velocities v), we can use a transformation of the Kramers-Henneberger type[22]

$$\phi=e^{-\frac{iZ^2e^2}{8Mc^2\hbar\omega}a_0^2(\sin 2\omega t-8\sin\omega t+6\omega t)}e^{\frac{iZe}{Mc\hbar\omega}(\sin\omega t-\omega t)\cdot\mathbf{a}_0\mathbf{P}}\chi \ , \qquad (9.57)$$

[21] J. Bergou, "Wavefunctions of a free electron in an external field and their application in intense field interactions. I. Non-relativistic treatment", J. Phys. **A13** 2817 (1980); J. Bergou and S. Varro, "Wavefunctions of a free electron in an external field and their application in intense field interactions. II. Relativistic treatment", J. Phys. **A13** 2823 (1980).

[22] W. C. Henneberger, "Perturbation method for atoms in intense light beams", Phys. Rev. Lett. **21** 838-840 (1968); H. A. Kramers, "Non-relativistic quantum-electrodynamics and the correspondence principle", Solvay Congress 1948, in *Collected Scientific Papers*, North-Holland, Amsterdam (1956), p. 865; N. G. van Kampen, "Contribution to the quantum theory of light scattering", Kgl. Danske Videnskab. Selskab, Mat.-fys. Medd. **26** (15) (1951); W. Pauli and M. Fierz, "Zur Theorie der Emission langwelliger Lichtquanten", Nuovo Cimento **15** 167 (1938).

where χ satisfies the Schroedinger equation

$$i\hbar\frac{\partial\chi}{\partial t}=\left[\frac{1}{2M}P^2+\widetilde{V}(\mathbf{r})\right]\chi \tag{9.58}$$

with the radiation-dressed potential

$$\widetilde{V}(\mathbf{r})=e^{-\frac{Ze}{M\omega^2}(\sin\omega t-\omega t)\cdot\mathbf{E}_0 grad}V(\mathbf{r}) \ ; \tag{9.59}$$

we can see that in a very short time (given by $(ZeE_0/M\omega^2a)(\omega t)^3/6$, where a is the range of the potential $V(\mathbf{r})$) the potential $V(\mathbf{r})$ is practically removed from the problem, and the wavefunction χ is a free-particle wavefunction;[23] since the time-dependent exponents in the wavefunction ϕ (equation (9.57)) are immaterial for our problem, we arrive again at the conclusion that the effect of the electric field is included mainly in the transform given by equation (9.55). This is an interesting effect, which shows that in high-intensity radiation the ion scattering on short-range nuclear potentials is obliterated by the scattering by the radiation field, which has a large cross-section, as we see below (for Coulomb potentials the situation is different).

For reasonably high ion velocity we may consider the incoming ion wavefunction $e^{-\frac{i}{\hbar}\mathcal{E}t+i\mathbf{K}\mathbf{r}}$ and the outgoing wavefunction $e^{-\frac{i}{\hbar}\mathcal{E}^{'}+i\mathbf{K}^{'}\mathbf{r}}$ as χ-type wavefunctions and write the radiation scattering amplitude as

$$T=\frac{1}{v}\int d\mathbf{r}e^{-\frac{iZe}{\hbar\omega}\mathbf{E}_0\mathbf{r}(\cos\omega t-1)}e^{i\mathbf{q}\mathbf{r}}e^{-\frac{i}{\hbar}(\mathcal{E}-\mathcal{E}^{'})t} \quad , \tag{9.60}$$

where $\mathcal{E}=\hbar^2K^2/2M$ and $\mathcal{E}^{'}=\hbar^2K^{'2}/2M$ are the energies of the incoming and outgoing particles and $v\simeq d^3$ is the volume of the region F containing the radiation. This quantity is a (time-dependent) tunneling coefficient; it is easy to see that the transition probability, *i.e.* the number of scattered particles with momentum $\hbar K^{'}$ and energy $\mathcal{E}^{'}$ is given by

$$|T|^2=\frac{1}{v}\int d\mathbf{r}e^{-\frac{iZe}{\hbar\omega}\mathbf{E}_0\mathbf{r}(\cos\omega t-1)}e^{i\mathbf{q}\mathbf{r}} \tag{9.61}$$

(for a sufficiently large volume v). The integration in equation (9.61) gives a momentum transfer $\hbar\mathbf{q}$ in the direction of the electric field $\mathbf{E}_0$;

[23]This effect is highlighted by M. Apostol, "Fast ionization in strong electromagnetic radiation", Z. Naturforsch. **73a** 461 (2018).

we denote it by $\hbar q_0$; the integration along the coordinate in the field direction gives

$$|T|^2 = \frac{2\pi}{d}\delta\left[q_0 - \frac{ZeE_0}{\hbar\omega}(\cos\omega t - 1)\right] \ , \tag{9.62}$$

or

$$|T|^2 = \frac{2\pi}{d}\delta\left[q_0 - \frac{\xi v}{2\omega d}K(\cos\omega t - 1)\right] \ . \tag{9.63}$$

Usually, the cross-section is obtained as the ratio of the transition probability per unit time to the current density v/d^3; thereafter, the differential cross-section is obtained by multiplying with the number of available scattered states, which in our case is $d\mathbf{K}^{'}d^3/(2\pi)^3$, *i.e.*, $K^{'2}d^3dK^{'}\sin\theta d\theta/2\pi^2$ (it is worth noting that the volume available for the transition process to $\mathbf{K}^{'}$-states is d^3). The time derivative $\partial|T|^2/\partial t$ gives the probability per unit time for scattering both to and from the state $\mathbf{K}^{'}$, since it may be both positive and negative. Consequently, it is more convenient to compute the cross-section as the ratio of the number of scattered particles $|T|^2$ to the incoming particle flux $1/d^2$; the differential cross-section is given by

$$d^2\sigma = \frac{d^4}{\pi}K^{'2}\delta\left[q_0 - \frac{\xi v}{2\omega d}K(\cos\omega t - 1)\right]\sin\theta\cdot dK^{'}d\theta \ . \tag{9.64}$$

We can see in equation (9.64) that the scattering implies in this case absorption and emission of special "photons" which change the particle momentum along the direction of the electric field by the time-dependent amount $(\hbar\xi vK/2\omega d)(\cos\omega t - 1)$ (and the energy by a corresponding amount). In order to integrate with respect to $K^{'}$ it is convenient to take the z-axis along the direction of the electric field $\mathbf{E}_0$; then $q_0 = K - K^{'}\cos\theta$ and

$$d\sigma = \frac{d^4K^2}{\pi}\left[1 - \frac{\xi v}{2\omega d}(\cos\omega t - 1)\right]^2\frac{\sin\theta}{\cos^3\theta}d\theta \ , \ \ 0 < \theta < \pi/2 \ ; \tag{9.65}$$

for $\pi/2 < \theta$ the cross-section is zero. We may take the time average in equation (9.65) and get

$$d\sigma \simeq \frac{3d^4K^2}{8\pi}\left(\frac{\xi v}{\omega d}\right)^2\frac{\sin\theta}{\cos^3\theta}d\theta = \frac{3d^2}{2\pi}\left(\frac{ZeE_0d}{\hbar\omega}\right)^2\frac{\sin\theta}{\cos^3\theta}d\theta \ ; \tag{9.66}$$

we can see that the cross-section corresponding to the scattering by the high-intensity radiation field is large, with a singularity at right angle ($\theta = \pi/2$) with respect to the direction of the electric field. We note the coefficient $ZeE_0d/\hbar\omega$ entering the cross-section, which counts the number of "photons" $\hbar\omega$ included in the amount of energy ZeE_0d of the ion in the radiation field.[24]

It is worth discussing the case of electron projectiles. First, in order to preserve the non-relativistic character of the motion the radiation intensity must be limited to $I < 10^{15}w/cm^2$. Second, the critical value $\xi = 1$ for $I = 10^{15}w/cm^2$ restricts the electron energy to $\mathcal{E} < 5MeV$; it follows that for all allowed energies the parameter ξ is larger than unity and, consequently, the electron suffers an inelastic scattering.

---

[24]It is worth computing the usual Born scattering cross-section according to the method used here (*i.e.*, normalizing to the particle flux, not to the current density). The Born scattered wavefunction is given by

$$\psi_1 = -\frac{M}{2\pi\hbar^2}\int d\mathbf{r}' V(\mathbf{r}')e^{i\mathbf{k}\mathbf{r}'}\frac{e^{ik|\mathbf{r}-\mathbf{r}'|}}{|\mathbf{r}-\mathbf{r}'|} \simeq f\frac{e^{ikr}}{r} \;\; ;$$

the tunneling coefficient is

$$T_B = f\int d\mathbf{r}e^{-i\mathbf{k}'\mathbf{r}}\frac{e^{ikr}}{r} \;\; ,$$

where $\mathbf{k}'$ is directed along $\mathbf{r}$; the evaluation of this integral (over the scattering region) gives $T_B \simeq f/d$ (normalized to the scattering volume); the cross-section obtained as the ratio to the incoming particle flux is $d^2\left|T_B\right|^2 = |f|^2$, *i.e.* the Born cross-section. The density of states in the elastic scattering is the solid angle $d\Omega$, while the density of states in the inelastic scattering is $K'^2d^3dK'd\Omega/(2\pi)^3$. In the elastic scattering the energy is conserved, while in the inelastic scattering the energy is not conserved; the amplitudes of the two types of scattering cannot be added, such that cross-sections of the type $|Tf|$ (arising from $|T+f|^2$) are meaningless. If we use radiation-dressed wavefunctions in the Born formula we may get an amplitude $\sim J_nf$, where J_n is a Bessel function, by artificially selecting an n-photon exchange process; this procedure leads to the Kroll-Watson formula $d\sigma \sim J_n^2d\sigma_{el}$; but it corresponds both to absorption of n photons and emission of n photons. The energy of the emitted photons and the energy of the absorbed photons inside the radiation region go to charges which suffer a continuous energy exchange with the radiation in that region.

In scattering problems in the presence of the radiation the potential $V(\mathbf{r})$ contributes to diffraction, which appears at low- or moderate-intensity radiation, while in high-intensity radiation the scattering on the potential $V(\mathbf{r})$ is blurred by the large cross-section of the inelastic scattering by radiation.

For usual sources of light the parameter ξ is much smaller than unity, and the electrons suffer diffraction.

In conclusion, one may say that non-relativistic quasi-classical charged projectiles may suffer a diffraction effect during their scattering by a low- or moderate-intensity radiation field, while in higher-intensity radiation the scattering becomes inelastic. The behaviour of a non-relativistic charged particle in the field of electromagnetic radiation raises an interesting problem, related to the dipole hamiltonian given by equation (9.42) and the standard form of the non-relativistic hamiltonian given by equation (9.56). Classically, the two hamiltonians lead to equivalent equations of motion, while, quantum-mechanically, they are related through the Goeppert-Mayer unitary transform and, thus, are considered equivalent. In spite of this formal equivalence, there exist differences between the two hamiltonians, depending on the problem they are applied to. In scattering problems the radiation is confined to a finite region (denoted by F herein) and the asymptotic wavefunctions are eigenfunctions of the momentum $\mathbf{P} = M\mathbf{v}$; consequently, the appropriate starting point in these problems is the dipole hamiltonian given by equation (9.42). This has relevant effects in scattering problems, as shown herein, distinct from the results obtained by starting directly with the standard form of the non-relativistic hamiltonian (equation (9.56)). These effects, which are related to the phase factor given by equation (9.55), are lost when starting with the hamiltonian given by equation (9.56). However, apart from other peculiar points (like the selection of transition amplitudes associated with n-photon exchange), the standard non-relativistic hamiltonian is still the starting point. On the contrary, for problems related to quantum-mechanical bound-states in radiation field it is convenient to start with the standard form of the non-relativistic hamiltonian (equation (9.56)), which includes the radiation effects in the momentum $\mathbf{P} = M\mathbf{v} + (Ze/c)\mathbf{a}$ (the phase factor given in equation (9.55) has little effect in this case).

As a general comment, the quasi-classical approximation used here is not appropriate, since the projectile has still a quantum-mechanical behaviour along its trajectory; the energy is not resolved in equation (9.65).

10 Electron scattering

Electron scattering in matter proceeds mainly by scattering by atoms (molecules). As a first approximation we may use the screened potential seen by an electron on approaching an atom, as derived in the Scattering, Collisions chapter. The cross-section depends on the energy of the electron. The usual experiments involve electron energies in the range $1keV$ to $10MeV$. Most of these experiments are related to electron microscopy. The electron scattering in matter is accompanied by electronic excitations of the atoms, transitions in atoms, resonant scattering, inelastic scattering, absorption and, at high energies, relativistic effects. For low energies the atomic cross-section of an electron is of the order of the square of the Bohr radius ($a_H = \hbar^2/me^2 \simeq 0.53\text{Å}$, where $\hbar \simeq 10^{-27}erg \cdot s$ is Planck constant, $m \simeq 10^{-27}g$ is the electron mass and $e = -4.8 \times 10^{-10}esu$ is the electron charge; $1\text{Å} = 10^{-8}cm$). In order to have a well-defined situation we adopt a single-atom cross-section $\sigma_0 = 10^{-17}cm^2$ and an electron energy $E = 1keV$ (wavelength $\lambda \simeq 0.1\text{Å}$, momentum $p \simeq 10^{-18}g \cdot cm/s$). Usual electron flux densities are of the order $S = \Phi A = 10^{12}electrons/s$, where A is the cross-section of the sample.[1]

In crystals, electrons may suffer a coherent diffraction, similar to X-rays and neutrons.[2] The diffraction appears on very small crystals, or even on the surface layer, or on a few layers parallel to the surface layer. We adopt a number $N = 10^{16}$ atoms in the sample, which corresponds to a cross-section of the sample $A \simeq 10^{-4}cm^2$, with a mean inter-atomic separation distance $a = 4\text{Å}$. It follows a flux density $\Phi = S/A = 10^{16}/cm^2 \cdot s$.

We assume first an atomic collision. Consequently, the energy transfer

[1] R. Pike and P. Sabatier, eds., *Scattering*, Academic, NY (2002).

[2] C. J. Davisson and L. J. Germer, "The scattering of electrons by atoms", Nature **119** 558 (1927); H. A. Bethe, "Theorie der Beugung von Elektronen an Kristallen", Ann. Physik **87** 55 (1928).

to the atom is of the order $(v+p/M)p$, where v is the thermal velocity of the atom, p is the momentum of the incident electron and M is the atomic mass (see the Neutron Scattering chapter). For $v = p/M$ we get the energy of the incident electron $E = p^2/m = (Mv)^2/m$. By using $M = 10^5 m$ and $v = 10^4 cm/s$ (room temperature), we get an energy $E \simeq 1keV$. It follows that we may use $\Delta E = p^2/M = \frac{m}{M}E$ for $E > 1keV$. For $E = 1keV$ the energy transfer is of the order $\Delta E \simeq 10^{-14} erg$.

Let us examine first the conditions for a macroscopic coherent quantum scattering. The cross-section is $\overline{\sigma} = N\sigma_0/f$ and the collision time is $\tau = 1/\Phi\overline{\sigma} = f/\Phi N\sigma_0$. The conditions $\overline{\sigma} < A$ and $\tau > \Delta t_{eq} = \hbar/\Delta E$ read

$$f > \frac{N\sigma_0}{A} \;,\; f > \frac{\hbar\Phi}{\Delta E} N\sigma_0 \;. \tag{10.1}$$

The first inequality gives $f > 10^3$; since $1/ < f < 1$, it follows that $f = 1$ and $\overline{\sigma} = A$. The number of atoms in each domain is $N_d = 1/f = 1$. We have an incoherent scattering due to phonons. According to the Neutron Scattering chapter, the incoherent cross-section is $\sigma_0^{ph} = 10^{-21} cm^2$ and the total incoherent cross-section is $\sigma_{incoh} = 10^{-5} cm^2$. The force acting upon the sample is $F = p\Phi\sigma_{incoh} \simeq 10^{-7} dyn$ for an energy $E = 1keV$. For higher energy the force is larger and it may be used in electron force microscopy.

Let us examine now the conditions of a coherent diffraction. The cross-section is $\overline{\sigma} = N\sigma_0/f^{1/3}$ and the collision time is $\tau = 1/\Phi\overline{\sigma} = f^{1/3}/\Phi N\sigma_0$. The conditions $\overline{\sigma} < A$ and $\tau < \Delta t_{eq} = \hbar/\Delta E$ read

$$\frac{N\sigma_0}{A} < f^{1/3} < \frac{\hbar\Phi}{\Delta E} N\sigma_0 \;. \tag{10.2}$$

The principle of maximum cross-section requires $f^{1/3} = N\sigma_0/A = 10^3$, *i.e.* $f = 1$ (since $1/N < f < 1$). It follows that a coherent diffraction is not possible in the whole sample.

Very likely, the electron scattering proceeds in a few superficial layers of the sample. We take b such layers, each with $N_s = 10^{10}$ atoms. The conditions for a coherent quantum scattering are

$$f > \frac{b\sigma_0}{a^2} \;,\; f > \frac{\hbar\Phi}{\Delta E} N_s b\sigma_0 \;. \tag{10.3}$$

These two inequalities lead to $f > 10^{-2}b$ and $f > 10^{-4}b$ (for $E = 1keV$). The number of atoms in each domain is $N_d = 1/f < 10^2/b$. This number is much smaller than the critical value $\simeq 10^6$ of N_d, required by the continuity of the phase domains. Therefore, N_d tends to unity and we have an incoherent scattering.

The conditions for a coherent diffraction are

$$\frac{b\sigma_0}{a^2} < f^{1/3} < \frac{\hbar\Phi}{\Delta E} N_s b\sigma_0 \ . \tag{10.4}$$

With our numerical data the first inequality leads to $f > 10^{-6}b^3$ and $N_d < 10^6/b^3$. This number indicates that b is of the order of unity (for N_d to be comparable to the critical number of atoms in three dimensions $N_c \simeq 10^6$). In two dimensions the critical number of atoms for phase continuity differs from that in three dimensions (which is $\simeq 10^6$). Indeed, the variation of the area A due to atomic discontinuities is $\delta A/A = (a_0/a)^2$, where a_0 is the atomic dimension and a is the mean separation distance between the atoms. It follows $\delta A = (a_0/a)^2 a^2 N_c$. On the other hand, $\delta A = L_c a = N_c^{1/2} a^2$, where L_c is the length of the border of the area A, which encloses a critical number of atoms N_c. From the equality of these two formulae we get $N_c = (a/a_0)^4$ (compare to $N_c = (a/a_0)^9$ in three dimensions). For $a = 4a_0$ we get $N_c \simeq 10^3$ (slightly overestimated, in comparison to $N_c \simeq 10^6$ in three dimensions). Therefore, the coherence domains may exist in electron scattering in superficial layers. However, the second inequality (10.4) is only marginally satisfied. In this respect we should note that an atomic collision is not a quite good assumption for electron scattering. In an electron-atom collision the atomic electrons are perturbed, because they are lighter; they transfer the energy to the entire atom, by interaction, being left with a small amount of energy. It is this energy ΔE which enters the equilibrium time, such that the equilibrium time is much longer, and the second inequality (10.4) may be satisfied.

We conclude that the electron scattering is mostly incoherent, with a weak coherent diffraction. However, we should note that this result depends very much on the input data used herein.

11 Appendix

11.1 Center of mass

For two particles with masses $m_{1,2}$ and positions $\boldsymbol{r}_{1,2}$ the relative position is

$$\boldsymbol{r} = \boldsymbol{r}_1 - \boldsymbol{r}_2 \tag{11.1}$$

and the position of the center of mass is

$$\boldsymbol{R} = \frac{m_1\boldsymbol{r}_1 + m_2\boldsymbol{r}_2}{m_1 + m_2} \ , \tag{11.2}$$

such that

$$\boldsymbol{r}_1 = \boldsymbol{R} + \frac{m_2}{M}\boldsymbol{r} \ , \ \boldsymbol{r}_2 = \boldsymbol{R} - \frac{m_1}{M}\boldsymbol{r} \ , \tag{11.3}$$

where $M = m_1 + m_2$ is the total mass. The energy

$$E = \frac{1}{2}m_1v_1^2 + \frac{1}{2}m_2v_2^2 \ , \tag{11.4}$$

where $\boldsymbol{v}_{1,2} = \dot{\boldsymbol{r}}_{1,2}$ are the velocities, becomes

$$E = \frac{1}{2}M\dot{\boldsymbol{R}}^2 + \frac{1}{2}\mu\dot{\boldsymbol{r}}^2 \ , \tag{11.5}$$

where $\mu = m_1m_2/(M_1 + M_2)$ is the reduced mass. For an interaction $U(\boldsymbol{r}_1 - \boldsymbol{r}_2) = U(\boldsymbol{r})$, which depends only on the relative position, the center of mass is free, such that its momentum

$$\boldsymbol{P} = M\dot{\boldsymbol{R}} \tag{11.6}$$

is conserved.

From equation (11.5) the relative momentum is $\boldsymbol{p} = \mu\dot{\boldsymbol{r}}$, such that, from equations (11.3) we get the momenta

$$\boldsymbol{p}_1 = \frac{m_1}{M}\boldsymbol{P} + \boldsymbol{p} \ , \ \boldsymbol{p}_2 = \frac{m_2}{M}\boldsymbol{P} - \boldsymbol{p} \tag{11.7}$$

and

$$\boldsymbol{P} = \boldsymbol{p}_1 + \boldsymbol{p}_2 \ , \ \boldsymbol{p} = \frac{m_2}{M}\boldsymbol{p}_1 - \frac{m_1}{M}\boldsymbol{p}_2 \ . \tag{11.8}$$

Since the interaction depends only on the relative coordinate the center-of-mass coordinates are useful in scattering. In the laboratory frame the projectile has the momentum $\boldsymbol{p}_1$ and the target has momentum zero ($\boldsymbol{p}_2 = 0$). Therefore, $\boldsymbol{P} = \boldsymbol{p}_1$ and $\boldsymbol{p} = \frac{m_2}{M}\boldsymbol{p}_1$. The scattering amplitude is expressed by means of the relative momenta $\boldsymbol{p}$ and $\boldsymbol{p}'$, where the prime means the quantities after collision. The angle between $\boldsymbol{p}$ and $\boldsymbol{p}'$, or between $\boldsymbol{p}_1$ and $\boldsymbol{p}'$, is the scattering angle θ; we write $(\boldsymbol{p}, \boldsymbol{p}') = (\boldsymbol{p}_1, \boldsymbol{p}') = \theta$. According to equations (11.7) the momenta after collision are

$$\begin{aligned} \boldsymbol{p}_1^{'} &= \tfrac{m_1}{M}\boldsymbol{P}' + \boldsymbol{p}' = \tfrac{m_1}{M}\boldsymbol{P} + \boldsymbol{p}' = \tfrac{m_1}{M}\boldsymbol{p}_1 + \boldsymbol{p}' \ , \\ \boldsymbol{p}_2^{'} &= \tfrac{m_2}{M}\boldsymbol{P}' - \boldsymbol{p}' = \tfrac{m_2}{M}\boldsymbol{P} - \boldsymbol{p}' = \tfrac{m_2}{M}\boldsymbol{p}_1 - \boldsymbol{p}' \ . \end{aligned} \tag{11.9}$$

We multiply these equations by $\boldsymbol{p}_1$ in order to get the scattering angles $(\boldsymbol{p}_1^{'}, \boldsymbol{p}_1) = \theta_1$, $(\boldsymbol{p}_2^{'}, \boldsymbol{p}_1) = \theta_2$ after collision in the laboratory frame. Taking into account that $p = p'$ in the elastic collision, we get

$$\begin{aligned} p_1^{'} \cos\theta_1 &= \tfrac{1}{M}\left(m_1 + m_2 \cos\theta\right) p_1 \ , \\ p_2^{'} \cos\theta_2 &= \tfrac{m_2}{M}\left(1 - \cos\theta\right) p_1 \ . \end{aligned} \tag{11.10}$$

Also, from equations (11.9) we get

$$\begin{aligned} p_1^{'2} &= \left[1 - \tfrac{2m_1 m_2}{M^2}\left(1 - \cos\theta\right)\right] p_1^2 \ , \\ p_2^{'2} &= \tfrac{2m_2^2}{M^2}\left(1 - \cos\theta\right) p_1^2 \ , \end{aligned} \tag{11.11}$$

which, introduced in equations (11.10), give

$$\begin{aligned} \tan\theta_1 &= \tfrac{m_2 \sin\theta}{m_1 + m_2 \cos\theta} \ , \\ \theta_2 &= \tfrac{1}{2}\left(\pi - \theta\right) \ . \end{aligned} \tag{11.12}$$

The cross-section in the laboratory frame is obtained from the cross-section in the center-of-mass frame by expressing $\boldsymbol{k}$ and $\boldsymbol{k}'$ ($\boldsymbol{p} = \hbar\boldsymbol{k}$ and $\boldsymbol{p}' = \hbar\boldsymbol{k}'$) in terms of $\boldsymbol{k}_1$ ($= \boldsymbol{p}_1/\hbar$) and $\boldsymbol{k}_1^{'}$ ($= \boldsymbol{p}_1^{'}/\hbar$) by using the above formulae.

11.2 δ-function and the Fourier transform

Let x be a continuous variable in the range $-L/2 < x < L/2$ and k a parameter. The integral

$$I(k) = \int_{L/2}^{L/2} dx e^{ikx} = \frac{\sin(kL/2)}{k/2} \ , \tag{11.13}$$

viewed as a function of k, is L for $k \to 0$ and decreases with increasing k, being localized in the range $\Delta k \simeq 2\pi/L$ around $k = 0$. In addition,

$$\int_{-\infty}^{+\infty} dk I(k) = 2\pi \ . \tag{11.14}$$

Therefore, for $L \to \infty$ we may set $I(k) = 2\pi\delta(k)$

$$I(k) = \int dx e^{ikx} = 2\pi\delta(k) \ . \tag{11.15}$$

Obviously,

$$\int dk e^{ikx} = 2\pi\delta(x) \ . \tag{11.16}$$

Consequently, we may use the Fourier (integral) representation

$$f(x) = \int f(k) e^{ikx} \tag{11.17}$$

for any function $f(x)$, because the coefficients $f(k)$ are well determined by

$$f(k) = \frac{1}{2\pi} \int dx f(x) e^{-ikx} \ ; \tag{11.18}$$

in addition

$$\int dx \left| f(x) \right|^2 = 2\pi \int dk \left| f(k) \right|^2 \ . \tag{11.19}$$

With certain limitations, the same results can be obtained by discretization, *i.e.* by assuming $x \to x_n = a \cdot n$, $n = 0, 1, 2 ... N-1$ and $k = \frac{2\pi}{Na} \cdot m$, m any integer. Indeed,

$$\begin{aligned} S(k_m) &= a \sum_{n=0}^{N-1} e^{ik_m x_n} = a \frac{1 - e^{i2\pi N m}}{1 - e^{i\frac{2\pi}{N} m}} = \\ &= \begin{cases} aN = L \ , \ m = 0 \ , \\ 0 \ , \ a \neq 0 \end{cases} = L\delta_{k_m,0} \ ; \end{aligned} \tag{11.20}$$

if $k_m a \ll 1$, *i.e.* $N \gg 1$, k_m can be viewed as a continuous variable k and the function e^{ikx_n} varies slowly with n, such that

$$S(k_m) \rightarrow S(k) = e^{ikL/2} \int_{-L/2}^{L/2} dx e^{ikx} \qquad (11.21)$$

and

$$2\pi\delta(k) = L\delta_{k,0} \ . \qquad (11.22)$$

We note a certain asymmetry between x, which lies in the large, but finite, interval $(-L/2, L/2)$ (or $(0, L)$, the so-called thermodynamic, or macroscopic limit) and k, which takes all the values in the interval $(-\infty, +\infty)$. Also, in the limit $L \rightarrow \infty$ the variable k is continuous (since $\Delta k = \frac{2\pi}{L} \rightarrow 0$), while x remains discrete, since $\Delta x = a \neq 0$. The latter can be viewed as a continuous variable only in integrals of the form

$$\sum_n f(x_n) = \frac{1}{a} \int dx f(x) \ , \qquad (11.23)$$

where $f(x)$ is a slowly varying function. In contrast,

$$\sum_m f(k_m) = \frac{L}{2\pi} \int dk f(k) \qquad (11.24)$$

for any function $f(k)$, since k is a continuous variable in the limit $L \rightarrow \infty$. In addition, dx cannot be replaced by $\Delta x = a$, nor dk by $\Delta k = \frac{2\pi}{L}$, because both Δx and Δk remain finite, while dx and dk go to zero, in the sense of slowly varying integrands.

11.3 Electromagnetic field and polarization

The polarization $\boldsymbol{P} = nq\boldsymbol{u}$ is the diplole moment of the unit volume, generated by pointlike charges q with density n, displaced by $\boldsymbol{u}$. The induced charge density is $\rho = -qdiv(n\boldsymbol{u}) = -div\boldsymbol{P}$ and the induced current density is $\boldsymbol{j} = nq\dot{\boldsymbol{u}} = \dot{\boldsymbol{P}}$ (where the overdot means the time derivative $\partial/\partial t$). These induced charge and current densities generate an electric field $\boldsymbol{E}$ and a magnetic field $\boldsymbol{H}$. They are given by

$\boldsymbol{E} = -\frac{1}{c}\frac{\partial \boldsymbol{A}}{\partial t} - grad\Phi$ and $\boldsymbol{H} = curl\boldsymbol{A}$, where Φ and $\boldsymbol{A}$ are the electromagnetic potentials, with the Lorenz gauge $\frac{1}{c}\frac{\partial \Phi}{\partial t} + div\boldsymbol{A} = 0$. These potentials satisfy the equations

$$\begin{aligned} \frac{1}{c^2}\frac{\partial^2 \Phi}{\partial t^2} - \Delta\Phi = 4\pi\rho = -4\pi div\boldsymbol{P} \ , \\ \frac{1}{c^2}\frac{\partial^2 \boldsymbol{A}}{\partial t^2} - \Delta\boldsymbol{A} = \frac{4\pi}{c}\boldsymbol{j} = \frac{4\pi}{c}\frac{\partial \boldsymbol{P}}{\partial t} \ ; \end{aligned} \tag{11.25}$$

hence, the electric field satisfies the equation

$$\frac{1}{c^2}\frac{\partial^2 \boldsymbol{E}}{\partial t^2} - \Delta\boldsymbol{E} = -4\pi\left(\frac{1}{c^2}\frac{\partial^2 \boldsymbol{P}}{\partial t^2} - grad\, div\boldsymbol{P}\right) \ , \tag{11.26}$$

or

$$\frac{1}{c^2}\frac{\partial^2 \left(\boldsymbol{E} + \boldsymbol{4\pi} P\right)}{\partial t^2} - \Delta\left(\boldsymbol{E} + 4\pi\boldsymbol{P}\right) = 4\pi curl\, curl\boldsymbol{P} \ . \tag{11.27}$$

We add the equation of the (free) external electric field $\boldsymbol{E}_0$, and get

$$\frac{1}{c^2}\frac{\partial^2 \boldsymbol{D}}{\partial t^2} - \Delta\boldsymbol{D} = 4\pi curl\, curl\boldsymbol{P} \ , \tag{11.28}$$

where $\boldsymbol{D} = \boldsymbol{E}_t + 4\pi\boldsymbol{P}$ is the electric displacement (induction) and $\boldsymbol{E}_t = \boldsymbol{E} + \boldsymbol{E}_0$ is the total electric field. Since $\boldsymbol{D}$ is generated by external charges, this equation can be used outside the polarized body, for $\boldsymbol{D}$ equal to the electric field generated by polarization.

11.4 Maxwell equations in matter

Maxwell equations in matter are the two Gauss equations

$$div\boldsymbol{D} = 4\pi\rho_e \ , \ div\boldsymbol{B} = 0 \ , \tag{11.29}$$

where $\boldsymbol{D}$ is the electric induction (displacement), $\boldsymbol{B}$ is the magnetic induction and ρ_e is the external charge density; the Faraday equation

$$curl\boldsymbol{E} = -\frac{1}{c}\frac{\partial \boldsymbol{B}}{\partial t} \ , \tag{11.30}$$

where $\boldsymbol{E} = \boldsymbol{D} - 4\pi\boldsymbol{P}$ is the electric field and $\boldsymbol{P}$ is the polarization, and the Maxwell-Ampere equation

$$curl\boldsymbol{H} = \frac{1}{c}\frac{\partial \boldsymbol{D}}{\partial t} + \frac{4\pi}{c}\boldsymbol{j}_e \ , \tag{11.31}$$

where $\boldsymbol{H} = \boldsymbol{B} - 4\pi\boldsymbol{M}$ is the magnetic field, $\boldsymbol{M}$ is the magnetization and $\boldsymbol{j}_e$ is the external current density.[1] By integrating over infinitesimal boxes and loops at the interface of two bodies, we get the jumps (boundary conditions) for the normal components of the electric induction and the magnetic induction

$$\delta(D_n) = 4\pi\rho_s \ , \ \delta(B_n) = 0 \tag{11.32}$$

and the jumps for the tangential components of the electric field and the magnetic field

$$\delta(E_t) = 0 \ , \ \delta(H_t) = \frac{4\pi}{c}j_s \ , \tag{11.33}$$

where ρ_s and j_s are surface charge and current densities. In conductors the electric field and the magnetic induction are zero; surface charges and currents may exist.

[1]See M. Apostol, *Essays in Electromagnetic Field and Matter*, Lambert, Saarbruecken (2013).

Index

A

atom, 55

B

beam, 18
Born scattering, 12
Brillouin, Mandelstam, 198

C

center of mass, 57
charge distribution, 52
charge scattering, 207
charge-atom scattering, 13
charge-field interaction, 51
classical scattering, 117, 167
coherence domains, 132, 153, 165, 178
coherence/incoherence, 127
coherent cross-section, 162
coherent diffraction, 168
coherent scattering, 144
crystal field, 139

D

damping force, 32
Debye-Waller, 109, 125
delta potential, 14
diffraction, 90
diffuse scattering, 103
dipole field, 31
Doppler effect, 71

E

electron scattering, 231
energy transfer, 25, 115, 159
energy, force, momentum, 42
energy, momentum, force, 99
extinction, 82

F

Fermi's golden rule, 22
force upon target, 17, 120
form-factor, 92, 121

I

incoherent cross-section, 163
inelastic collisions, 24
interference, 69
ion diffraction, 221

L

laser pulse, 207
lattice distortion, 104, 110
Laue, Bragg, Debye-Scherer, 101
Laue, Bragg, Ewald, 89
line breadth, 33
longitudinal field, 47

M

maximal cross-section, 166
Maxwell equations, 39
Moessbauer effect, 186
momentum transfer, 16

N
neutrino coherence, 129
neutrino scattering, 150, 175
Neutrinos, 113
neutrinos, 170
neutron scattering, 157

O
other projectiles, 176

P
phase continuity, 132, 164, 233
phase defects, 106
phonon excitations, 184
photon emission/absorption, 65, 141
photons, 49
polarization, 85

Q
quantum scattering, 136, 160, 173

R
radiation absorption, 37
radiation quantum scattering, 62
radiation scattering, 74
Raman scattering, 75
Rayleigh cross-section, 35
resonance fluorescence, 35, 183

S
scattering by sound, 198
scattering length, 157
sound, 195
surface distortion, 111

T
Thomson cross-section, 34

W
Weber scattering, 150, 175

X
X-ray scattering, 87
X-rays, 80